Transport and Structural Formation in Plasmas

Plasma Physics Series

Series Editors:

Professor Peter Stott, JET
Professor Hans Wilhelmsson, Chalmers University of Technology

Other books in the series

Plasma Physics via Computer Simulation
C K Birdsall and A B Langdon

Radiofrequency Heating of Plasmas
R A Cairns

An Introduction to Alfvén Waves
R Cross

Transition Radiation and Transition Scattering
V L Ginzburg and V N Tsytovich

Tokamak Plasma: a Complex Physical System
B B Kadomtsev

MHD and Microinstabilities in Confined Plasmas
W M Manheimer and C N Lashmore-Davies

Electromagnetic Instabilities in Inhomogeneous Plasma
A B Mikhailovskii

Physics of Intense Beams in Plasma
M V Nezlin

Instabilities in a Confined Plasma
A B Mikhailovskii

Forthcoming titles in the series

Particle Transport in Magnetically Confined Plasmas
G L Fussmann

Collective Modes in Inhomogeneous Plasma and Advanced Fluid Theory
J Wieland

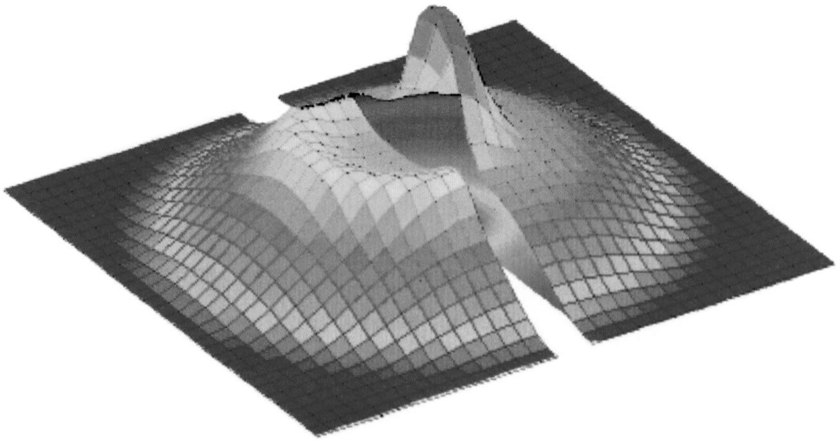

Plasma Physics Series

Transport and Structural Formation in Plasmas

K Itoh

National Institute for Fusion Science
Toki

S-I Itoh

Research Institute for Applied Mechanics
Kyushu University, Kasuga

A Fukuyama

Faculty of Engineering
Kyoto University, Kyoto

Contributor: M Yagi

Institute of Physics Publishing
Bristol and Philadelphia

British Library Cataloguing-in-Publication Data

A catalogue record for this book is available from the British Library.

ISBN 0 7503 0449 9 hbk

Library of Congress Cataloging-in-Publication Data are available

Published by Institute of Physics Publishing, wholly owned by The Institute of Physics, London

Institute of Physics Publishing, Dirac House, Temple Back, Bristol BS1 6BE, UK

US Office: Institute of Physics Publishing, The Public Ledger Building, Suite 1035, 150 South Independence Mall West, Philadelphia, PA 19106, USA

Typeset in TEX using the IOP Bookmaker Macros
Printed in the UK by Bookcraft (Bath) Ltd, Midsomer Norton

This monograph is dedicated to the parents of the authors

Contents

Preface

$\Pi\alpha\nu\tau\alpha$ $\rho\varepsilon\iota$—All things flow
(Heraclitus)

Plasma is a state of matter where the majority of electrons are not bound to ions to form neutral atoms. Electrons and ions move either independently or when forming a structure through the long range interactions of electromagnetic fields. Almost all matter in the universe, of which we have empirical knowledge, is in a plasma state. This means that an understanding of plasma, in principle, is basic to our knowledge of and insight into nature itself. To what extent, in comparison with the abundance of plasma in nature, does our knowledge of plasmas constitute the realm of physics?

A branch of science can be said to have a cultural impact if it finds new ways of recognizing structures in our perception of reality. Such attempts have been of historical interest to mankind in understanding nature. What is the nature of the elements that constitute matter? In addition to this fundamental question, that has been a driving force of research for more than 2000 years, there is another historical enigma: all things flow. What is the law that governs the flow of *all things*? We have been faced with this question again and again whenever new observations of nature have been added to human knowledge.

In this monograph the phenomena of turbulence and turbulent transport in confined plasmas are discussed, particularly in relation to plasma inhomogeneities. Two characteristic features of this area are, first, that the gradient is the order parameter of transport processes, and, secondly, that the long range interactions of the fields play essential roles. Special attention is devoted to structural formation and transitions between various plasma states.

It is intended to show that one central thread of ideas could consistently describe turbulence, turbulent transport and structural transitions. Through this description, a theme is developed, i.e., the physics of plasmas has provided progress in the area of nonlinear nonequilibrium transport. Encyclopaedic description is beyond the scope of this monograph. This naturally limits the subject. Many essential processes are not covered, and important references may be missed. The authors would be happy if this monograph could succeed in illustrating a number of interesting aspects of one of the most active research areas in modern physics.

K Itoh, S-I Itoh, A Fukuyama
September 1998

Acknowledgments

The authors owe much of this monograph to daily communication with their colleagues worldwide. They acknowledge Dr M Yagi for his valuable contributions to the study of turbulence and structural formation. They cordially thank Professor F Wagner, Professor T Ohkawa, Professor S Yoshikawa and Professor A J Lichtenberg for elucidating discussion and guidance during the course of the research, and Professor A Yoshizawa and Dr J W Connor for valuable discussion of the general problems. They are also grateful to Professor K Nishikawa, Professor K Lackner, Professor G H Wolf and the late Academician B B Kadomtsev for helpful suggestions, Dr M Azumi, Dr K C Shaing and Dr H Sanuki for stimulating discussion of the subject and Dr O J F Kardaun for comments on the manuscript. They appreciate the continuous encouragement of Professor A Iiyoshi and Professor Y Furutani. They would also like to thank the generous authors and institutions who have permitted them to reproduce the figures.

The authors wish to acknowledge discussion of experimental observations with the ASDEX Team, the ASDEX-U Group, Dr K Ida, Dr A Fujisawa and the CHS Group, Dr V S Chan and the D-III D Group, Professor M Keilhacker, Dr J A Wesson and the JET Team, Dr Y Miura, the late Professor H Maeda and the JFT-2M Group, the JIPP T-IIU Group, the JT-60U Team, the Heliotron-E Group, the START Group, the TEXTOR Group, Dr K L Wong and the TFTR Group, the TRIAM-1M Group and the Wendelstein 7AS Group. It is a pleasure to express our gratitude to the editorial office of Institute of Physics Publishing for continued cooperation.

This work is partially supported by a grant-in-aid for scientific research from the Ministry of Education, Science, Sports and Culture, Japan, by the collaboration programme of the National Institute for Fusion Science and by the collaboration programme of the Advanced Fusion Research Centre, RIAM, of Kyushu University. The hospitality of the Max-Planck-Institut für Plasmaphysik is also acknowledged.

Chapter 1

Introduction

The structure and dynamics of high temperature plasmas have played a main role for our understanding the universe. The process of nuclear fusion, which takes place in the plasmas of stars and the sun, is the origin of the energy that generates light. The spectacular dynamics in this universe, such as formation of stars and galaxies (some of which is shown in plates 1.1–1.3), are mainly observed by mankind through the observation of light of various wavelengths or of particles travelling across the space. They are formed by the varieties of plasma dynamics under celestial circumstances. Another fundamental element of the source of energy—gravity—is also essential for evolution of the universe and can also be recognized mainly by observing the plasma interactions from a distance. Plasma physics, the subject of which covers the study of structural formation of charged particles and electromagnetic fields, provides the key to understanding how the present universe is formed and our observations of real phenomena from far distances in space and time.

In laboratory experiments, high temperature plasmas are confined by magnetic fields in toroidal devices (plate 1.4). Confinement research is aimed at realizing thermonuclear fusion, which tries to utilize, in a controlled manner, the energy-generating process in the sun. In this field of endeavour, the understanding of transport and structural formation of plasmas is of critical importance. Detailed observations of the solar surface show that (semi-) toroidal plasmas are naturally formed (figure 1.1). Plates 1.1–1.4 and figure 1.1 illustrate a variety of plasma structures in the universe or the laboratory. The understanding of structural formation in hot plasmas is itself a challenging branch of modern physics, and is a key issue to understanding the phenomena occurring in the cosmos. The physics of toroidal plasmas is prototypical for plasmas under general circumstances.

The plasma, consisting of positively charged and negatively charged particles, has an intrinsic property: structural formation through interaction with the electromagnetic field. The mutual interaction between many particles is an essential aspect of plasmas. Collective effects must be taken into account.

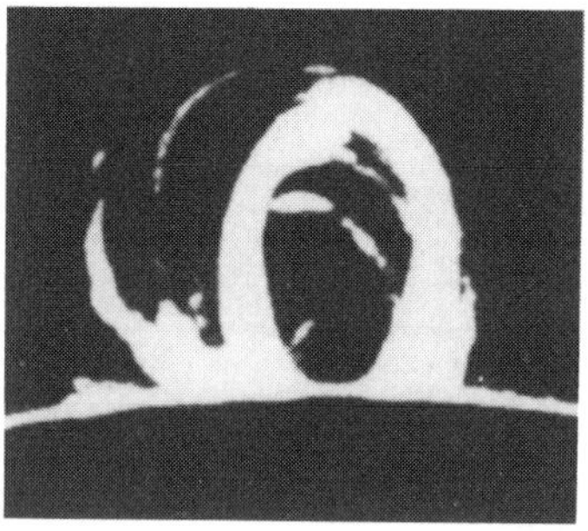

Figure 1.1. (Semi-) toroidal plasma is also observed in nature, the sun. (Courtesy of Dr M Keilhacker [1.5].)

One key concept is that of screening. If the charged particles are located individually as well as independently, then the mutual interaction can be tremendously large and such a state would require a very large free energy. It is well known that the charged particles (e.g., with charges e and e_1 at a mutual distance r_1) interact with the Coulomb potential. The potential energy has the dependence

$$e\phi = \frac{ee_1}{4\pi\varepsilon_0 r_1}. \tag{1.1}$$

The number of interacting particles at distance $r_1 < r < r_1 + \Delta$ for a certain solid angle increases, on average, as r_1^2. Therefore, the potential energy which is influenced by the particles in this region,

$$e\phi = e\left(\sum_{r_1<r<r_1+\Delta} \frac{e_1}{4\pi\varepsilon_0 r}\right) \tag{1.2}$$

could increase in the far distance, if the particles are distributed arbitrarily. The total interaction potential can be tremendously large. In reality, this is not the case. This can be understood from the fact that the presence of a test particle modifies the possible distribution of other particles. If ions are located at the origin of figure 1.2, then the electrons are attracted, while other ions are repelled. As a result of this screening, called Debye screening, the influence of this test charge on the potential can be written in the form of the Yukawa potential [1.6]

$$\phi_{eff}(r) = \frac{e}{4\pi\varepsilon_0 r}\exp\left(-\frac{r}{\lambda_D}\right) \tag{1.3}$$

and the cutoff distance (called the Debye length) is given by

$$\lambda_D = \frac{v_{the}}{\omega_p}$$

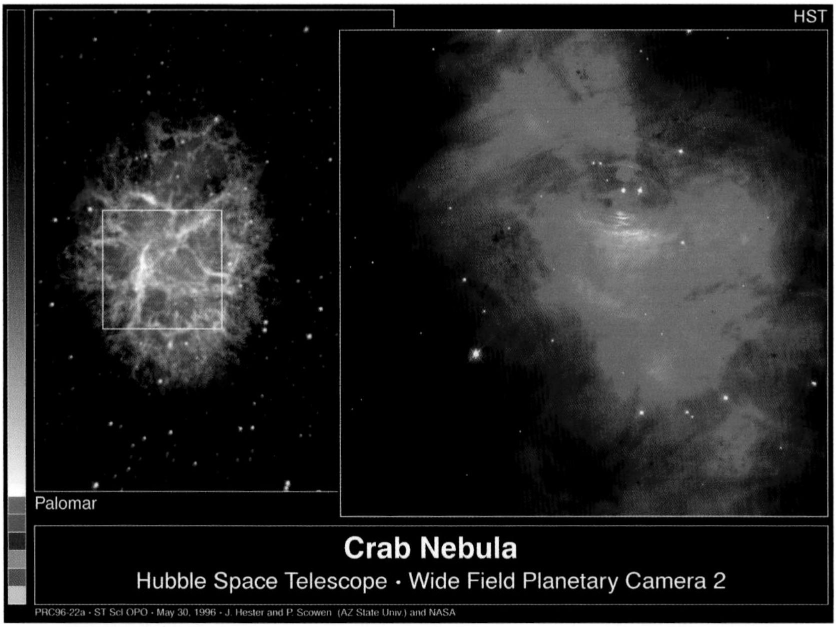

Plate 1.1. Plasmas in the Crab Nebula, recorded by Teika Fujiwara, observed by the Hubble Space Telescope. (Courtesy of AURA/STScI [1.1].)

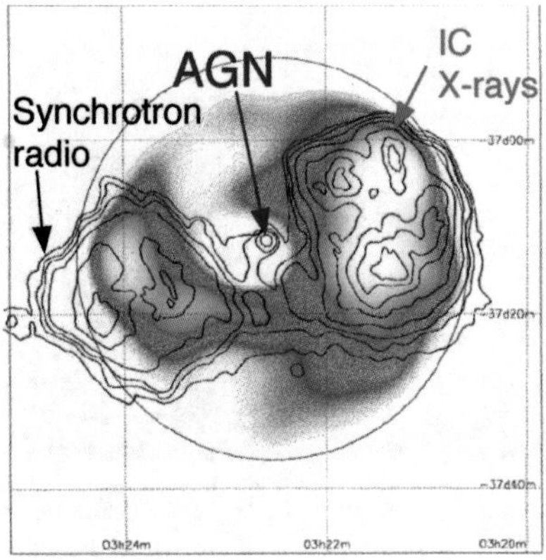

Plate 1.2. Plasmas surrounding an active galactic nucleus. (Courtesy of Dr H Kaneda [1.2].)

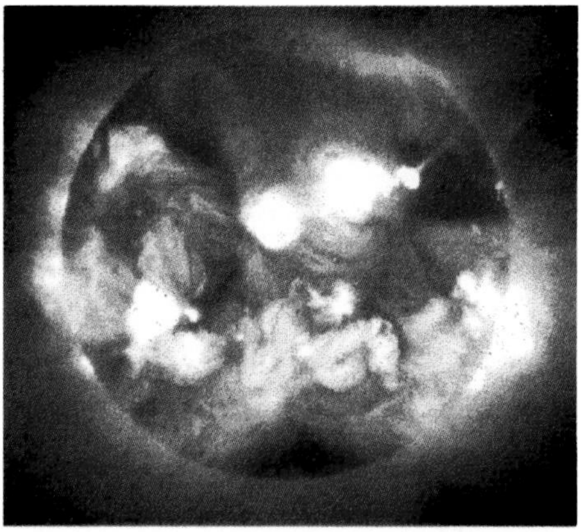

Plate 1.3. High temperature plasmas surrounding the sun. (Courtesy of Dr S Tsuneta [1.3]; copyright The Japan Society of Plasma Science and Nuclear Fusion Research.)

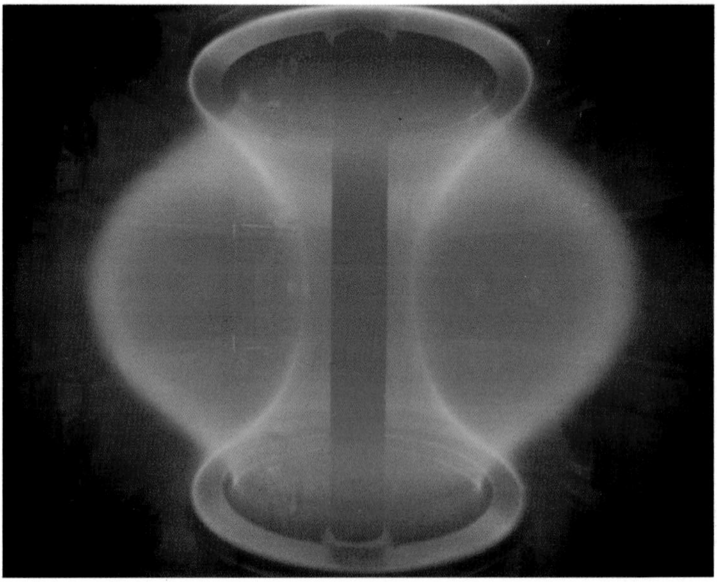

Plate 1.4. Plasmas in laboratory experiments. A toroidal (doughnut-shaped) plasma is confined by the magnetic field. (Central rod and top and bottom rings are current-carrying conductors.) (Courtesy of Dr A Sykes [1.4].)

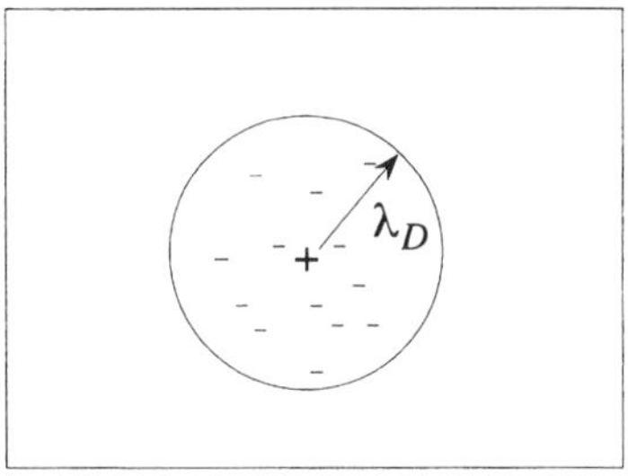

Figure 1.2. Debye screening.

where ω_p is the plasma frequency, which is defined by $\omega_p^2 = n_e e^2/\varepsilon_0 m_e$, v_{the} is the thermal velocity of electrons, n_e is the electron density and m_e is the electron mass. The test particle interacts with the particles mainly within this cutoff distance. The number of interacting particles, $n_e\lambda_D^3$, is a key parameter. If the inverse of this parameter

$$\frac{1}{n_e\lambda_D^3} \tag{1.4}$$

which is often called the *plasma parameter*, is small, the test particle interacts with many particles and the total interaction would be averaged out. When this parameter becomes large, the interaction depends strongly on the mutual configuration of the interacting pair. The plasma with $n_e^{-1}\lambda_D^{-3} > 1$ is called strongly coupled. In such plasmas, processes such as crystallization can take place. At large values of the parameter $n_e^{-1}\lambda_D^{-3}$, varieties of plasma activity and states may appear.

The other key aspect of plasmas is the inhomogeneity and the presence of collective (oscillating) modes. Clusters of charged particles respond to the long range electromagnetic field, and their response generates a contribution to the electromagnetic field. This mutual interaction of charged particles and the large scale electromagnetic field causes the dynamical nature of plasmas. As illustrated in plates 1.1–1.4 and figure 1.1, plasmas exist in inhomogeneous states. Gradients of density, temperature, flow velocity and electric current can exist. Associated with the inhomogeneity, flows appear. These flows are usually accompanied by fluctuating structures of various length scales. The fluctuations occur either in the parameters of the plasma or in the electromagnetic fields. Figure 1.3 illustrates two directions that characterize the structural formation in plasmas. The laws of physics that govern the global structural formation in toroidal plasmas are the subject of this monograph.

The collective interactions, which have much longer spatial scale-length than the range of the potential of the individual particle λ_D, are also the essence of the free energy that sustains the plasma dynamics. An analogy is seen in

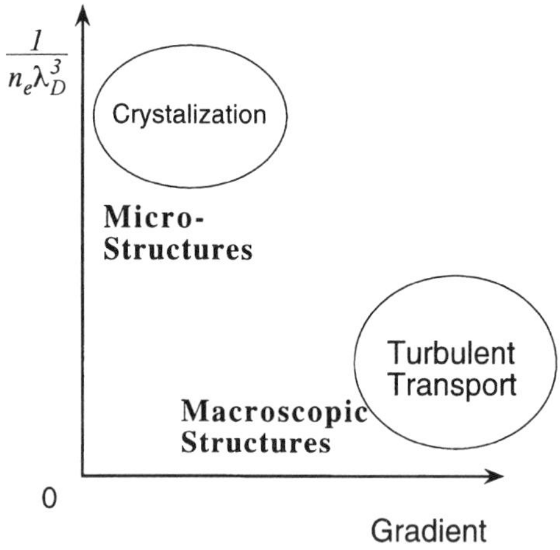

Figure 1.3. Two directions in the structural formation.

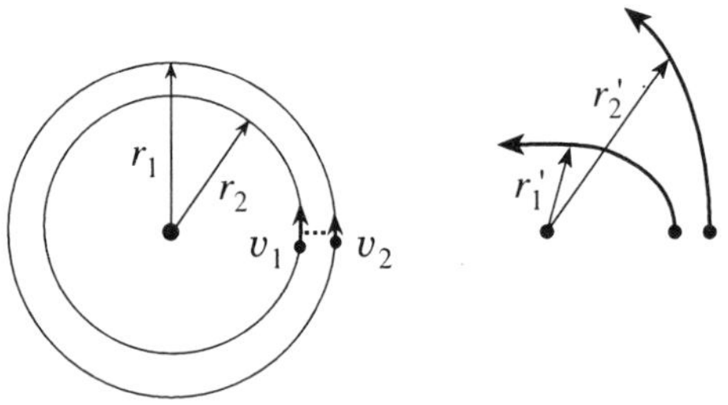

Figure 1.4. Keplerian motion of two elements around the strong gravity centre. The dotted line indicates the interaction that equilibrates the velocities. The inner element falls into the centre, releasing the potential energy.

motion under the influence of gravitational forces. For example, consider two elements which are rotating around a strong gravitational centre as shown in figure 1.4. The two elements move in Keplerian orbits, if they are mutually independent. The angular momentum is conserved for each element, and they stay on stable orbits. The Keplerian motion is characterized by the differential rotation,

$$\Omega_{rot} \propto r^{-3/2}.$$

If there is an interaction between two elements, such as viscosity, their angular velocities tend to be equalized. By this interaction, the angular velocity of the inner element v_1/r_1 is reduced and the one for the outer element v_2/r_2 is increased. Body 1 is subject to a reduced centrifugal force and starts to fall towards the centre. On the other hand, body 2 moves from the centre. One element gains potential energy and the other one loses. The net potential energy is reduced, and free energy is released by the viscous interactions. The released energy could be an origin of further dynamical evolution.

Similar phenomena are found, albeit in a much more complicated manner, in inhomogeneous plasmas. The breakdown of symmetry (and the associated conservation law) within a limited element leads to a much stronger release of free energy. The released energy increases the level of fluctuations, i.e., the strength of violation of the symmetry, and accelerates the evolution of plasmas towards a new state, with a different structure. The global structural formation, the released energy and the fluctuations are self-sustaining in plasmas. For this reason, the study of confined plasmas will have an impact on the present and future research in physics. This monograph intends to show the characteristic features of turbulent transport and structural formation in confined plasmas. The physics of the global structure formation, the released energy and the fluctuations is introduced. It is intended to show that the structural formation in hot plasmas is an attractive and challenging branch of modern physics, and that it is the key to understanding the phenomena in the cosmos as well as in the early universe.

The construction of this monograph is as follows. This monograph has four branches.

In chapter 2, characteristic features of experimental observations, which identify a state of the confined plasma, are illustrated. This represents the subject of research on far nonequilibrium systems.

In chapters 3–6, elementary processes that are related to the confinement, fluctuation-driven transport and instabilities are surveyed. These problems have been well documented in literature, and some important elements are stated.

In chapters 7–15, an example of analysis of turbulence and turbulent transport is presented. In order to analyse the turbulence in inhomogeneous and nonequilibrium plasmas, we introduce the method of the dressed test mode. Along this line of approach, the characteristic properties of fluctuations (i.e., fluctuation level, correlation time and length) and turbulent transport coefficients are discussed. In this course of considerations, it is shown that the gradients of plasma profiles are the order parameter to govern the turbulence and turbulent transport and that fluctuations could be subject to subcritical excitations. In addition to an analytic estimate, a direct nonlinear simulation and discussion based on a scale-invariant property are developed.

In chapters 16–22, structural formation and transition in confined plasmas are explained. The turbulence and plasma structure are shown to be controlling each other. In addition, the fields (electric fields and magnetic fields), which have long range interactions through excitation of the collective modes in plasmas, play a role in the structural formation. Owing to these mutual couplings, various characteristic natures of structural formation in plasmas appear. Emphasis is placed on the role of the radial electric field that causes a bifurcation and structural transitions in confined plasmas.

A summary is given in chapter 23. An annex, chapter 24, is attached at the end. In the annex, transport simulation is briefly described, in which transport properties in inhomogeneous plasmas are taken into account. Structural formation in experimental circumstances are shown and experimental observations are briefly commented on.

REFERENCES

[1.1] This figure was created with support from the Space Telescope Science Institute from NASA, and is reproduced with permission from AURA/STScI (STScI-PRC96-22a).

[1.2] Modified from Kaneda H *et al* 1995 *Astrophys. J. Lett.* **453** L13, courtesy of Dr Hidehiro Kaneda.

[1.3] For details of experimental setup, see, e.g., Tsuneta S *et al* 1991 *Solar Phys.* **136** 37

[1.4] For details of experimental setup, see, e.g., Sykes A *et al* 1997 *Phys. Plasmas* **4** 1665

[1.5] Keilhacker M, Keen B E and Watkins M L 1993 *JET Report* JET-P(93)56

[1.6] Spitzer L Jr 1958 *Physics of Fully Ionized Gases* (New York: Interscience). The cut-off distance λ_D is called the Debye length after Debye who has investigated this type of screening for ionic solutions.

Chapter 2

Transport Phenomena in Toroidal Plasmas

We first overview the experimental observations of confined plasmas. The confinement of plasmas has been pursued motivated by the research for the nuclear fusion. The majority of the data has been arranged so as to illustrate the possibility of realizing nuclear fusion. However, the accumulation of data has also revealed the enchanting feature of inhomogeneous plasmas, in the light of the general physics interest in nonlinear and nonequilibrium systems. Basic plasma properties and concepts have been described in textbooks [2.1–2.13]. In order to clarify the challenging problem of this subject and to show the focus of this monograph, we first try to extract the basic and generic feature of plasmas under the condition of magnetic confinement. A few characteristic features are summarized in this chapter, for which theoretical analysis is devoted in the following chapters.

2.1 Confinement and Plasma Profile

Plasmas for which the transport and structural formation are discussed in this monograph, in general, are confined in the nested *magnetic surfaces*. Charged particles are subject to the gyro-motion in the perpendicular direction to the strong magnetic field. The motion along the magnetic field lines is less affected by the magnetic field, and charged particles move almost freely along the magnetic field line. Therefore, magnetic field lines are arranged so as to form a toroidal surface, in order to confine energetic charged particles within a laboratory system size (figure 2.1). The gyro-radius is given as

$$\rho_j = \left| \frac{v_j}{e_j B} \right| \tag{2.1}$$

where v_j and e_j are the velocity and charge of particles ($j = e, i$), respectively, and B is the magnitude of the magnetic field. The condition

$$\rho_j \ll a \tag{2.2}$$

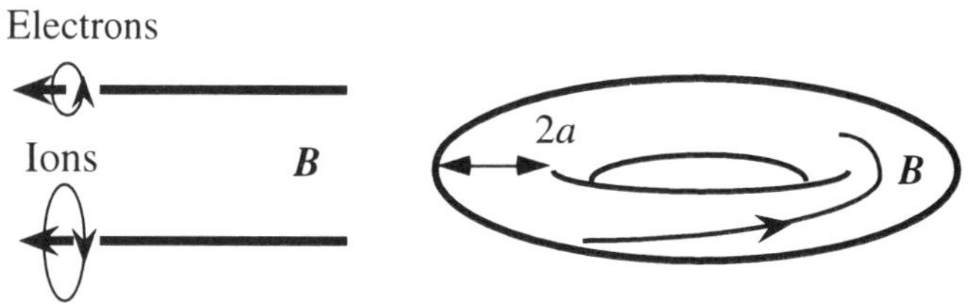

Figure 2.1. Gyro-motion of charged particle across the magnetic field (a) and toroidal magnetic surface (b).

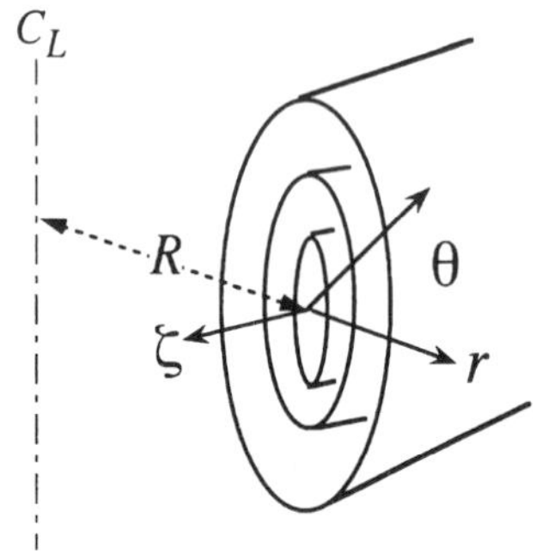

Figure 2.2. Nested magnetic surfaces; coordinates (r, θ, ζ) indicate the minor radius, poloidal angle and toroidal angle, respectively.

is prerequisite to the confinement, where a is the minor radius of the plasma column.

When toroidal surfaces are nested, as illustrated in figure 2.2, trajectories of particles, moving along the field line, on one inner surface are not mixed up with those on the outer surfaces. In the zeroth order approximation, plasmas on one magnetic surface are separated from those on the other surfaces. High temperature plasmas may be confined in the central region, and are separated from the colder plasmas near the edge, so that the hot parts can be separated from the wall materials which are usually close to the room temperature. The method to construct such magnetic field structure is explained in text books [2.4, 2.7].

The state of the plasma is characterized by the number density, n_j, velocity $\boldsymbol{V}_j$ and temperature T_j, where the suffix j stands for the species (e.g., electrons, various kinds of ion). Owing to the insulation capability of the nested magnetic surfaces, plasma parameters (such as temperature or pressure) gradually decrease from the centre to the edge (figure 2.3).

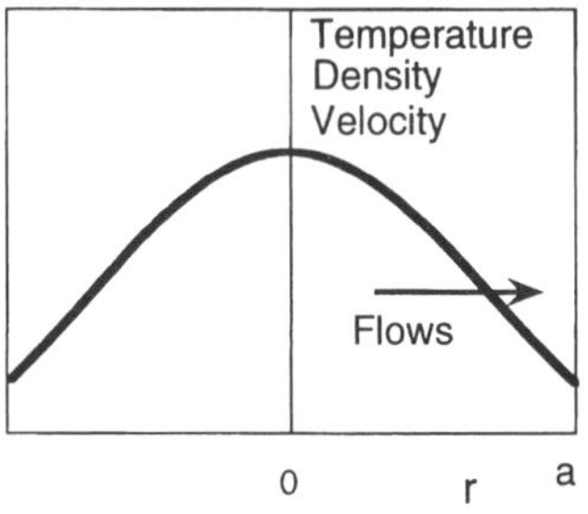

Figure 2.3. Schematic drawing of the profiles of plasma parameters.

One of the main problems of plasma confinement is to find a rule that governs the profile of inhomogeneous plasmas. Since the plasma temperature is much higher than surrounding materials, high temperature (and the ionized state as well) can be realized and sustained by the injection of energy from an external supply, which one calls 'heating'. The energy flows from the hotter region to the colder one, and the temperature difference between the centre and the edge in figure 2.3 is realized associated with this heat flow. The heat flows from the centre to the edge within a finite time.

This time, the energy confinement time τ_E, has been thought to be of great importance for fusion research. Controlled fusion will be realized if the energy confinement time τ_E is long enough at high temperature which is relevant for thermonuclear reaction [2.1]. The energy confinement time is usually defined via the energy balance equation as

$$\frac{\mathrm{d}}{\mathrm{d}t} W_p = P_{in} - \frac{W_p}{\tau_E} \tag{2.3}$$

where W_p is the total plasma energy (i.e., volume integral of plasma internal energy) and P_{in} is the power supply to the plasma. High temperature plasmas have been generated by use of the intense heating. Experiments have revealed that the energy confinement time turns out to be strongly dependent on the plasma temperature. Moreover, contrary to the prediction from collisional transport, the nature of the dependence varies as the plasma temperature increases, as schematically reproduced in figure 2.4. The reduction of the confinement time which appears according to the increment of the temperature has been often referred to as 'degradation of confinement time', because this prohibits the rise of the temperature in proportion to the heating power. This feature has been the main obstacle in fusion research, but, at the same time, shows a characteristic feature of the plasma of a nonlinear–nonequilibrium medium. The confinement time is a scalar variable, which represents the integral of total plasma energy. The dependence of τ_E on various parameters will be understood if the mechanisms that govern the plasma profile are clarified.

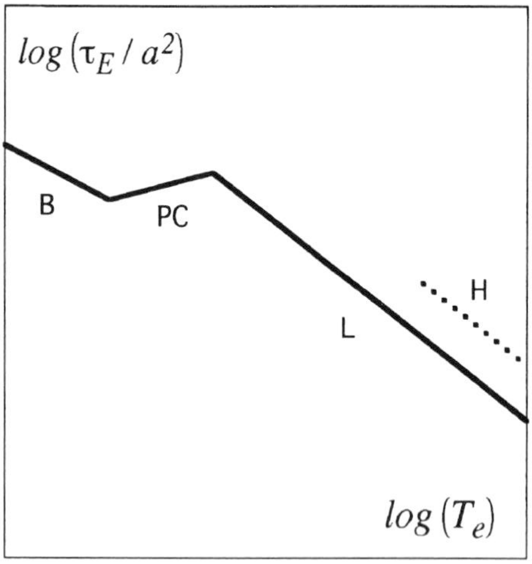

Figure 2.4. Temperature dependence of the energy confinement time. Symbols B (Bohm confinement), PC (pseudo-classical), L (L-mode) and H (H-mode) denote the historical evolution of the confinement time and temperature.

2.2 Picture of Transport Phenomena

The quantity τ_E represents only a part of the nature of the confined plasma. Much fruitful knowledge has emerged from the experiments of confined plasmas. The observations on confined plasmas can be summarized as a few generic features of inhomogeneous plasmas. Four aspects of the transport nature are shown here.

2.2.1 Gradient–Flux Relation

The relationship between the flows and gradients of parameters is of fundamental importance. In the first approximation, it is assumed that plasma parameters are constant on each magnetic surface. This approximation is useful for the first step of an analysis, because an equilibration on the magnetic surface may be realized in a very short time as a result of the fast motion of particles along magnetic field line. This equilibration time is much shorter than τ_E. Thus the flow across the magnetic surface is a quantity which is subject to the analysis. The temperature gradient and the energy flux are chosen as typical examples. The ratio between the energy flux per particle, q_r/n, and the temperature gradient, $-\nabla T$, is called thermal conductivity χ. One may write

$$q_r = -n\chi\nabla T \tag{2.4}$$

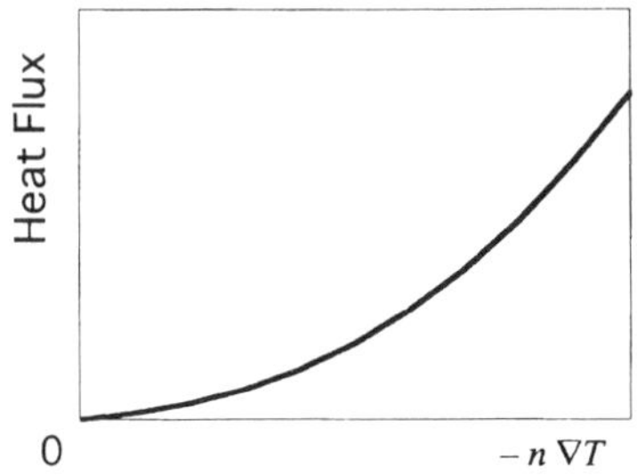

Figure 2.5. Gradient–flux relation for the energy flow across the magnetic surface.

(the suffix r indicates that the flow is directed in the radial direction, and the coordinates (r, θ, ζ) are shown in figure 2.2). The transport equation of the plasma energy,

$$n\frac{\partial}{\partial t}T = \nabla \cdot n\chi\nabla T + P \tag{2.5}$$

where P is the density of power deposition, determines the energy confinement time. The confinement time τ_E in equation (2.3) has a (dimensional) relation

$$\frac{1}{\tau_E} = \frac{\langle\chi\rangle}{a^2} \tag{2.6}$$

where the bracket $\langle\,\rangle$ indicates an average over plasma volume.

The relation (2.4) is called the Fourier relation of heat transport, if χ is constant. The relation between q_r/n and $-\nabla T$ has been intensively studied experimentally. Figure 2.5 shows a typical relation which was observed in experiments [2.14]. It is clearly seen that the flux is not a linear function of the gradient; it nonlinearly increases as the gradient becomes stronger.

This result has two important implications. First, this relation identifies the fundamental difficulty in the fusion research, because the increase of the temperature difference requires a larger heat flow as the temperature gradient becomes higher. Increment of χ is equivalent to the decrease of τ_E. The dependence in figure 2.4 reflects the more complex physical nature of the transport process in confined plasmas.

Second, which is of physics interest, the confined plasma is really the target medium of the nonlinear–nonequilibrium physics. Confined plasmas belong to the family of nonequilibrium systems. The high temperature plasmas are open and sustained only if they are associated with the energy flow. When χ is independent of the gradient, the linear nonequilibrium thermodynamics may apply. For such systems, the theory of the linear response has been established [2.15, 2.16]. Studies on the linear response of the plasma have long been developed (see, e.g. [2.1–2.12]). The transport properties, caused by the thermal fluctuations, could be calculated by using the linear susceptibility [2.3]. In contrast, the experimental study, which is shown in figure 2.5, clearly demonstrates that the plasma is a medium which belongs to nonlinear nonequilibrium thermodynamics. The transports are driven by the enhanced fluctuations (not by the thermal fluctuations) associated with the release of the free energy of plasma inhomogeneity.

Figure 2.5 stimulates us to study the important role of *inhomogeneity* in the non-equilibrium transport. In addition to inhomogeneity, a temporal change is another important element that defines the nonequilibrium. Maxwell has suggested a form such as

$$\tau \dot{q}_r + q_r = -n\chi \nabla T \tag{2.7}$$

in order to study the effect of nonstationarity (τ being some constant). There is a certain amount of evidence of nonstationarity in confined plasmas [2.17]: if the observation is fitted to the model formula of equation (2.7), the constant τ may not be positive definite.

2.2.2 Interference of Fluxes

The gradient–flux relation in the plasma is more complicated than a simple expression such as equation (2.4). Fluxes and gradients are subject to mutual interferences. Figure 2.6 illustrates typical examples of the plasma density, velocity and temperature under various heating conditions [2.18]. The profile that is shown by the dashed line is realized by supplying the particle, momentum and energy to the central region, by injecting energetic particles in the central region. When an rf wave is additionally supplied, the profile changes as shown by the solid lines. In this process, the supply of the energy alone is added to the central plasma. (The supply of particles is not changed much.) The result shows, however, not only the increase of the temperature but also the increments of the particle density and velocity are realized. This suggests that there are interferences between the energy flow and flows of particles and momentum as well. Figure 2.7 illustrates the flow of the toroidal momentum in the radial direction as a function of the gradient of the velocity. The gradient of the velocity drives the flow of the momentum: however, the flow of the momentum remains finite even in the absence of the gradient of the flow velocity.

The interference is described by the presence of the off-diagonal element in the transport matrix $\mathbf{M}$ which may be defined by

$$\begin{pmatrix} \Gamma_r \\ P_{\theta r} \\ P_{\zeta r} \\ q_r \end{pmatrix} = -\mathbf{M} \begin{pmatrix} \nabla n \\ \nabla V_\theta \\ \nabla V_\zeta \\ \nabla T \end{pmatrix}. \tag{2.8}$$

This is also the characteristic nature of confined plasmas. In classical thermodynamics, the interference of fluxes is known (e.g., Joule–Thomson effect,

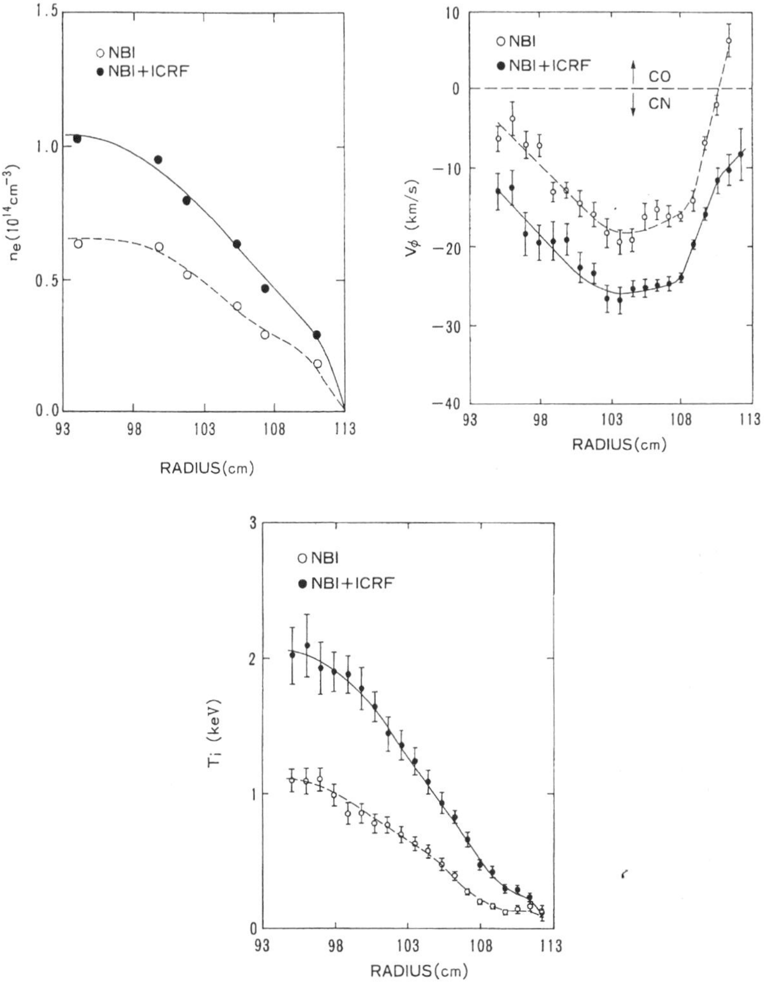

Figure 2.6. Profiles of the density, velocity and temperature in toroidal plasmas. Dashed lines show the case where the plasma is sustained by the injection of the energetic particles. In the case of the solid lines, an additional energy is supplied by rf waves. (From [2.18].)

thermo-electric effect etc [2.20]). What is important for such a classical system is the *Curie principle*, i.e., the flow of a scalar quantity does not interfere with that of a vector quantity. This is the consequence of the symmetry imposed on the systems. In contrast, the interference between the fluxes of momentum and energy appears in confined plasmas. The interference suggests that the

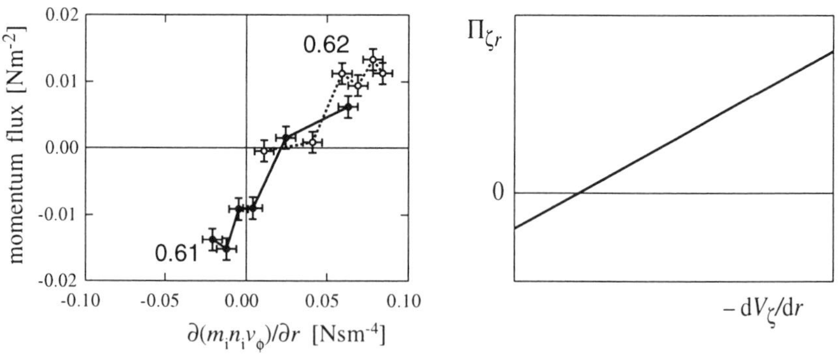

Figure 2.7. Flow of the toroidal momentum in radial direction $\Pi_{\zeta r}$ as a function of the gradient of the toroidal flow velocity $\mathrm{d}V_\zeta/\mathrm{d}r$. The observation on the JFT-2M tokamak (a) and a schematic illustration (b). (From [2.19].)

symmetry is violated although the global plasma profile, on average, is subject to symmetry. In reality, the symmetry is violated when the microscopic structure is taken into considerations. The understanding of the interference is another key for the plasma transport.

The interference plays an important role for structural formations of plasmas in wider circumstances. For instance, the flow of the angular momentum is of vital importance of stellar plasmas. The dissipation of angular momentum is essential for the release of gravitational energy in an accretion disc, which is the origin of the variety in structures.

2.2.3 Bifurcation and Transition

The gradient–flux relation has more variety in the plasma. The flux can be a decreasing function of the gradient, or be even multi-valued. The bifurcation, i.e., a substantial change of the flux caused by an infinitesimal small variation of gradient (or vice versa), has been observed. Such a change appears abruptly, i.e., the transition occurs.

Figure 2.8 illustrates the multiple plasma profiles realized in tokamaks. For a fixed power supply to the plasma, the plasma gradient near the edge is either weak or very sharp [2.21]. In the change from a weak gradient profile (dashed line) to a sharp gradient one (solid line), the heat flux across the relevant magnetic surface instantaneously drops. The drop happens within a very short time, and is considered as a transition (figure 2.9).

Historically, this state of the plasma with the sharp gradient has been called the 'H-mode'. (Correspondingly, the state with weak gradient was named the 'L-mode'. The word 'mode' indicates not an oscillation patterns but a global

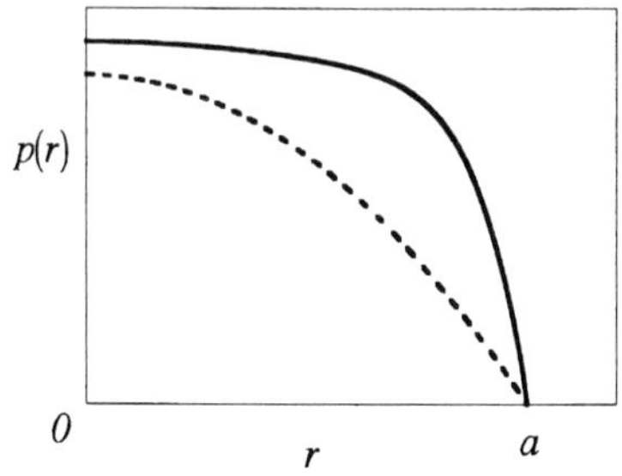

Figure 2.8. Multiple plasma profiles realized for the same external supply.

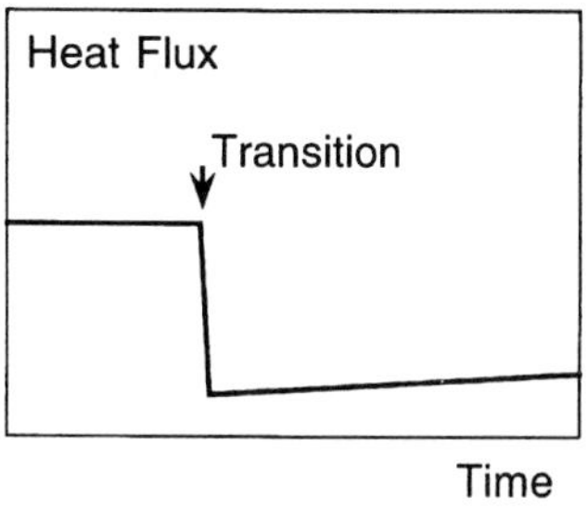

Figure 2.9. Sudden change of the heat flux across the outermost surface of the plasma when the transition from the L-mode structure to the H-mode structure happens.

trend of a discharge (such as 'Paris-mode', etc).) Various profiles have been identified in confined plasmas [2.22]. More than one mechanism that can cause bifurcations are considered to play roles in plasmas.

The existence of the multiple states and the rapid change of the flux suggest a nonlinear and nonmonotonic relation between the gradient and flux, as illustrated in figure 2.10. The study of such a relation is the fundamental issue to understand the confined plasmas.

Bifurcation and transition are unambiguously observed in evolution of electrostatic potential. Plasmas are composed of electrons and ions being associated with electromagnetic field. Small but finite difference of the electron flow and ion flow, electromagnetic fields evolve within a plasma. Plasmas are often charged positively or negatively, and the profile of static potential varies across a plasma radius. The polarity of the static potential shows variety depending on plasma parameters. It is found that the potential profile is subject to a very rapid change. Figure 2.11 shows an example of a radial profile of electrostatic potential, which was obtained in the CHS (Compact Helical

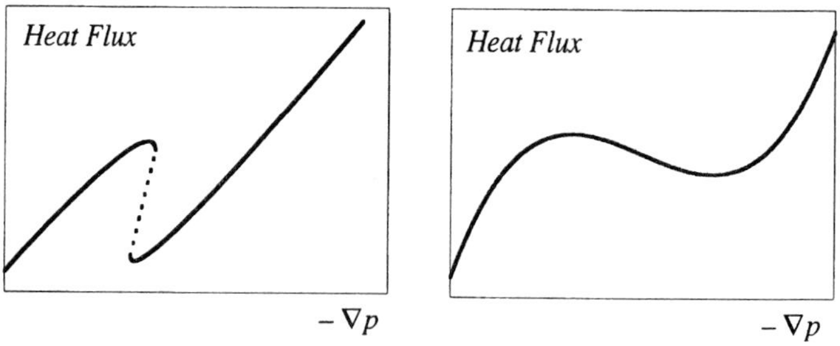

Figure 2.10. Nonmonotonic relation between the gradient and flux. The hard bifurcation type (left) and soft bifurcation type (right) are shown.

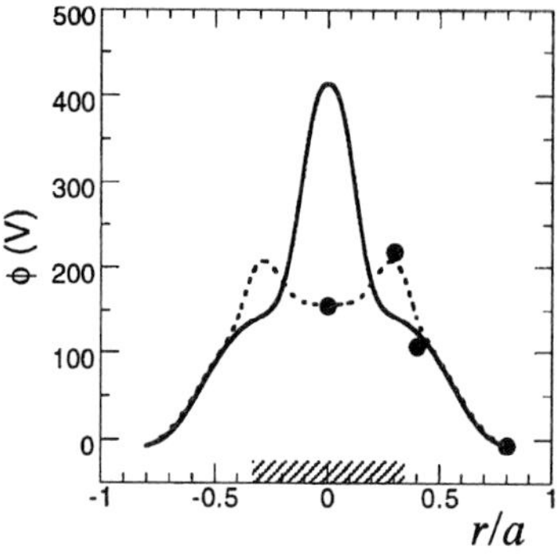

Figure 2.11. Radial profile of electrostatic potential. The peaked profile (solid line) is realized. A rapid transition to a hollow profile (dashed line) happens. Thick dots show measured data just after the bifurcation occurs. (Based on an observation on the CHS device [2.23].)

System) device [2.23]. This observation is made on a plasma in which electrons are heated by an external supply of power. As is schematically illustrated in figure 2.3, electron temperature is high in the centre. Plasma is positively charged, and an electrostatic potential is found high in the centre. A peaked profile is seen. This profile is subject to a sudden bifurcation, and the second type of profile, which is shown by a dotted line in figure 2.11, is also realized under the same circumstances of external supply of energy. It is found that the potential takes two profiles, and bifurcation between them takes place.

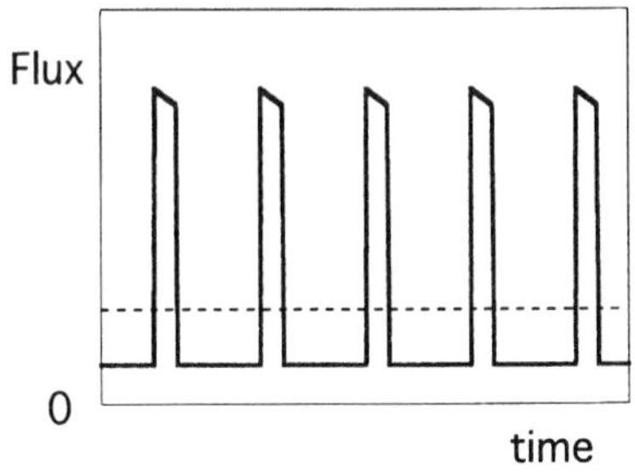

Figure 2.12. Burst in the plasma outflux across the outermost magnetic surface. The dotted line indicates the constant supply from the inside.

2.2.4 Bursts, Periodicity and Intermittence

The nonlinear and bifurcation characteristics give rise to various dynamics of flows in confined plasmas. Bursts of the flow, or self-generated oscillations, are widely observed.

Figure 2.12 illustrates the periodic bursts of energy expelled from the confined plasma across the surface. The external supply of the energy in the central region is kept constant in time, but the energy outflow could be in the form of successive bursts. The time average of the bursts of flux is of course equal to that of the constant external supply. There exists a mechanism to cause a self-regulating oscillatory temporal change of transport coefficients.

The sequence of such bursts is often periodic, but is also more or less irregular. In the plasma-parameter space, the region without bursts, that with periodic bursts and that with irregular ones have not been identified yet. However, it is empirically known that bursts could appear sensitively to the parameters and to the operation conditions; this also indicates that the bifurcation mechanisms, if they exist in the plasma, could be hard bifurcations.

The abruptness of the occurrence of each burst is also noticeable. When a burst like that in figure 2.12 appears, there arises a strong (high level) fluctuation associated with it. When the time-expanded view at the occurrence of the burst is shown, the level of fluctuations also jumps in a very short time, as is schematically drawn in figure 2.13. This phenomenon is called a *trigger* phenomenon [2.24].

2.2.5 Other Complex Processes

Before closing this chapter, we point out some other processes that are influential in the structural formation of the plasma.

The mixing of ion species is important. The ion species of experimental plasmas are to be strictly controlled. Particularly in fusion-oriented plasmas, the

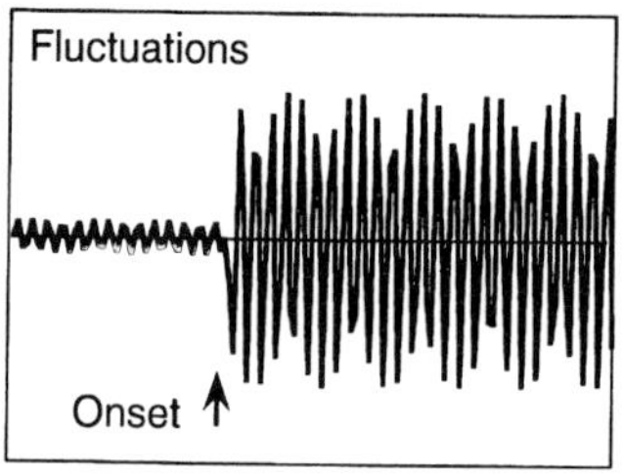

Figure 2.13. Schematic drawing of the change of fluctuations at the onset of the burst.

fuel ions (an isotope of hydrogen) are confined for the fusion reaction, and others (such as the elements that construct the wall) are considered as 'impurities'. The mixing of ion species, nevertheless, occurs in reality, and is also a source of free energy from the thermodynamical point of view. The mixing ratio is dictated by a chemical potential in the thermodynamics near the equilibrium. In addition to this, when atoms of high charge numbers move into the plasma, they absorb the plasma energy through their ionization. As is the role of the latent heat in the dynamics of cloud formation, this exchange of energy could provide an additional process of structural formation.

Another example is the role of neutral particles. Neutral particles are not restricted by the magnetic field, and interact with plasma particles through atomic processes such as the charge exchange or ionization processes. These processes give rise to a long range (remote) transfer of momentum or energy, because the mean free path of neutral particles in the plasma can be long. It can often be longer than the scale-length associated with the sharp gradient near the edge (in figure 2.8). Therefore they play a role in the mechanisms to dictate the structural formation in plasmas.

The majority of this monograph discusses the dynamics in the system of electrons and main ions. These additional processes are also briefly described.

REFERENCES

There are many textbooks that cover a related area of this monograph. Some of them are listed as examples. Certain basic elements of the turbulent transport and structural formation have been discussed in the following.

[2.1] Yoshikawa S and Iiyoshi A 1972 *Introduction to Controlled Thermonuclear Reaction* (Tokyo: Kyoritu) (in Japanese)

[2.2] Kadomtsev B B 1992 *Tokamak Plasma: a Complex Physical System* (Bristol: Institute of Physics)

Kadomtsev B B 1965 *Plasma Turbulence* (New York: Academic)

The basic property of plasma media is explained in the following.

[2.3] Ichimaru S 1973 *Basic Principles of Plasma Physics* (Reading, MA: Benjamin)

[2.4] Miyamoto K 1976 *Plasma Physics for Nuclear Fusion* (Cambridge, MA: MIT Press)

[2.5] Krall N A and Trivelpiece A W 1973 *Principles of Plasma Physics* (New York: McGraw-Hill)

[2.6] Nishikawa K and Fukuyama A 1998 *Plasma Kinetic Theory* (Tokyo: Baihukan) (in Japanese) to be published

Basic concepts of the toroidal plasmas are given in the following.

[2.7] Wesson J 1987 *Tokamaks* (Oxford: Oxford University Press)

[2.8] Hazeltine R D and Meiss J D 1992 *Plasma Confinement* (Addison-Wesley, Redwood City)

[2.9] Goldston R J and Rutherford P H 1995 *Introduction to Plasma Physics* (Bristol: Institute of Physics)

Varieties of plasma instabilities are explained in [2.10]–[2.13].

[2.10] Mikailowski A B 1992 *Electromagnetic Instabilities in an Inhomogeneous Plasma* (Bristol: Institute of Physics)
Mikailowski A B 1974 *Theory of Plasma Instabilities* transl. J B Barbour (New York: Consultants Bureau)

[2.11] White R 1989 *Theory of Tokamak Plasmas* (New York: North-Holland)

[2.12] Hasegawa A 1975 *Plasma Instabilities and Nonlinear Effects* (Berlin: Springer)

[2.13] Biskamp D 1993 *Nonlinear Magnetohydrodynamics* (Cambridge: Cambridge University Press)

[2.14] JET Team 1992 *Plasma Physics and Controlled Nuclear Fusion Research 1992* vol 1 (Vienna: IAEA) p 15

[2.15] Kubo R 1957 *J. Phys. Soc. Japan* **12** 570

[2.16] An alternative approach to the linear-nonequilibrium thermodynamics is given by Prigogine I 1961 *Introduction to Thermodynamics of Irreversible Processes* 2nd edn (New York: Interscience)

[2.17] Wagner F and Stroth U 1993 *Plasma Phys. Control. Fusion* **35** 1321

[2.18] Ida K, Kawahata K, Toi K *et al* 1991 *Nucl. Fusion* **31** 943

[2.19] Ida K, Miura Y, Matsuda T *et al* 1995 *Phys. Rev. Lett.* **74** 1990

[2.20] Landau L D and Lifshitz E M 1984 *Electrodynamics of Continuous Media* 2nd edn, transl. J B Sykes *et al* (Oxford: Pergamon). Section 27 describes thermogalvanomagnetic effects (e.g., Nernst effect, Leduc Righi effect, Ettingshausen effect).
Landau L D and Lifshitz E M 1980 *Statistical Physics, Part 1* 3rd edn, transl. J B Sykes *et al* (Oxford: Pergamon) section 18

[2.21] ASDEX Team 1989 *Nucl. Fusion* **29** 1959

[2.22] For a review, see, e.g., Itoh S-I, Itoh K and Fukuyama A 1995 *J. Nucl. Mater.* **220–222** 117

[2.23] Fujisawa A, Iguchi H, Sanuki H *et al* 1997 *Phys. Rev. Lett.* **79** 1054
Fujisawa A, Iguchi H, Idei H *et al* 1997 *Phys. Rev. Lett.* **81** 2256

[2.24] A review of trigger and collapse events is given in, e.g., Itoh S-I, Itoh K, Zushi H, Fukuyama A 1998 *Plasma Phys. Control. Fusion* **40** 879

Chapter 3

Mechanical Equilibrium and Absolute Trapping of Particles

When inhomogeneous plasmas are confined stationarily, plasma parameters (e.g., density, velocity etc) remain almost constant in time. Such a state is called a mechanical equilibrium (sometimes, it is abbreviated simply as 'equilibrium'). Force balance, on each small part of the plasma, is satisfied. Various symmetries often hold in a coarse-grained scale length. The property of mechanical equilibrium and motion of plasma particles in this idealized condition is surveyed first.

3.1 Mechanical Equilibrium

In studying the transport processes, we impose the condition that the force balance of the plasma is satisfied. In confined plasmas, the pressure is usually high in the centre and low at the edge. The pressure gradient may cause the expansion of the plasma, if the pressure is not balanced by a certain force. In a current-carrying medium, the Lorentz force $J \times B$ exists. The simplified force balance condition is given as

$$J \times B = \nabla p \tag{3.1}$$

provided that the plasma pressure is isotopic and that the viscous friction is small. The current density J and the pressure p are expressed as

$$J = e_i n_i V_i - e n_e V_e \tag{3.2}$$

and

$$p = \sum_j n_j T_j \tag{3.3}$$

respectively. If the plasma pressure p is inhomogeneous, the current J flows on the iso-pressure surface, on which the magnetic field lines lie, i.e., the magnetic

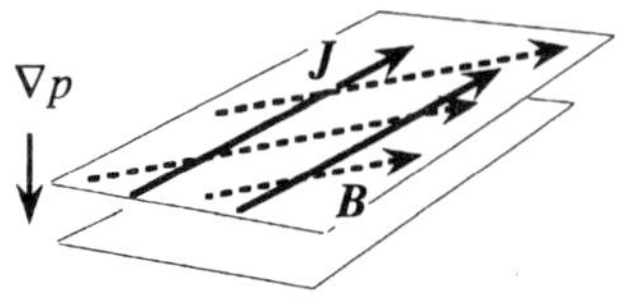

Figure 3.1. Magnetic surfaces and pressure gradient.

surface (figure 3.1). When the plasma structure is characterized in terms of six profiles, $\{n_j, V_j, T_j\}$ $(j = e, i)$, one constraint, equation (3.1), is required for the force balance.

The order of accuracy to which equation (3.1) is satisfied is very high. When the force balance of the form (3.1) does not hold, the plasma is subject to acceleration. Complicated plasma dynamics appear. Some typical examples are described in this monograph.

The other constraint, which is the electric property, is the charge neutrality. The (quasi-) charge neutrality condition holds as

$$n_e \simeq \sum_I Z_I n_I \tag{3.4}$$

where the suffix I indicates the ion species, and Z is the charge number. This approximate condition is valid when the scale-length of the gradient is longer than the Debye length. The accuracy of the approximate relation (3.4) is estimated from the Poisson equation for static plasmas

$$\varepsilon_0 \nabla \cdot \boldsymbol{E} = e\left(\sum_I Z_I n_I - n_e\right). \tag{3.5}$$

By the help of the order estimate $\nabla \cdot \boldsymbol{E} \sim -\Delta\phi/a^2$, equation (3.5) yields the relation

$$\frac{\left(\sum_I Z_I n_I - n_e\right)}{n_e} \sim \frac{\lambda_D^2}{a^2}\frac{e\Delta\phi}{T_e} \tag{3.6}$$

where $\Delta\phi$ is the potential difference across the distance a, and λ_D is the Debye length

$$\lambda_D = \frac{v_{the}}{\omega_p} \qquad \omega_p^2 = \frac{n_e e^2}{\varepsilon_0 m_e} \tag{3.7}$$

(ω_p being the electron plasma oscillation frequency).

Equation (3.6) shows that the charge neutrality approximately holds if $(\lambda_D^2/a^2)(e\Delta\phi/T_e)$ is small. The potential difference of the order $e\Delta\phi/T_e \sim$ O(1) is observed in experiments of the magnetic confinement. This indicates that the charge neutrality is satisfied, as an average, to the accuracy of the order $\lambda_D^2 a^{-2}$. This value may be in the range of 10^{-4}–10^{-7} for the magnetic confinement experiments, if one takes a to be the system size. Another property is that the potential difference can exist without causing large charge separation

$$\left|\frac{e\Delta\phi}{T_e}\right| \gg \frac{\left|\sum_I Z_I n_I - n_e\right|}{n_e}$$

so long as the length of interest is much longer than the Debye length. The charge neutrality condition (3.4) is often employed as a simple approximation.

It should be stressed, as is discussed later, that a potential difference of the order of $e\Delta\phi/T_e \sim \mathrm{O}(1)$ is enough to cause variety in structural formation processes. Care is required when one employs the charge neutrality condition for simplicity. (An extreme case is the non-neutral plasma, for which interesting physics has been developed [3.1]. However, this topic is beyond the scope of this monograph.)

3.2 Magnetic Surface

The force balance relation (3.1), which is satisfied to a high accuracy, indicates that the pressure is constant on the field line, and that the magnetic field line lies on the constant pressure surface. In order to sustain the pressure gradient in toroidal systems, the magnetic field lines constitute nested tori. The surface of nested tori is called a magnetic surface.

The existence of magnetic surfaces has long been studied intensively [2.8, 3.2]. The solution in general circumstances has not yet been completely obtained. However, the case in symmetric systems is easily understood. Assume that the magnetic field is static and has axial symmetry. Examples are the straight cylinder or symmetric torus as shown in figure 3.2. Cylindrical coordinates (R, φ, z) are used here. Owing to the axial symmetry,

$$\partial/\partial\varphi = 0 \tag{3.8}$$

the vector potential A_φ depends only on (R, z). We use the function

$$\Psi(R, z) \equiv RA_\varphi \tag{3.9}$$

which has a property of the stream function for the magnetic field, i.e.,

$$B_R = -\frac{1}{R}\frac{\partial}{\partial z}\Psi \tag{3.10-1}$$

$$B_z = \frac{1}{R}\frac{\partial}{\partial R}\Psi. \tag{3.10-2}$$

The relation (3.10) indicates that

$$\boldsymbol{B}\cdot\nabla\Psi = 0 \tag{3.11}$$

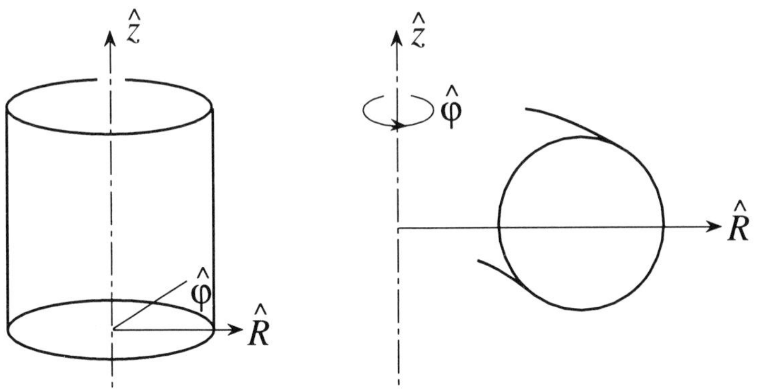

Figure 3.2. Axis-symmetric systems. In the cylinder and torus, axial symmetry, $\partial/\partial\varphi = 0$, could be satisfied.

i.e., the magnetic field line is on the constant-Ψ surface. Equation (3.1) gives

$$\boldsymbol{B} \cdot \boldsymbol{\nabla} p = 0. \tag{3.12}$$

Comparing equations (3.11) and (3.12), we see that the pressure is constant on the surface $\Psi =$ constant. Figure 3.3 shows an example of the nested magnetic surfaces.

Magnetic surfaces are slender at the inside of torus. At the centre, the toroidal surface is degenerated as a line. This is called the *magnetic axis*. On the magnetic surface, the condition $\partial\Psi/\partial R = \partial\Psi/\partial z = 0$ is satisfied, i.e., Ψ takes an optimum value.

This identity of constant-pressure surfaces and constant-Ψ surfaces clarifies the fundamental importance of the nature of magnetic surfaces. There are several quantities which characterize the topology of magnetic surfaces.

First is the pitch of the field line, which has been described by the *safety factor*

$$q = \frac{r}{R}\frac{B_t}{B_p}. \tag{3.13}$$

(The suffixes t and p indicate the toroidal (ζ) and poloidal (θ) directions, respectively. Toroidal coordinates (r, θ, ζ) are shown in figure 2.2.) If the safety factor q is a rational number, such a magnetic surface is called a 'rational surface'. The field line closes itself on the surface. If q is not a rational number, one field line covers the magnetic surface.

Another important concept is *magnetic shear*. The pitch of the field lines may have a difference between the neighbouring magnetic surfaces. The rate of

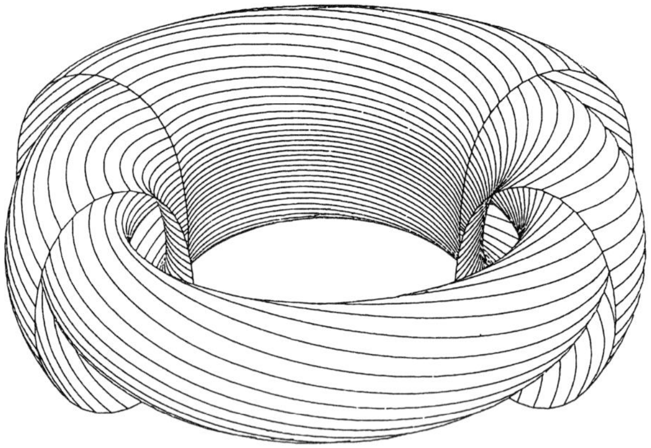

Figure 3.3. Calculation of nested magnetic surfaces [3.3]. Lines indicate the magnetic field lines.

change of the pitch is the magnetic shear, and

$$s = \frac{r}{q} \frac{\mathrm{d}q}{\mathrm{d}r} \tag{3.14}$$

is called a shear parameter. The example in figure 3.3 consists of sheared magnetic surfaces. In the case of figure 3.3, the safety factor is smaller inside and is larger outside.

A shift of magnetic surface is important as well. A poloidal cross-section is illustrated in figure 3.4. Cross-sections of nested magnetic surfaces are shown. The geometrical centre of one magnetic surface is not always identical to the magnetic axis, but is shifted by an amount Δ. This shift is called the *Shafranov shift* [3.4]. If the Shafranov shift increases ($\mathrm{d}\Delta/\mathrm{d}r < 0$), the poloidal magnetic field becomes stronger on the outside of the torus. The shift Δ has an influence on plasma instabilities.

3.3 Absolute Trapping of Particles

In a static magnetic field, it is possible to arrange the configuration for plasma particles to move on isolated surfaces, so that they are confined in space. This is called 'absolute trapping' of particles.

When one takes a plasma particle and follows its trajectory, the orbit could be integrable under the situations which are relevant for the existence of the magnetic surfaces. In a zeroth order argument, particles move almost freely along the static magnetic field lines. If the magnetic field line constitutes a magnetic surface, the trajectory of the particle may also form the surface.

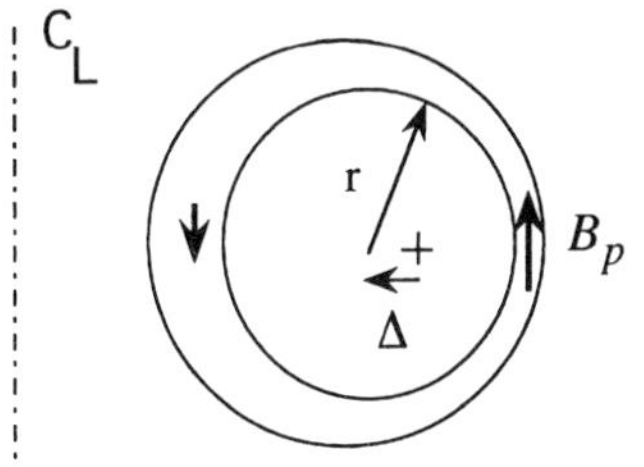

Figure 3.4. Poloidal cross-section of toroidal plasma. A magnetic surface (minor radius r) and the magnetic axis (denoted by the symbol $+$). The geometrical centre of the surface is shifted by an amount Δ.

This idea could be accurately quantified, and the characteristic of the orbits is clarified. Within an ordering (equation 2.2), the equation of motion is sometimes integrable. (If an integral of motion exists, it is invariant along the particle orbit. Some, e.g., the energy, are exact constants, and some, e.g., the magnetic moment, $v_\perp^2/B$, are approximately constant. The latter is called the *adiabatic invariance*, because its change is of the order of $\exp(-\ell/\rho)$, where ℓ is the scale-length of the variation of the magnetic field and ρ is the gyroradius [3.5].)

The Hamiltonian of the particle is given as

$$H = \frac{1}{2m}\left\{P_R^2 + \left(\frac{P_\varphi}{R} - eA_\varphi\right)^2 + (P_z - eA_z)^2\right\} \tag{3.15}$$

where $\boldsymbol{P}$ and $\boldsymbol{A}$ are the canonical angular momentum and the vector potential, respectively. In the presence of the axial symmetry,

$$\partial/\partial\varphi = 0$$

the canonical angular momentum

$$P_\varphi \equiv R(mv_\varphi + eA_\varphi) \tag{3.16}$$

is constant. The trajectory of particle is described by two constants, i.e., energy and canonical angular momentum,

$$H = W \text{ and } P_\varphi = P_0 \tag{3.17}$$

respectively. Substituting these constants in equation (3.15), and by using the magnetic flux function Ψ of equation (3.9), one has the relation

$$W = \frac{1}{2}m(v_R^2 + v_z^2) + \frac{1}{2mR^2}(P_0 - e\Psi)^2. \tag{3.18}$$

Owing to the axial symmetry, $\partial/\partial\varphi = 0$, the flux function Ψ depends only on (R, z). Equation (3.18) describes the particle orbit projected on the (R, z) plane.

From equation (3.18), it is possible to show that the particle orbit is restricted in a narrow region in space. Equation (3.18) provides a bound

$$\frac{e^2}{2mR^2}\left(\frac{P_0}{e} - \Psi\right)^2 < W. \tag{3.19}$$

The value of Ψ varies along the particle orbit, but is also bounded for equation (3.19) to be satisfied. Equation (3.19) gives the upper bound of the variation of Ψ along the orbit as

$$|\Delta\Psi| \leq \frac{\sqrt{2mW}R}{e} = \frac{mv}{e}R. \tag{3.20}$$

The deviation of Ψ along the orbit indicates that the particle orbit is slightly deviated from the magnetic surface, because Ψ is constant on the magnetic surface. Taylor expanding the flux function as

$$\Psi = \Psi(R_0, z_0) + \Delta_{orbit}[\nabla\Psi]_{R_0, z_0} + \cdots \tag{3.21}$$

where Δ_{orbit} is the deviation of the particle orbit relative to the magnetic surface, the shift of orbit Δ_{orbit} is estimated as

$$\Delta_{orbit} \simeq |\Delta\Psi|\left(\frac{\partial\Psi}{\partial r}\right)^{-1}. \tag{3.22}$$

From equations (3.10-2) and (3.22), we have an estimate

$$\Delta_{orbit} \sim \frac{mv}{eB_z}. \tag{3.23}$$

In the static magnetic field, the axisymmetry provides a basis that particles are trapped in a localized space.

In the case of a straight cylinder, figure 3.5, the main axial magnetic field is effective in restricting the particle orbit near the initial magnetic field line. The estimate equation (3.23) shows that the gyro-radius

$$\rho_j = \frac{m_j v_j}{e_j B} \tag{3.24}$$

is the scale of deviation of the particle orbit.

In the toroidal geometry, the deviation is not restricted by the main magnetic field, but by the poloidal magnetic field. The shift is characterized by the poloidal gyro-radius ρ_p

$$\rho_p = \left|\frac{mv}{eB_p}\right| \tag{3.25}$$

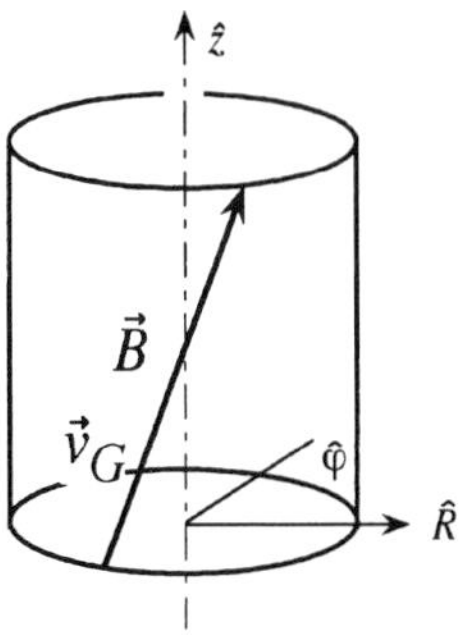

Figure 3.5. Guiding centre motion (v_G) is very close to the magnetic field direction in the cylindrical plasma.

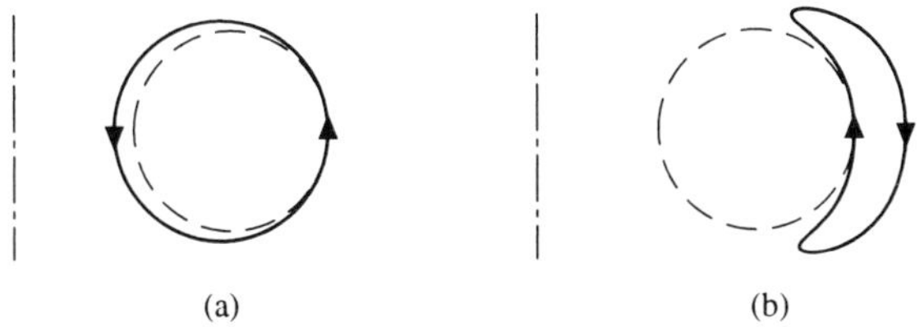

Figure 3.6. The guiding centre orbit (solid line) can be deviated from the magnetic surface (dashed line) in toroidal plasmas. Projection of the orbit on the poloidal cross-section is shown for transit particles (a) and for trapped particles (b).

in toroidal plasmas. The deviation in the cylindrical system is the gyro-radius ρ. The ratio between them, $\rho_p/\rho = B/B_p$, can be of the order of ten in real systems. Figure 3.6 illustrates the trajectory of the guiding centre of a charged particle in axisymmetric toroidal plasmas. Some particles circumnavigate in poloidal directions (transit particles, figure 3.6(a)), but some are bounded in the outside of the torus (trapped particles, figure 3.6(b)).

The fact that the toroidality gives varieties in the particle orbit is one of the origins of complexity in the analysis of confined plasmas. Owing to the toroidicity, the strength of the magnetic field is stronger inside the torus. Hence, the particles can be reflected on the inside (figure 3.7). There are two classes of orbits confined in an axisymmetric torus; a passing (or transit) orbit and a trapped orbit.

Noting that the magnetic moment is (quasi-) invariant, the perpendicular kinetic energy $v_\perp^2$ varies as

$$v_\perp^2 = \frac{B(\theta)}{B(0)} v_{\perp 0}^2 \qquad (3.26)$$

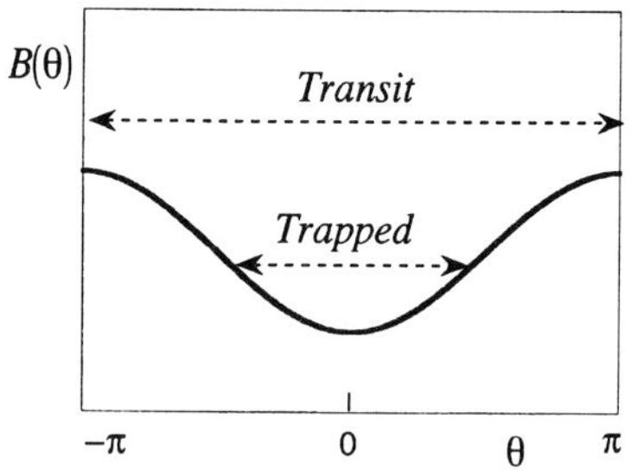

Figure 3.7. Variation of the magnetic field. The region where trapped particles are confined and the one in which transit particles move are also indicated.

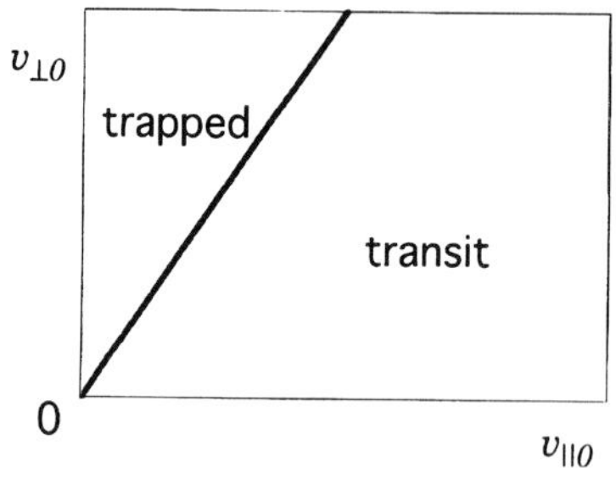

Figure 3.8. The region of trapped (transit) particles in the velocity space.

where 0 indicates the poloidal angle at the weak magnetic field side. Particles with small initial parallel velocity are reflected at some poloidal angle. Such particles are called 'trapped particles'. If the parallel velocity is large enough,

$$v_{\parallel 0}^2 \geq \left[\frac{B(\max)}{B(0)} - 1 \right] v_{\perp 0}^2 \tag{3.27}$$

particles continue to move in one direction. Trapped and transit particles are separated by the condition

$$\left| \frac{v_{\parallel 0}}{v_{\perp 0}} \right| \geq \sqrt{\frac{B(\max)}{B(0)} - 1} \tag{3.28}$$

as shown in figure 3.8. Trapped particles turn into transit particles if the pitch angle varies by a small amount $\sqrt{B(\max)/B(0) - 1}$. In the simple geometry of figure 3.4, magnetic field is approximately expressed as

$$B(\theta) = \frac{1 + r/R}{1 + (r/R)\cos\theta} B(0). \tag{3.29}$$

The boundary of equation (3.28) is estimated as $|v_{\parallel 0}/v_{\perp 0}| \geq \sqrt{2r/R}$.

The argument of the orbit can be generalized in the presence of static electric field, some of which will be discussed later, in relation to the structural formation and transition.

The formation of a toroidal configuration loses one symmetry compared to the straight cylinder. (Comparing figure 3.2(b) with figure 3.2(a), an axial symmetry is preserved, but the translational symmetry in the $\hat{z}$ direction is lost.) It should be noted that an analogy between the stretched straight cylinder and the toroidal plasma cannot always be drawn. The lack of poloidal symmetry provides a change of the character of particle orbits. When the axial symmetry is violated (either intentionally or due to the technological limitation in constructing devices), a much more complex orbit appears as is discussed in chapter 17. In this monograph, we do not intend to discuss thoroughly the effect of geometry. Rather, emphasis will be placed on the generic feature which is common for various configurations.

REFERENCES

[3.1] Recent progress in the physics related to nonneutral plasmas is reported by O'Neil T M and Dubin D H E 1998 *Phys. Plasmas* **5** 2163

Mohri A *et al* 1997 *Buturi* **52** 585 (in Japanese)

The elementary process is explained in, e.g., Davidson R 1990 *Physics of Nonneutral Plasmas* (Redwood City: Addison-Wesley). Structural formation in dense and strongly coupled plasmas is explained in, e.g., Ichimaru S 1991 *Statistical Plasma Physics II* (Reading, MA: Addison-Wesley)

[3.2] Hamada S 1962 *Nucl. Fusion* **2** 23

[3.3] Courtesy of Professor D Düchs 1997

[3.4] Shafranov V D 1966 *Reviews of Plasma Physics* vol 2, ed M A Leontovich (New York: Consultants Bureau) p 103

[3.5] Husimi K 1964 *Classical Mechanics* (Tokyo: Iwanami) (in Japanese)

Sivukhin D V *Review of Plasma Physics* vol 1, ed M A Leontovich (New York: Consultants Bureau) p 1

Chapter 4

Fluctuation-Driven Flux

Although a particle trajectory is slightly deviated from a magnetic surface, the orbit could be integrable in an axisymmetric system when the magnetic field is static. Particles remain on the integral surfaces of the orbit. When this integral disappears, the cross-field plasma fluxes are generated. A heuristic discussion is given on a mechanism which causes a loss of the integral of motion.

4.1 Elementary Processes

4.1.1 Collision

One of the processes which cause a change of orbit is coulomb collision with other particles. The particle trajectory lies on one magnetic surface as an average, but it deviates in and out by a quantity Δ_{orbit}. The orbit of the particle intersects with those of particles on a neighbouring magnetic surface. When coulombic collision takes place between two particles, the particle can move by the amount of Δ_{orbit}. When collisions occur independently with the frequency ν_{col}, particles are subject to a diffusion process with the diffusivity

$$D_c \simeq \nu_{col} \Delta_{orbit}^2 \tag{4.1}$$

where the suffix c denotes coulombic collision.

Coulombic collision occurs between two discrete particles. When a test particle is subject to accumulated interactions with many particles, a statistical average must be taken into account to evaluate the collision frequency. A collective interaction of plasma particles causes the Debye screening, and the test particle has an effective range of interaction. This range is given by the Debye length λ_D [2.3, 2.4], and the test particle does not interact with particles which are separated in space more than λ_D. The number of particles in this sphere,

$$N_D = (4\pi/3) n_e \lambda_D^3$$

is considered to be a large number in this monograph. (The system of $N_D \gtrsim 1$ is another extremum case, and provides interesting physics as shown by the vertical axis of figure 1.3 [3.1].)

In the presence of thermal noise, the number of interacting particles fluctuates around N_D with the deviation δN_D, where δN_D is estimated as

$$\frac{\delta N_D}{N_D} \simeq \frac{1}{\sqrt{N_D}} \tag{4.2}$$

after the central limiting theorem. If the limit $N_D \to \infty$ is taken, the discreteness of particles disappears and only the mean value is of relevance. In such a case, the relative variance $\delta N_D/N_D$ disappears. The Coulombic collision frequency between likely electrons is given as (apart from a numerical factor of order unity)

$$\nu_c \simeq \frac{1}{N_D}\omega_p. \tag{4.3}$$

In the limit of $N_D \to \infty$ with ω_p being fixed, the collisional transport vanishes.

4.1.2 $\boldsymbol{E} \times \boldsymbol{B}$ Convection

In plasmas, various kinds of electromagnetic fluctuation can be generated. From the impact on transport processes, fluctuations in the frequency range lower than the ion cyclotron frequency are important. When the temporal variation of the electric field is slower than the ion cyclotron frequency, the particle motion is approximately described by the sum of gyromotion and $\boldsymbol{E} \times \boldsymbol{B}$ motion of the guiding centre. An averaging of the fast gyration around the gyrocentre gives the expression of the motion of the gyrocentre in terms of the $\boldsymbol{E} \times \boldsymbol{B}$ drift as

$$\tilde{\boldsymbol{V}}_{E\times B} = \frac{\tilde{\boldsymbol{E}} \times \boldsymbol{B}}{B^2}. \tag{4.4}$$

The symbol $\tilde{}$ indicates fluctuating electric field and the response. Figure 4.1 shows schematically the $\boldsymbol{E} \times \boldsymbol{B}$ motion by the oscillatory electric field.

The diffusion of a fluid element under the given electric field fluctuations is evaluated from the averaged rate of its motion. In a random walk process, the average vanishes,

$$\langle \tilde{x}(t) - \tilde{x}(0) \rangle = 0 \tag{4.5-1}$$

but the variance increases in time

$$\langle \{\tilde{x}(t) - \tilde{x}(0)\}^2 \rangle = Dt. \tag{4.5-2}$$

When the fluctuating electric field is Fourier decomposed and is written as

$$\tilde{E}(x) = \sum_{\boldsymbol{k}.\omega} E_{\boldsymbol{k}.\omega} \exp\{\mathrm{i}\boldsymbol{k} \cdot \boldsymbol{x} - \mathrm{i}\omega\} \tag{4.6}$$

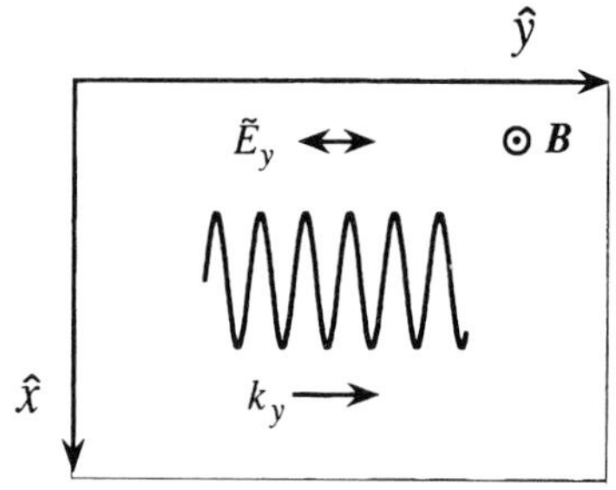

Figure 4.1. Plasma is subject to the $\boldsymbol{E} \times \boldsymbol{B}$ motion in the presence of a fluctuating field.

the motion of the plasma in response to one (k, ω) Fourier component is given according to equation (4.4) as

$$\tilde{x}(t) - \tilde{x}(0) \simeq \frac{\mathrm{i}E_y}{(\omega - k_z v_z + \mathrm{i}\nu)B} \exp\{\mathrm{i}\boldsymbol{k} \cdot \boldsymbol{x} - \mathrm{i}\omega\}. \tag{4.7}$$

In deriving equation (4.7) approximations are made that the motion along the field line is unaltered, and that the perpendicular wavelength is much longer than $\tilde{x}(t) - \tilde{x}(0)$. A small but finite value of ν indicates some weak decorrelation around the resonance at $\omega - k_z v_z = 0$.

Under the stationary oscillating field, the plasma is also vibrated with the amplitude $\Delta_{\boldsymbol{E}\times\boldsymbol{B}} = E_y B^{-1}(\omega - k_z v_z + \mathrm{i}\nu)^{-1}$. However, the lifetime of electric fluctuations is often finite. The motion due to the $\boldsymbol{E} \times \boldsymbol{B}$ drift gives rise to the diffusion coefficient of the fluid elements, which is of the order of

$$D \simeq \frac{1}{\tau_{dec}} \Delta^2_{\boldsymbol{E}\times\boldsymbol{B}} \tag{4.8}$$

where τ_{dec} is the decorrelation time of the oscillating motion. The diffusivity is a quadratic form of the electric fluctuation amplitude, when other characteristic coefficients are constant.

When the fluctuation amplitude becomes large, the relation $D \propto \tilde{E}^2$ does not hold. The shift due to the $\boldsymbol{E} \times \boldsymbol{B}$ motion, $\Delta_{\boldsymbol{E}\times\boldsymbol{B}}$, during the time of τ_{dec} can become comparable to the perpendicular wavelength $k_\perp^{-1}$. In such a condition, $\Delta_{\boldsymbol{E}\times\boldsymbol{B}} \simeq k_\perp^{-1}$, equation (4.8) provides the relation $D \simeq \tau_{dec}^{-1} k_\perp^{-2}$. In other words, the decorrelation rate is to be limited by this scale-length, and becomes short as

$$\tau_{dec}^{-1} \simeq D k_\perp^2. \tag{4.9}$$

In this limiting case, the decorrelation time of the fluctuating motion, τ_{dec}, is determined by the diffusion due to the fluctuations. The response of the plasma of equation (4.7) is dictated by this decorrelation rate, and we have an estimate

$\nu \sim \tau_{dec}^{-1}$. The step size is evaluated from equation (4.7), with the help of condition $\tau_{dec}^{-1} \sim \nu \gg |\omega - k_z v_z|$, as

$$\Delta_{E\times B} = E_y B^{-1} \tau_{dec}.$$

Substituting equation (4.9) and this estimate of $\Delta_{E\times B}$ into equation (4.8), equation (4.8) turns out to be

$$D \simeq \frac{1}{Dk_\perp^2}\left(\frac{E_y}{B}\right)^2. \tag{4.10}$$

This is a strong turbulence limit. The form

$$D \simeq \frac{1}{k_\perp}\frac{\tilde{E}_y}{B} \tag{4.11}$$

is derived from equation (4.10). In the strong turbulence limit, the diffusivity is linearly proportional to the fluctuation amplitude. A more rigorous derivation is discussed later.

4.1.3 Magnetic Perturbation

Since plasma parameters tend to equilibrate along the magnetic field line owing to the rapid motion, it is expected that the small perturbation of the magnetic field can easily enhance the plasma transport across the magnetic field. The effect of the perturbed magnetic field $\tilde{\boldsymbol{B}}$, which is superimposed on the strong and static magnetic field $\boldsymbol{B}_0$, has been investigated.

Let us study the motion of the guiding centre along the magnetic field line. A simplified limit is assumed, where the velocity parallel to the magnetic field is constant and the perturbation of the magnetic field varies slowly in time and affects only the direction of motion. In this simplified situation, the deviation in the x-direction due to the magnetic perturbation is given as

$$\tilde{x}(t) - \tilde{x}(0) = \int_0^t \left(\frac{\tilde{B}_x}{B_0}\right) v_z \,\mathrm{d}t. \tag{4.12}$$

A sum of this motion and the $\boldsymbol{E} \times \boldsymbol{B}$ motion, equation (4.7), gives the diffusion coefficient $D_p = \lim_{t\to\infty}\langle\{\tilde{x}(t) - \tilde{x}(0)\}^2\rangle t^{-1}$. An expansion in powers of $1/B_0$ gives D_p as

$$D_p = \int_{-\infty}^{\infty} \mathrm{d}\tau \langle \tilde{A}(r(\tau), \tau)\tilde{A}(r(0), 0)\rangle \tag{4.13}$$

where

$$\tilde{A}(r,t) \equiv \frac{v_z}{B_0}\tilde{B}_x(r,t) + \frac{1}{B_0}\tilde{E}_y(r,t). \tag{4.14}$$

The first term on the right-hand side of equation (4.14) causes diffusion by magnetic perturbation, and the second one yields equation (4.8). In equation (4.13), the bracket $\langle\rangle$ indicates the auto-correlation of the fluctuating quantity $\tilde{A}$.

The contribution of the magnetic perturbation, i.e., the first term of equation (4.14), to D_p is given by a simple estimate. Since the particle velocity along the field line is assumed constant, and the magnetic field perturbation is considered to change slowly in time, the time integral in equation (4.13) is replaced by the spatial integration

$$\mathrm{d}\ell = v_z(0)\,\mathrm{d}\tau$$

and the diffusion coefficient is given as

$$D_p = v_z(0) D_M \tag{4.15-1}$$

and

$$D_M = \int_{-\infty}^{\infty} \mathrm{d}\ell \langle \tilde{b}(r(\ell), \ell)\tilde{b}(r(0), 0)\rangle \tag{4.15-2}$$

where

$$\tilde{b} = \tilde{B}_x(r, t)/B_0.$$

The coefficient D_M is dependent only on the geometrical configuration of magnetic field lines. It is often called the diffusion coefficient of the field line, because it becomes finite if the field line no longer lies on one magnetic surface.

4.1.4 Magnetic Island

A topology of magnetic surfaces changes near rational surfaces, if a resonant perturbation to the corresponding rational number is superimposed. The magnetic surface with the pitch of rational number m/n can be vulnerable to the perturbations of the form

$$\tilde{B}_r = \tilde{B}\cos(m\theta - n\zeta). \tag{4.16}$$

This perturbation is resonant at the surface of the rational pitch number

$$r = r_{s,(m,n)} \tag{4.17-1}$$

where the condition

$$q = \frac{m}{n} \tag{4.17-2}$$

holds. In the presence of the resonant magnetic perturbation, the $q = m/n$ rational surface is destroyed, and a magnetic *island* is formed nearby (figure 4.2). The magnetic field lines encircle in the vicinity of the resonant surface and small and localized ellipses appear in the Poincaré plot; these are called *magnetic islands*.

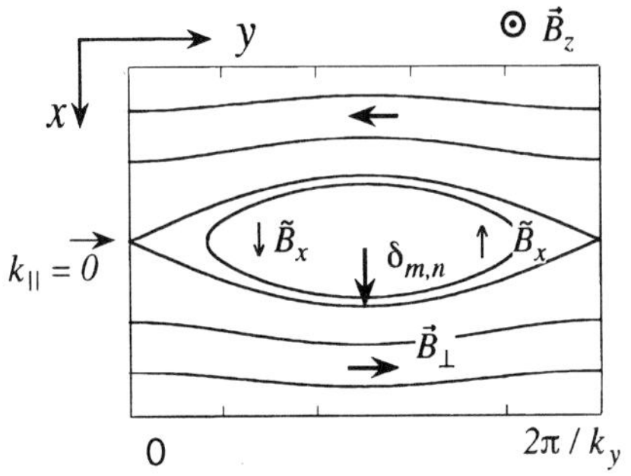

Figure 4.2. A magnetic island is formed at the mode rational surface. Local Cartesian coordinates (x, y) correspond to radial and poloidal directions, respectively.

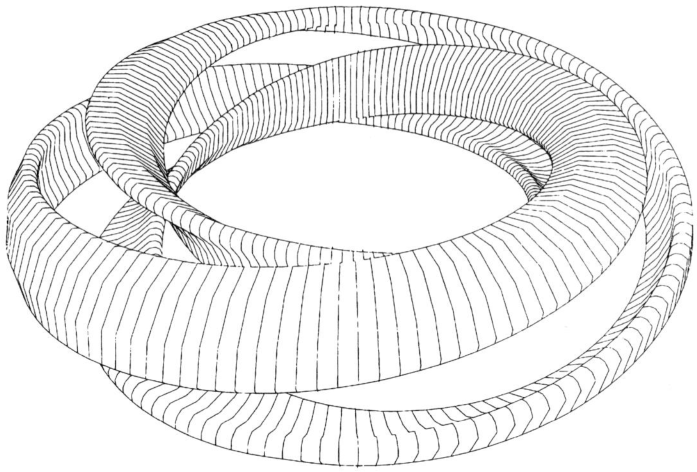

Figure 4.3. The $(m = 3, n = 1)$ island in toroidal plasmas [3.3].

In a toroidal geometry, the island on the $r = r_{s(m,n)}$ surface helically winds around the toroidal axis. Figure 4.3 shows the $(m = 3, n = 1)$ island in toroidal plasmas.

The width of the magnetic island, $\delta_{m,n}$, is expressed as

$$\delta_{m,n} \sim \sqrt{\frac{rR}{ms} \frac{\tilde{B}_r}{B_t}}. \tag{4.18}$$

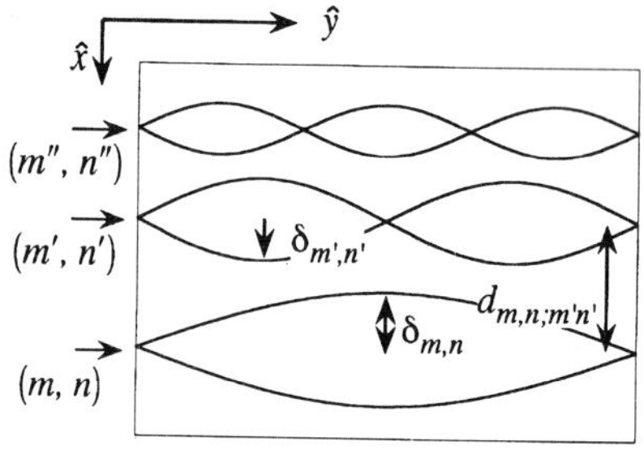

Figure 4.4. Island chains could appear when perturbations of various mode numbers coexist.

4.1.5 Magnetic Braiding and Stochasticity

When the fluctuation includes various Fourier components, layered magnetic islands appear as is illustrated in figure 4.4. If the perturbation amplitude is small, adjacent islands are completely separated by magnetic surfaces. When the island size becomes comparable to or greater than the separation distance between islands, the island overlapping condition is satisfied as

$$\delta_{m,n} + \delta_{m',n'} > d_{m,n;m'n'} \tag{4.19-1}$$

$$d_{m,n;m'n'} = |r_{s(m,n)} - r_{s(m',n')}| \tag{4.19-2}$$

the separatrix is destroyed, and some field lines across the island are connected [4.1, 4.2]. This relation implies that the diffusion of magnetic field lines takes place if the perturbation exceeds a certain threshold amplitude. An example of the braided magnetic field is shown in figure 4.5.

The magnetic stochasticity gives rise to the diffusion of the magnetic field lines. The diffusion coefficient (4.15) is evaluated under the circumstance of the braided magnetic surfaces. When the magnetic field perturbation exceeds the threshold and the stochasticity sets in, then the stochastic average of $\langle\tilde{b}(r(\ell), \ell)\tilde{b}(r(0), 0)\rangle$ remains appreciable. The correlation $\langle\tilde{b}(r(\ell), \ell)\tilde{b}(r(0), 0)\rangle$ tends to disappear if the distance ℓ becomes large. There are two ways in which the correlation $\langle\tilde{b}(r(\ell), \ell)\tilde{b}(r(0), 0)\rangle$ disappears. Let us consider the case where the magnetic fluctuations have correlation length along the field line (auto-correlation length) L_{ac}, and that across the field line δ_{cor} (figure 4.6). Parameters L_{ac} and δ_{cor} characterize the magnetic field perturbations. The z-axis is taken in the direction of unperturbed magnetic field $\boldsymbol{B}_0$. The x-axis denotes the perpendicular direction to $\boldsymbol{B}_0$. The trajectory along the perturbed magnetic field is schematically represented by the thick curved arrows.

The integrand of equation (4.15-2), $\langle\tilde{b}(r(\ell), \ell)\tilde{b}(r(0), 0)\rangle$, is schematically illustrated by the shaded area in figure 4.7. The integral (4.15-2) is estimated

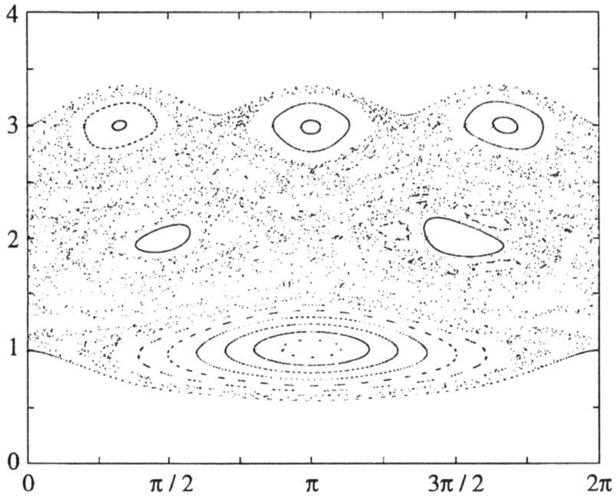

Figure 4.5. An example of braided magnetic field lines. A Poincaré plot on the (x, y)-plane is illustrated. (Same configuration as figure 4.4.).

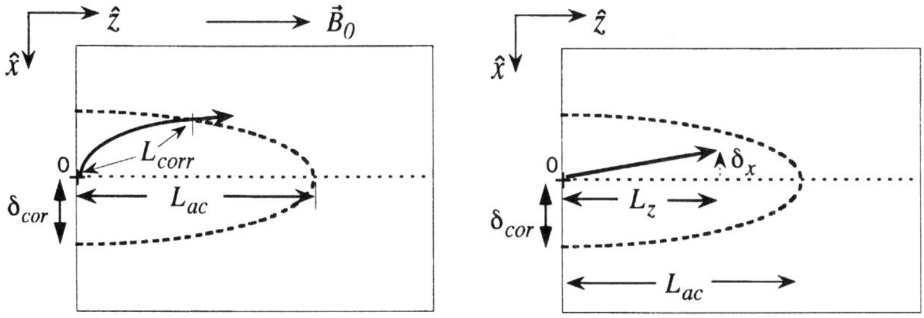

Figure 4.6. Correlation length along the unperturbed magnetic field, L_{ac}, and that across the field, δ_{cor}. The dashed line indicates the region where the correlation $\langle \tilde{b}(r(\ell), \ell)\tilde{b}(r(0), 0)\rangle$ remains finite. The solid arrow indicates a typical example of the trajectories. The arc length within the correlation region is denoted by L_{corr}.

by use of the typical value of the amplitude

$$\tilde{b}^2 \equiv \langle \tilde{b}(r(0), 0)\tilde{b}(r(0), 0)\rangle$$

and the length of the path within the correlation region, L_{corr}, as

$$D_M = \int_{-\infty}^{\infty} \mathrm{d}\ell \langle \tilde{b}(r(\ell), \ell)\tilde{b}(r(0), 0)\rangle \simeq \tilde{b}^2 L_{corr}. \qquad (4.20)$$

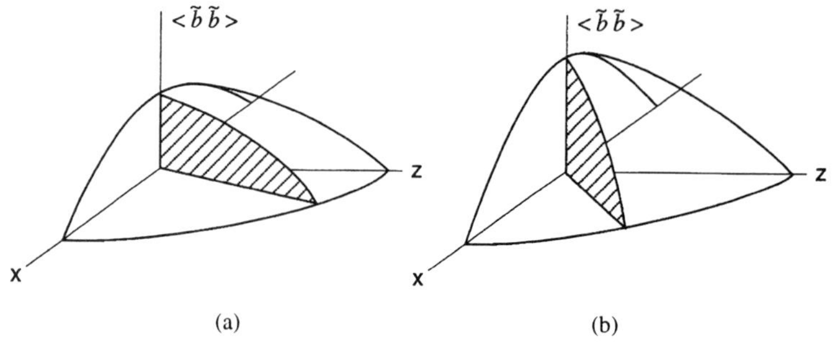

Figure 4.7. Correlation $\langle \tilde{b}(r(\ell), \ell)\tilde{b}(r(0), 0)\rangle$ is shown schematically. The integrand of equation (4.15-2), as a function along the path, is shown by the shading. The case of weak perturbation is shown in (a), and (b) represents the case of strong perturbation.

The length of the path within the correlation region, L_{corr}, depends on the perturbation amplitude. Correlation could be lost either by motion in the z-direction (case of figure 4.7(a)) or by deviation into the x-direction (case of figure 4.7(b)). The path length L_{corr} is calculated in the cases of weak and strong perturbations. The deviation in the x-direction, δ_x, is given in terms of the distance in the z-direction, L_z. (See figure 4.6(b).) During the time interval Δt, an average of δ_x is given by the diffusive formula as

$$\delta_x^2 = D_p \Delta t \tag{4.21}$$

and the motion in the z-direction amounts to $L_z = v_z(0)\Delta t$. The relation $\delta_x^2 = (D_p/v_z(0))L_z = D_M L_z$ holds.

The correlation $\langle \tilde{b}(r(\ell), \ell)\tilde{b}(r(0), 0)\rangle$ vanishes if one of the conditions

$$L_z > L_{ac} \tag{4.22-1}$$

or

$$\delta_x > \delta_{cor} \tag{4.22-2}$$

is satisfied. (In figure 4.7(a), the condition $L_z > L_{ac}$ is satisfied first; the condition $\delta_x > \delta_{cor}$ is fulfilled first in figure 4.7(b).) The deviation in the x-direction at the distance $L_z = L_{ac}$, δ_{xL}, is introduced as

$$\delta_{xL}^2 \equiv D_M L_{ac}. \tag{4.23}$$

In calculating δ_{xL}, we use a limiting condition that $\delta_x \ll L_z$.

The relative magnitude of δ_{xL} with respect to δ_{cor} determines which of equations (4.22-1) and (4.22-2) is satisfied first. If the condition

$$\delta_{xL} < \delta_{cor} \tag{4.24-1}$$

is satisfied, (i.e., $D_M L_{ac} < \delta_{cor}^2$), the condition (4.22-1) determines the path length L_{corr} as

$$L_{corr} = L_{ac}. \tag{4.25}$$

Combining equations (4.20) and (4.25), one has

$$D_M = \tilde{b}^2 L_{ac}. \tag{4.26}$$

On the other hand, when the opposite condition

$$\delta_{xL} > \delta_{cor} \tag{4.24-2}$$

is satisfied, (i.e., $D_M L_{ac} > \delta_{cor}^2$), the path length L_{corr} is dictated by the condition (4.22-2). The length L_{corr} is given by the relation $\delta_x = \delta_{cor}$ at $L_z = L_{corr}$, i.e., $D_M L_{corr} = \delta_{cor}^2$. We have

$$L_{corr} = \frac{\delta_{cor}^2}{D_M}. \tag{4.27}$$

Combining equations (4.20) and (4.27), one has $D_M = \tilde{b}^2 \delta_{cor}^2 / D_M$, or

$$D_M = \tilde{b} \delta_{cor}. \tag{4.28}$$

Summarizing equations (4.26) and (4.28), we have

$$D_M \simeq \begin{array}{ll} L_{ac} \tilde{b}^2 & (\tilde{b} < \delta_{cor} L_{ac}^{-1}) \\ \delta_{cor} \tilde{b} & (\tilde{b} > \delta_{cor} L_{ac}^{-1}) \end{array} \tag{4.29}$$

under the magnetic braiding condition. When the perturbation is small, the diffusivity is in proportion to the square of the perturbation amplitude (assuming that the other parameters of fluctuations are constant). If the fluctuation becomes stronger, the diffusivity linearly grows with the perturbation amplitude. (See figure 4.8.) The change from the quadratic dependence (weak turbulence) to the linear dependence (strong turbulence) is the same as that in the diffusion by the $\boldsymbol{E} \times \boldsymbol{B}$ motion. A more thorough classification is given in, e.g., [4.3] and [4.4].

4.2 Collisional Transport (to Thermal Equilibrium)

Even in thermal equilibrium, fluctuations are excited in the form of thermal fluctuations. Each mode can be pumped up to the level of an equilibrium temperature T. The influence of this fluctuating field on the transport is represented by the collisional relaxation process which is described in section 4.1.1.

The transport driven by the binary collision is essentially given in the form (4.1). As is discussed in section 3, the deviation of the particle orbit from the magnetic surface is strongly dependent on the geometry. Thus the main issue for the collisional transport is to account for the geometry effect carefully.

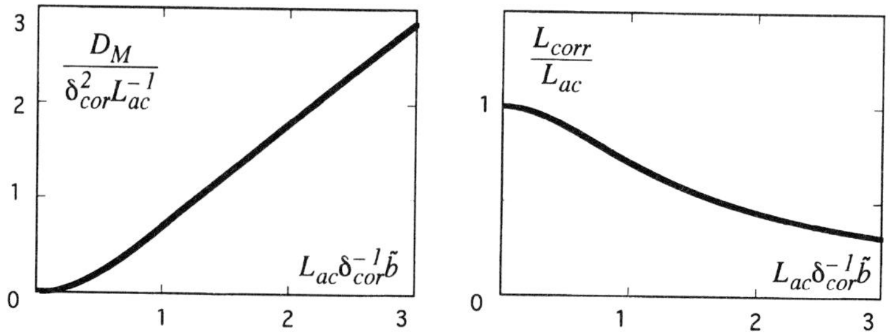

Figure 4.8. Schematic illustration of diffusion coefficient of magnetic field lines, D_M, and correlation length L_{corr} as a function of magnetic perturbation amplitude $\tilde{b}$. In a small amplitude limit, $D_M \propto \tilde{b}^2$ holds, while relations $D_M \propto \tilde{b}$ and $L_{corr} \propto \tilde{b}^{-1}$ are satisfied in a strong amplitude limit.

In the previous chapter, it is shown that the shift of particle orbit from the magnetic surface is enhanced by the toroidicity. The geometry not only affects the shift Δ_{orbit} but also is important in determining the collision frequency. Coulomb collision frequency ν_c is often expressed by the rate at which the pitch angle changes by an amount of 90° as an average. The coulombic collision in the plasma is treated as an accumulation of the small angle scatterings, and the time rate by which the pitch angle changes by the amount of $\delta\theta_v$, ν_{eff}, is given as

$$\nu_{eff} \simeq \nu_c \delta\theta_v^{-2}. \tag{4.30}$$

In a toroidal geometry, a small variation in the pitch angle can change the orbit characteristics. Namely, trapped particles turn into transit particles if the pitch angle varies by a small amount (see figure 3.7)

$$\delta\theta_v \sim \sqrt{B(\max)/B(0) - 1}.$$

The effective collision frequency, which is to be substituted into equation (4.1), is

$$\nu_{eff} \simeq \frac{\nu_c}{[B(\max)/B(0) - 1]}. \tag{4.31}$$

In the case of a tokamak with circular cross-section, an approximation relation $B(\max)/B(0) \simeq 1 + 2r/R$ holds. This gives an enhancement factor for ν_{eff} as $\nu_{eff} \simeq \nu_c R/2r$.

The collisional transport has been calculated based on the established procedure. Combining these geometrical influences, the transport matrix has been obtained [4.5]. For instance, the particle flux, conductive heat fluxes of electrons and ions are given in the limit of a low collision frequency as

$$\begin{pmatrix} \Gamma_r \\ q_{e.r} \\ q_{i.r} \end{pmatrix} = -D_b \begin{pmatrix} 1.12\left(1+\frac{T_i}{T_e}\right)n & -0.43n & -0.19n \\ -1.53\left(1+\frac{T_i}{T_e}\right)nT_e & 1.81nT_e & 0.27nT_e \\ M_{31} & M_{32} & 0.48\left(\frac{m_i T_e}{m_e T_i}\right)^{1/2} nT_i \end{pmatrix} \begin{pmatrix} \frac{n'}{n} \\ \frac{T'_e}{T_e} \\ \frac{T'_i}{T_i} \end{pmatrix}$$

$$+\sqrt{\frac{r}{R}} \begin{pmatrix} -2.44n \\ 1.75nT_e \\ W_3 \end{pmatrix} \frac{E_\parallel}{B_\theta} \tag{4.32-1}$$

$$E_\parallel = 0.51\frac{m_e}{ne^2}\nu_e\left(1-1.95\sqrt{\frac{r}{R}}\right)^{-1}(J_\parallel - J_{BS}) \tag{4.32-2}$$

$$J_{BS} = -\frac{nT_e}{B_\theta}\sqrt{\frac{r}{R}}\left\{2.44\left(1+\frac{T_i}{T_e}\right)\frac{n'}{n} + 0.69\frac{T'_e}{T_e} - 0.42\frac{T'_i}{T_i}\right\} \tag{4.32-3}$$

$$D_b = \sqrt{\frac{r}{R}}\left(\frac{qR}{r}\right)^2 \nu_e\rho_e^2 \tag{4.32-4}$$

where the electron collision frequency is defined as $\nu_e = (4\sqrt{2\pi}\ln\Lambda/3)m_e^{-1/2} \times T_e^{-3/2}n$. ($\Lambda$ is a Coulomb logarithm [2.3].) In this expression, terms M_{31}, M_{32} and W_3 give higher order corrections with respect to the ratio m_e/m_i for $q_{i.r}$ and are not explicitly written. In the formula of equation (4.32), fluxes are carried by trapped particles. The deviation of trapped-particle orbit from the magnetic surface is amplified as $\Delta_{orbit} \sim (q\sqrt{R/r})\rho_e$. Equation (4.31) shows that the rate of change of an orbit from a trapped one to a transit one is also amplified as $\nu_{eff} \sim (R/r)\nu_e$. Considering that the ratio of trapped particle density to plasma density is of the order $\sqrt{r/R}$, an estimate of $D \sim D_b$ is obtained from equation (4.1). Equation (4.32) describes mutual interferences between various gradients and fluxes. The last term of equation (4.32-1) shows that radial fluxes are caused by a parallel electric field. This term is called 'Ware pinch' [4.6]. Equation (4.32-2) indicates that a current along the magnetic field line is induced by gradients of plasma. This component of the current is called 'bootstrap current' [4.7].

The complication in the expression is due to a variation of the strength of magnetic field. From the physics point of view, apart from a form factor arising from the geometrical complexity, the collisional transport has two characteristic features. First, it is described by the local parameters of the plasma and electromagnetic field. The gradients of the plasma parameters (such as temperature gradient) are not included in the transport coefficient. In this sense, it describes the linear response of the flux against the gradient. Second, if the binary collision becomes less frequent, the transport becomes smaller. In the collisionless limit, it disappears. This latter fact has importance for fusion research. In controlled thermonuclear fusion, a high temperature (such as 10 keV) is required. The collisional transport coefficient predicts that the loss rate decreases, i.e., the confinement time becomes longer, as the temperature becomes higher. However, this is not the case in experiments. Only some elements in the transport matrix (4.32), which are related to the electric conductivity, can be explained by the collisional transport in toroidal plasmas [2.1, 2.4].

This collisional transport is also understood as a contribution of thermal fluctuations to the $\boldsymbol{E} \times \boldsymbol{B}$ transport. Let us consider a thermal fluctuation in a homogeneous, stationary and planar plasma. The plasma oscillation is excited by the thermal fluctuation in the thermal equilibrium state [2.3]. The time scale and spatial scale are given as ω_p^{-1} and λ_D, respectively. The number of modes is represented by $k^3 \simeq \lambda_D^{-3}$, and the amplitude of electric fluctuation is order estimated as

$$\tilde{E}^2 = \sum_k^{\infty} \tilde{E}_k^2 \simeq k^3 \tilde{E}_k^2. \tag{4.33}$$

If the energy of each mode is excited to the level of T_e, i.e.,

$$(\varepsilon_0 \tilde{E}_k^2 / 2) \simeq T_e \tag{4.34}$$

the fluctuation level is given as $\tilde{E}^2 \simeq \lambda_D^{-3} \varepsilon_0^{-1} T_e$, or

$$\tilde{E}^2 \simeq \frac{1}{n_e \lambda_D^3} \omega_p^2 \frac{m_e T_e}{e^2}. \tag{4.35}$$

A heuristic argument can be made based on equation (4.35). Substituting the relation

$$\tau_{cor} \simeq \omega_p^{-1} \tag{4.36}$$

and equation (4.35) into equation (4.8), the diffusivity is given as

$$D \simeq \frac{\omega_p}{N_D} \frac{m_e T_e}{e^2 B^2} \simeq \nu_{c.e} \rho_e^2 \tag{4.37}$$

where N_D is the particle number in a Debye sphere, $N_d \simeq n_e \lambda_D^3$. Equation (4.37) is essentially equal to equation (4.1) in a uniform and static magnetic field. The collisional transport process in the plasma thus belongs to the phenomena caused by thermal fluctuations in an equilibrium state.

4.3 Quasilinear Transport for Weak Turbulence (Close to Thermal Equilibrium)

4.3.1 Formulation in Terms of the Conductivity Tensor

As the perturbation of a space–time-varying electromagnetic field becomes stronger than the thermal fluctuations, the cross-field flux tends to overcome that driven by the collisional process. When the perturbation amplitude is small, the response of plasmas to such perturbations can be calculated by a linear theory. The resultant transport is called quasilinear transport. In this framework, each Fourier component of perturbations is treated separately, and their linear superposition is allowed.

Let us take an electrostatic perturbation and consider a linear response of plasma. We choose a *small region* of the plasma, and derive an averaged flux within this small volume.

The particle flux

$$\Gamma_x = nV_x = \sum_{k.\omega.k'.\omega'} \langle \tilde{n}_{k.\omega}(\tilde{E}_{y.k'.\omega'} B^{-1}) \rangle \tag{4.38}$$

is expressed in terms of the linear response function. By using the conductivity tensor for the oscillating field $\boldsymbol{\sigma}_{j.k\omega}$

$$\tilde{\boldsymbol{J}}_{j.k\omega} = \boldsymbol{\sigma}_{j.k\omega} \tilde{\boldsymbol{E}}_{k\omega} \tag{4.39}$$

where the suffix j denotes the species, the density perturbation is expressed as

$$\tilde{n}_{j.k\omega} = \frac{1}{e_j} \frac{1}{\omega} \boldsymbol{k} \cdot \boldsymbol{\sigma}_{j.k\omega} \tilde{\boldsymbol{E}}_{k\omega}. \tag{4.40}$$

In rewriting equation (4.40), the continuity equation is employed as

$$\omega \tilde{n}_{j.k\omega} = e_j^{-1} \boldsymbol{k} \cdot \tilde{\boldsymbol{J}}_{j.k\omega}. \tag{4.41}$$

Substituting equation (4.40) into equation (4.38), the cross-field flux is given as

$$\Gamma_{j.x} = \frac{1}{e_j B} \sum_{k.\omega} \left\langle \frac{1}{\omega} \tilde{E}_{y-k.-\omega} \boldsymbol{k} \cdot \boldsymbol{\sigma}_{j.k\omega} \tilde{\boldsymbol{E}}_{k\omega} \right\rangle \tag{4.42-1}$$

or

$$\Gamma_{j.x} = \frac{1}{e_j B} \sum_{k.\omega}^{\infty} \Re \left(\frac{k_y}{\omega} \boldsymbol{k} \cdot \boldsymbol{\sigma}_{j.k\omega} \boldsymbol{k} \right) |\tilde{\phi}_{k\omega}|^2 \tag{4.42-2}$$

for electrostatic perturbations, where ϕ is the electrostatic potential, $\boldsymbol{E} = -\boldsymbol{\nabla}\phi$.

The driving mechanism of flux in the direction of density gradient is illustrated in figure 4.9. Consider the case where a wave is excited by electrons and is absorbed by ions. Via the wave momentum, ions and electrons receive an action $\boldsymbol{F}_i$ and a reaction $\boldsymbol{F}_e$, respectively. The $\boldsymbol{F} \times \boldsymbol{B}$ drift causes the flux in the $\hat{x}$-direction.

The Hermitian part of the conductivity tensor is related to the rate of absorption/excitation of waves by the plasma. It is also related to the growth rate or the damping rate of electromagnetic perturbations. Maxwell equations are used to obtain

$$-\varepsilon_0 \nabla \cdot \frac{\partial}{\partial t} \tilde{\boldsymbol{E}}_{k\omega} = \sum_j \nabla \cdot \tilde{\boldsymbol{J}}_{j.k\omega}. \tag{4.43}$$

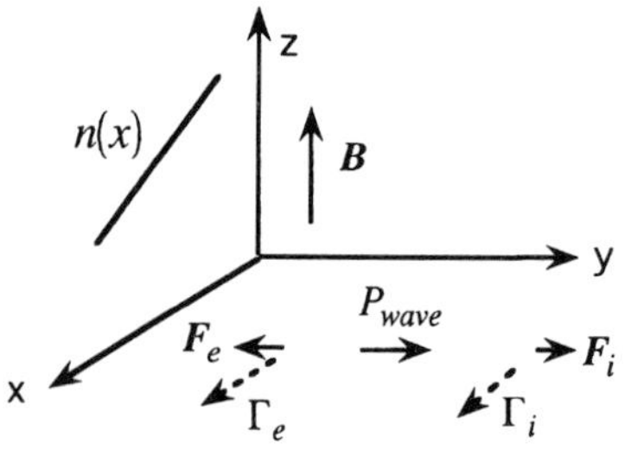

Figure 4.9. Schematic drawing of the mechanism to cause cross-field flux. Through excitation and absorption of waves, ions and electrons receive an action $\boldsymbol{F}_i$ and a reaction $\boldsymbol{F}_e$, respectively. The $\boldsymbol{F} \times \boldsymbol{B}$ drift causes the flux in the $\hat{x}$-direction. The drift fluxes of ions and electrons are in the same direction.

For electrostatic perturbations, this relation is rewritten in terms of the static potential, by use of the conductivity tensor, as

$$\frac{\partial}{\partial t}\tilde{\phi}_{k\omega} = -(\varepsilon_0 k^2)^{-1}\left(\sum_j \boldsymbol{k}\cdot\boldsymbol{\sigma}_{j.k\omega}\boldsymbol{k}\right)\tilde{\phi}_{k\omega}. \tag{4.44}$$

The absolute value of the perturbation obeys the rule

$$\frac{\partial}{\partial t}|\tilde{\phi}_{k\omega}|^2 = -\frac{2}{(\varepsilon_0 k^2)}\Re\left(\sum_j \boldsymbol{k}\cdot\boldsymbol{\sigma}_{j.k\omega}\boldsymbol{k}\right)|\tilde{\phi}_{k\omega}|^2. \tag{4.45}$$

The linear growth rate is expressed in terms of the Hermitian part of the conductivity tensor. The decorrelation rate of a linear mode is represented by the linear growth rate. The expression in the quasilinear theory is given in terms of the linear response function and the quadratic form of the perturbing fields [4.8]. The precise expression of the linear conductivity tensor has been obtained in literature (e.g., [2.1–2.6, 2.10, 4.9, 4.10]).

It is noted that the Hermitian part of the conductivity in the kinetic analysis comes sometimes from the resonant particles satisfying the condition

$$\omega - k_\parallel v_\parallel \simeq 0. \tag{4.46}$$

There are several kinds of particle trajectory in toroidal geometry as is discussed in the preceding chapter. Owing to this complexity, the linear response function becomes dependent on the plasma geometry and inhomogeneities. The majority of efforts has been directed to the research to find out the most dangerous mode in a plasma with complicated geometry [2.1–2.12]. Typical examples of the low frequency fluctuation are explained in the next chapter.

Equation (4.42) indicates that the cross-field transport could be enhanced if perturbations are excited more strongly than the thermal fluctuation. The treatment in the quasilinear transport has advanced. However, within this framework, there is no way to determine the level of turbulence. The quasilinear theory is not closed to describe the state of confined plasmas.

4.3.2 Other Elements of Fluxes

The particle flux, fluxes of momentum and energy are driven by fluctuations. Various flow elements are expressed by the plasma response function. The density n is expressed in terms of the distribution function as

$$n(\boldsymbol{r}, t) = \int f(\boldsymbol{r}, \boldsymbol{v}, t)\, \mathrm{d}\boldsymbol{v}. \tag{4.47}$$

The fluctuating component is separated as $n = n_0 + \tilde{n}$, and the suffix 0 indicates a global profile. Global density profile $n_0(\boldsymbol{r}, t)$ varies in a large spatial scale and a slow time scale, while $\tilde{n}(\boldsymbol{r}, t)$ changes rapidly in a short scale-length. Similar definitions are used for the velocity V_i, the pressure tensor P_{ij}, the heat flux tensor Q_{ijk}, the fourth moment R_{ijkl} etc:

$$V_i = \frac{1}{n} \int (v_i - V_{0i}) f(\boldsymbol{r}, \boldsymbol{v}, t)\, \mathrm{d}\boldsymbol{v} \equiv \frac{1}{n_0} \Gamma_i \tag{4.48}$$

$$P_{ij} = m \int (v_i - V_{0i})(v_j - V_{0j}) f(\boldsymbol{r}, \boldsymbol{v}, t)\, \mathrm{d}\boldsymbol{v} \tag{4.49}$$

$$Q_{ijk} = m \int (v_i - V_{0i})(v_j - V_{0j})(v_k - V_{0k}) f(\boldsymbol{r}, \boldsymbol{v}, t)\, \mathrm{d}\boldsymbol{v} \tag{4.50}$$

$$R_{ijkl} = m \int (v_i - V_{0i})(v_j - V_{0j})(v_k - V_{0k})(v_l - V_{0l}) f(\boldsymbol{r}, \boldsymbol{v}, t)\, \mathrm{d}\boldsymbol{v}. \tag{4.51}$$

The fluctuation parts are also divided as $V_i = V_{0i} + \tilde{V}_i$, $P_{ij} = P_{0ij} + \tilde{P}_{ij}$, $Q_{ijk} = Q_{0ijk} + \tilde{Q}_{ijk}$, $R_{ijkl} = R_{0ijkl} + \tilde{R}_{ijkl}$ and so on (i, j and k take x, y and z).

Average fluxes are formally given, in the leading terms of the ρ/L expansion (L being the gradient scale-length), as

$$\Gamma_x = \frac{1}{B} \langle (n\tilde{\boldsymbol{E}} + \tilde{\boldsymbol{\Gamma}} \times \tilde{\boldsymbol{B}})_y \rangle \tag{4.52}$$

$$\begin{aligned} P_{\parallel x} = {} & \frac{m}{B} \langle (\tilde{\Gamma}_z - \tilde{n} V_z)(\tilde{\boldsymbol{E}} + \boldsymbol{V} \times \tilde{\boldsymbol{B}})_y + (\tilde{\Gamma}_y - \tilde{n} V_y)(\tilde{\boldsymbol{E}} + \boldsymbol{V} \times \tilde{\boldsymbol{B}})_z \rangle \\ & + \frac{1}{B} \langle \tilde{B}_x (\tilde{P}_{zz} - \tilde{P}_{yy}) + \tilde{B}_y \tilde{P}_{yx} - \tilde{B}_z \tilde{P}_{zx} \rangle \end{aligned} \tag{4.53}$$

$$Q_{\perp x} = \frac{1}{2}\sum_{i=x,y}\left[\frac{1}{B}\left\langle \tilde{P}_{ii}(\tilde{\boldsymbol{E}}+\boldsymbol{V}\times\tilde{\boldsymbol{B}})_y + 2\tilde{P}_{iy}(\tilde{\boldsymbol{E}}+\boldsymbol{V}\times\tilde{\boldsymbol{B}})_i \right.\right.$$
$$\left.\left. + \frac{2P_{iy}}{n}(\tilde{n}\tilde{\boldsymbol{E}}+\tilde{\boldsymbol{\Gamma}}\times\tilde{\boldsymbol{B}})_i\right\rangle\right] + \frac{1}{2}\sum_{i=x,y}\left[-P_{ii}V_x + \frac{1}{B}\langle \tilde{B}_x\tilde{Q}_{zii} - \tilde{B}_z\tilde{Q}_{xii}\rangle\right]$$
$$- \frac{1}{B}\langle \tilde{B}_y\tilde{Q}_{xyz} - \tilde{B}_x\tilde{Q}_{yyz}\rangle \qquad (4.54)$$

$$Q_{\parallel x} = \frac{1}{B}\langle \tilde{P}_{zz}(\tilde{\boldsymbol{E}}+\boldsymbol{V}\times\tilde{\boldsymbol{B}})_y + 2\tilde{P}_{zy}(\tilde{\boldsymbol{E}}+\boldsymbol{V}\times\tilde{\boldsymbol{B}})_z\rangle - P_{\parallel}V_x$$
$$+ \frac{1}{B}\langle \tilde{B}_x\tilde{Q}_{zzz} - \tilde{B}_z\tilde{Q}_{zzx}\rangle + \frac{2}{B}\langle \tilde{B}_y\tilde{Q}_{xyz} - \tilde{B}_x\tilde{Q}_{yyz}\rangle. \qquad (4.55)$$

In formulae (4.52)–(4.55), the suffix 0 is suppressed to avoid a clumsy expression. Note that some of the fluctuating quantities could be shown by the conductivity tensor. By the definition $\tilde{\boldsymbol{J}}_j = e_j\tilde{\boldsymbol{\Gamma}}_j$ (j denotes particle species), the perturbed flux $\tilde{\boldsymbol{\Gamma}}$ is related to the conductivity tensor as $\tilde{\boldsymbol{\Gamma}}_j = e_j^{-1}\boldsymbol{\sigma}\tilde{\boldsymbol{E}}$.

The perturbed distribution function is calculated within a framework of linear response theory. The distribution function is written as the sum of the average and the fluctuating parts as

$$f_j(\boldsymbol{r},\boldsymbol{v},t) = f_{0j}(\boldsymbol{r},\boldsymbol{v}) + \tilde{f}_j(\boldsymbol{r},\boldsymbol{v},t) \qquad (4.56)$$

where the suffix j indicates the particle species, and the time dependence of slow variation is suppressed. The fluctuating part is given by a path integral of the Vlasov equation as

$$\tilde{f}_j(\boldsymbol{r},\boldsymbol{v},t) = \frac{e_j}{m_j}\int_{-\infty}^{t} \mathrm{d}t'(\tilde{\boldsymbol{E}}(\boldsymbol{r}',t') + v'\times\tilde{\boldsymbol{B}}(\boldsymbol{r}',t'))\frac{\partial}{\partial\boldsymbol{v}'}f_{0j}(\boldsymbol{r}',\boldsymbol{v}') \qquad (4.57)$$

where $\boldsymbol{r}'(t')$ and $\boldsymbol{v}'(t')$ are the quantities along the *unperturbed orbit* with the boundary condition $\boldsymbol{r} = \boldsymbol{r}'(t)$ and $\boldsymbol{v} = \boldsymbol{v}'(t)$ at $t = t'$. For the perturbation with the phase $\exp(\mathrm{i}\boldsymbol{k}\cdot\boldsymbol{r} - \mathrm{i}\omega t)$, the magnetic perturbation is expressed as

$$\boldsymbol{B}(\boldsymbol{r},t) = \omega^{-1}\boldsymbol{k}\times\boldsymbol{E}(\boldsymbol{r},t).$$

The $(\boldsymbol{k},\omega)$-Fourier component of the perturbed distribution function, $\tilde{f}_{k.\omega}(\boldsymbol{v})$ $\exp(\mathrm{i}\boldsymbol{k}\cdot\boldsymbol{r} - \mathrm{i}\omega t)$, is given as

$$\tilde{f}_{j.k\omega}(\boldsymbol{v}) = \frac{e_j}{m_j}\int_{-\infty}^{t} \mathrm{d}t' \exp\{\mathrm{i}\boldsymbol{k}\cdot(\boldsymbol{r}'-\boldsymbol{r}) - \mathrm{i}\omega(t'-t)\}$$
$$\times\left(\frac{\partial}{\partial\boldsymbol{v}'}f_{0j}(\boldsymbol{r}',\boldsymbol{v}')\right)\cdot\left(\boldsymbol{I} + \frac{1}{\omega}\boldsymbol{v}'\times\boldsymbol{k}\times\right)\tilde{\boldsymbol{E}}_{k\omega} \qquad (4.58)$$

where $\boldsymbol{I}$ denotes the unit tensor. This equation represents the linear response of the perturbed distribution function to the perturbed field.

The set of equations provides the method to calculate various fluxes. If one specifies the distribution function of the global plasma profile, $f_j(\boldsymbol{r}, \boldsymbol{v})$, then the perturbed function $\tilde{f}_j(\boldsymbol{r}, \boldsymbol{v}, t)$ is given by equation (4.58). Substituting this form of $\tilde{f}_j(\boldsymbol{r}, \boldsymbol{v}, t)$ into equations (4.47)–(4.51), the perturbed quantities

$$\begin{pmatrix} \tilde{n} \\ \tilde{V}_i \\ \tilde{P}_{ij} \\ \tilde{Q}_{ijk} \end{pmatrix}$$

are expressed in terms of linear functions of the perturbed electromagnetic field. Substituting them into equations (4.52)–(4.55), we obtain the fluxes as

$$\begin{pmatrix} \Gamma_i \\ P_{ix} \\ Q_{\perp x} \\ Q_{\parallel x} \end{pmatrix} = \text{quadratic form of } (\tilde{\boldsymbol{E}}, \tilde{\boldsymbol{B}}). \tag{4.59}$$

An explicit example of equation (4.59) is given in the appendix. One may write it in a matrix form as

$$\begin{pmatrix} \Gamma_i \\ P_{ix} \\ Q_{\perp x} \\ Q_{\parallel x} \end{pmatrix} = -\boldsymbol{M}\nabla \begin{pmatrix} n \\ V_i \\ T_\perp \\ T_\parallel \end{pmatrix}. \tag{4.60}$$

Each matrix element in $\boldsymbol{M}$ is now dependent on plasma inhomogeneities and has the form of a quadratic function of $(\tilde{\boldsymbol{E}}, \tilde{\boldsymbol{B}})$.

4.3.3 Energy Exchange via Fluctuations

Equation (4.45) describes an excitation of a wave by the jth element and an absorption of a wave by another element. This indicates the existence of energy flow from a jth element to another element via excited waves. Thus an energy exchange between plasma elements is also caused by fluctuations. The power partition is also influenced. For the simplicity, the case of electrostatic perturbation is shown in the following. The change of total electric field energy is expressed as

$$\frac{\partial}{\partial t}\left[\sum_{k\omega} \frac{(\varepsilon_0 k^2)}{2} |\tilde{\phi}_{k\omega}|^2\right] = -\sum_j \sum_{k\omega} \Re(\boldsymbol{k} \cdot \boldsymbol{\sigma}_{j,k\omega} \boldsymbol{k}) |\tilde{\phi}_{k\omega}|^2. \tag{4.61}$$

Contributions from various plasma species are separated, then equation (4.61) is written as

$$\frac{\partial}{\partial t}\left[\sum_{k\omega} \frac{(\varepsilon_0 k^2)}{2} |\tilde{\phi}_{k\omega}|^2\right] = \sum_j P_{j\to wave} \tag{4.62}$$

where the energy transfer from the jth species to the electric field energy is given as

$$P_{j\to wave} = -\sum_{k\omega} \Re(\boldsymbol{k}\cdot\boldsymbol{\sigma}_{j.k\omega}\boldsymbol{k})|\tilde{\phi}_{k\omega}|^2 \tag{4.63}$$

In a stationary state, the power $P_{e\to wave}$ and $P_{i\to wave}$ balance, i.e.,

$$P_{i\to wave} = -P_{e\to wave}. \tag{4.64}$$

This term $P_{e\to wave}$ expresses the rate of the energy exchange between plasma species, $P_{e\to i}$.

The energy exchange rate and the cross-field flux are related to each other. The energy exchange rate between electrons and waves, $P_{e\to wave}$, gives the rate of the energy transfer to ions in a stationary state. Comparing equations (4.42) and (4.63), one has

$$\langle V_p\rangle\Gamma_{e.x} = \frac{1}{eB}P_{e\to wave} \tag{4.65}$$

where the average phase velocity of the wave $\langle V_p\rangle$ is defined as

$$\langle V_p\rangle = \frac{\sum_{k\omega}\Re(\boldsymbol{k}\cdot\boldsymbol{\sigma}_{j.k\omega}\boldsymbol{k})|\tilde{\phi}_{k\omega}|^2}{\sum_{k\omega}\Re\left(\frac{k_y}{\omega}\boldsymbol{k}\cdot\boldsymbol{\sigma}_{j.k\omega}\boldsymbol{k}\right)|\tilde{\phi}_{k\omega}|^2}. \tag{4.66}$$

In a stationary state, we have

$$P_{e\to i} = eB\langle V_p\rangle\Gamma_{e.x}. \tag{4.67}$$

The result shows that the energy is transferred from electrons to ions, if the wave propagates in the direction of electron diamagnetic velocity. If, on the contrary, the wave propagates in the ion diamagnetic direction, the energy is given from ions to electrons. For the given propagation direction, the direction of energy flow is independent of the temperature difference between ions and electrons. An energy transfer can occur from a colder component to a hotter component.

Equation (4.67) holds as a general relation in the presence of fluctuations. As a specific example, let us apply it to a wave in the range of drift wave frequency, $\omega \sim \omega_* = k_y T_e/(eBL_n)$. ($L_n$ is the density gradient scale-length, $L_n^{-1} = |\nabla n|/n$.) The drift wave belongs to the modified ion acoustic branch, and is excited in the presence of the density gradient. The phase velocity is characterized by the drift velocity,

$$\langle V_p\rangle = C\frac{T_e}{eBL_n} \tag{4.68}$$

where C denotes the propagation direction and magnitude, and is of order unity. By substituting this into equation (4.65), we have

$$P_{e\to i} = C\frac{T_e}{L_n}\Gamma_{e.x}. \tag{4.69}$$

4.3.4 Momentum Conservation and Ambipolarity of the Flux

It is worth noting the momentum conservation relation associated with the perturbation-driven flux. Substituting equation (4.42-2) into (4.45), one has the relation

$$\frac{\partial}{\partial t}\sum_{k,\omega}\frac{k_y}{\omega}\frac{(\varepsilon_0 k^2)}{2}|\tilde{\phi}_{k\omega}|^2 = \sum_j e_j B_z \Gamma_{j.x} = -(\boldsymbol{J}\times\boldsymbol{B})_y. \tag{4.70}$$

This relation shows that an increasing change in the wave momentum (left-hand side) is equal to the momentum reduction rate of the plasma, i.e.,

$$\frac{\partial}{\partial t}P_{wave,y} = -(\boldsymbol{J}\times\boldsymbol{B})_y. \tag{4.71}$$

In the stationary state, if it exists, the relation $\partial P_{wave,y}/\partial t = 0$ holds. Then the momentum balance equation leads to an ambipolar condition

$$\sum_j e_j \Gamma_{j.x} = 0 \tag{4.72}$$

so long as an *average in a certain region* is treated. The basis and validity of ambipolarity is discussed in [4.4] and [4.11].

It should be noted that the ambipolarity does not necessarily hold as the local balance. As is noted in the beginning of section 4.3.1, the flux is calculated as an *average within a small plasma volume*. In general, the wave can propagate in the direction of the gradient ($\hat{x}$-direction), which is shown in figure 4.10. If the averaged forces $F(x)$ and $F(x')$ are given on thin layers near $x = x$ and $x = x'$, respectively, the momentum balance does not hold in each region. The momentum exchange between the different magnetic surfaces can take place as is discussed in the following chapters. In such a case, *local* charge neutrality does not hold.

4.4 Strong Turbulence (Far from Thermal Equilibrium)

The main issue in obtaining the fluctuation-driven flux is the determination of the fluctuation amplitude and the correlation time. Efforts to describe a strong turbulence are sketched.

In determining the level of fluctuations, one must derive the equation which dictates the temporal evolution of the wave amplitude. In the zero wave amplitude limit, it may be written as

$$\frac{\partial}{\partial t}\tilde{\phi} = \gamma_L\tilde{\phi} \tag{4.73}$$

where γ_L is the linear growth rate and $\tilde{\phi}$ is the fluctuation amplitude. When the amplitude grows, the growth of the wave would be affected by the other

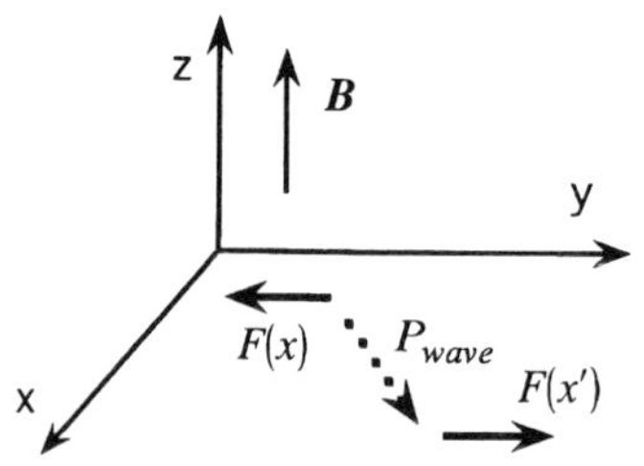

Figure 4.10. Wave is excited at x and is absorbed on a different magnetic surface, x'. Opposite flows are generated at x and x'. On each surface, x and x', the flux can be bipolar.

fluctuations. If there should exist a stationary state, the competing mechanisms by which modes are destabilized or damped must balance with each other. The nonlinear damping rate is expressed as $\gamma_{N.damp}$, and the growth of the mode may be written as

$$\frac{\partial}{\partial t}\tilde{\phi} = (\gamma_g - \gamma_{N.damp})\tilde{\phi}. \tag{4.74}$$

The growing rate γ_g may or may not be equal to the linear growth rate γ_L.

Typical examples in modelling the rate coefficients, γ_g and $\gamma_{N.damp}$, are illustrated as follows.

A widely employed *ansatz* is that the mode of interest is also subject to the diffusion which is caused by the scattering of other waves [4.12]. One further assumes that *the diffusion damping rate on the test mode is equal to the diffusion coefficient of the global structure*. This kind of *ansatz* has been used in the field of statistical physics like the Weiss approximation for the magnetization, or the Onsager *ansatz* on the decorrelation rate [4.13]. With the help of this simplicity, the damping rate on the test mode is given by the dimensional argument as

$$\gamma_{N.damp} \simeq Dk_\perp^2. \tag{4.75}$$

(A more elaborate discussion will be given later where it is necessary.) A further simplification is that the driving mechanism is unaffected, i.e., $\gamma_g = \gamma_L$. Substituting the relation $\gamma_g = \gamma_L$ and equation (4.75) into (4.74), one obtains the relation $\gamma_L = Dk_\perp^2$ for the stationary state. An order estimate

$$D \simeq \frac{\gamma_L}{k_\perp^2} \tag{4.76}$$

is obtained. This formula is sometimes referred to as the *mixing-length estimate* (or Kadomtsev formula [2.2]) because the essential assumption originates from the Prandtl mixing length estimate in fluid turbulence [4.14], aside from the estimate of nonlinear decorrelation time.

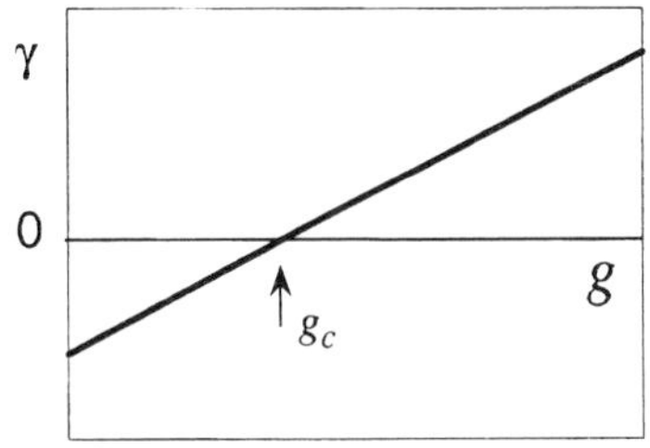

Figure 4.11. Schematic drawing of the linear growth rate of the instability as a function of a parameter.

When one employs this framework of analysis, a necessary step is the calculation of linear growth rate. A variety of instability modes have been theoretically found [2.10, 4.10]. However, this approach may not be sufficient in understanding the experimental observations. The comparison of the experimental data with the prediction has also been in progress [4.15].

A fundamental theoretical question is whether the mechanisms that causes instabilities are unaltered in strongly turbulent plasmas. The relation $\gamma_g = \gamma_L$ does not always hold. The variation of γ_g is considered in the following. The linear growth rate of the plasma instability often appears when some equilibrium parameter (say, g) reaches a criterion ($g = g_c$), as is illustrated in figure 4.11.

However, there have been several analyses that have pointed out a possibility of nonlinear destabilization in confined plasmas [4.16, 4.17]. Figure 4.12 shows schematically the growth rate as a function of the turbulence level. When the amplitude is small but finite, the growth rate can be an increasing function of the perturbation amplitude. The nonlinear growth rate, $\gamma_{N.grow}$, dominates over the linear growth rate and γ_g is replaced by $\gamma_{N.grow}$ as

$$\frac{\partial}{\partial t}\tilde{\phi} = (\gamma_{N.grow} - \gamma_{N.damp})\tilde{\phi}. \tag{4.77}$$

If this is the case, the memory of the linear growth rate does not directly reflect on the transport coefficient. The stationary state, if it exists, is expressed by the condition $\gamma_{N.grow} = \gamma_{N.damp}$.

The nonlinear instability is also related to the issue of investigating whether the plasma turbulence belongs to a family of *supercritical turbulence* or of *subcritical turbulence*. Consider the situation that this parameter g gradually changes and reaches the critical condition $g = g_c$ (figure 4.11). When the excitation of the turbulence is soft, the fluctuations are enhanced if $g \geq g_c$ (figure 4.13(a)). However, it could also be possible that the finite amplitude fluctuation is sustained below the linear stability criterion $g = g_c$, as is demonstrated in figure 4.13(b). The former is called *supercritical*, and the latter *subcritical*.

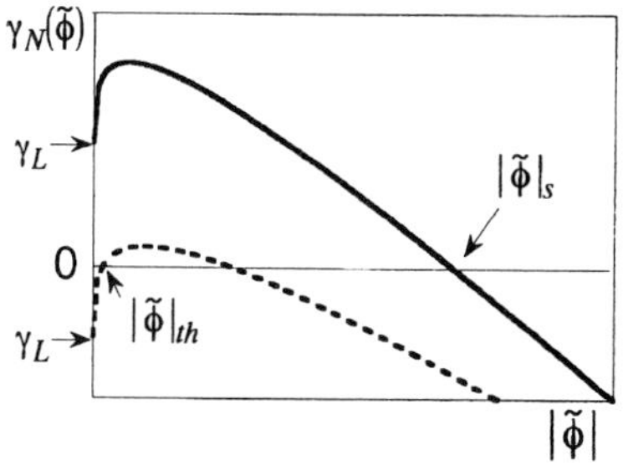

Figure 4.12. Nonlinear growth rate for subcritical turbulence as a function of fluctuation level. Linearly unstable case ($g > g_c$, solid line) and linearly stable case ($g < g_c$, dashed line) are shown. If linearly stable, it becomes unstable above the threshold amplitude $|\tilde{\phi}|_{th}$. In the stationary state, the fluctuation level $|\tilde{\phi}| = |\tilde{\phi}|_s$ is reached.

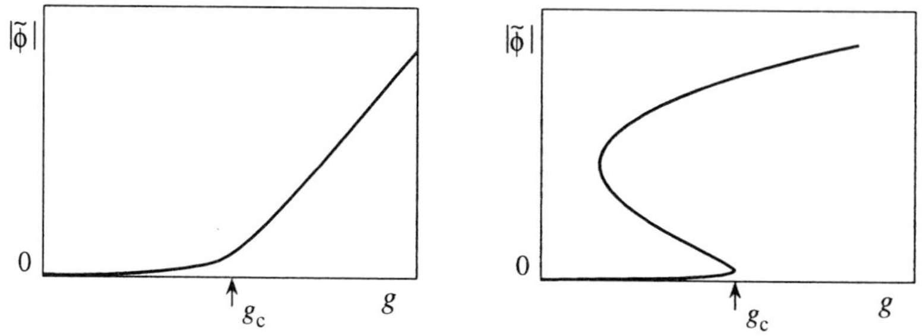

Figure 4.13. Fluctuation level as a function of the controlling parameter g. Linear instability appears for $g > g_c$. The case of supercritical turbulence (left) and subcritical turbulence (right).

For a subcritical excitation, the growth rate has a nature such that nonlinear interaction among the finite amplitude fluctuations *destabilizes* the mode. The theme of this monograph is to present a view based on the subcritical excitation of plasma turbulence.

Appendix 4A Formula for Cross-Field Flux in a Quasilinear Theory

Consider the case where the distribution function of the plasma is given by a shifted Maxwellian,

$$f(v) = \frac{n}{(2\pi/m)^{3/2} T_\perp \sqrt{T_\parallel}} \exp\left\{ -\frac{mv_\perp^2}{T_\perp} - \frac{m(v_\parallel - V_\parallel)^2}{T_\parallel} \right\}. \tag{4A.1}$$

The density, n, parallel flow velocity, $V_\parallel$, and temperatures, $T_\perp$ and $T_\parallel$, are considered to be inhomogeneous in the x-direction. Substituting this form into equation (4.58), the perturbed distribution function $\tilde{f}$ is calculated. Combining $\tilde{f}$ with equations (4.52)–(4.55), we obtain the cross-field fluxes explicitly as follows [4.18]

$$\Gamma_x = \frac{1}{B}\sum_k \int \frac{\mathrm{d}\omega}{2\pi}\left(\frac{-\mathrm{i}e_j}{k_y T_\perp}\right)\Upsilon \tag{4A.2}$$

$$\begin{aligned}P_{zx} = {} & \frac{m}{B}\sum_k \int \frac{\mathrm{d}\omega}{2\pi}\left(\frac{-\mathrm{i}e_j}{k_y T_\perp}\right)\left(\frac{\omega}{k_\parallel} - V_\parallel\right)\Upsilon \\ & + \frac{m}{B}\sum_k \int \frac{\mathrm{d}\omega}{2\pi}\left(\frac{-\mathrm{i}e_j}{k_y T_\perp}\right)(I_{12}+I_{22})\frac{k_\parallel}{\omega}\frac{\Omega^2}{k_\perp^2}\left(\frac{T_\perp}{T_\parallel} - 1 + \mathcal{D}\right) b\Lambda_0' n \\ & \times \frac{m}{B}\sum_k \int \frac{\mathrm{d}\omega}{2\pi}\left(\frac{-\mathrm{i}e_j}{k_y T_\perp}\right) I_{23}\frac{\Omega}{\Omega_i}\frac{k_\parallel}{\omega}\frac{\Omega^2}{k_\perp^2}\left(\frac{T_\perp}{T_\parallel} - 1 + \mathcal{D}^*\right)(-2b^2\Lambda_0')n\end{aligned} \tag{4A.3}$$

$$\begin{aligned}Q_{x\parallel} = {} & -T_\parallel\Gamma_x + \frac{m}{B}\sum_k \int \frac{\mathrm{d}\omega}{2\pi}\left(\frac{-\mathrm{i}e_j}{k_y T_\perp}\right)\left(\frac{\omega}{k_\parallel} - V_\parallel\right)^2\Upsilon \\ & + \sum_k \int \frac{\mathrm{d}\omega}{2\pi}\left(\frac{-\mathrm{i}me_j}{Bk_y T_\perp}\right)(I_{12}+I_{22})\frac{\Omega^2}{k_\perp^2} \\ & \times \left(\frac{\omega - k_\parallel V_\parallel}{\omega}\left(\frac{T_\perp}{T_\parallel} + \mathcal{D}\right) - \frac{T_\perp k_y}{\Omega m\omega}\frac{\partial}{\partial x}\right) 2b^2\Lambda_0' n \\ & + \sum_k \int \frac{\mathrm{d}\omega}{2\pi}\left(\frac{-\mathrm{i}me_j}{Bk_y T_\perp}\right) I_{23}\left(\frac{\Omega}{\Omega_i}\right)^2\frac{\Omega^2}{k_\perp^2} \\ & \times \left(\frac{\omega - k_\parallel V_\parallel}{\omega}\left(\frac{T_\perp}{T_\parallel} + \mathcal{D}\right) - \frac{T_\perp k_y}{\Omega m\omega}\frac{\partial}{\partial x}\right)(-4b^2\Lambda_0')n\end{aligned} \tag{4A.4}$$

$$\begin{aligned}Q_{x\perp} = {} & -2T_\perp\Gamma_x + \sum_k \int \frac{\mathrm{d}\omega}{2\pi}\left(\frac{-\mathrm{i}me_j}{Bk_y T_\perp}\right)\left(\frac{\Omega}{k_\perp}\right)^2\frac{1}{2}(I_{11}+2I_{12}+I_{22}) \\ & \times (\mathcal{D}-\mathcal{D}^*)b(\Lambda_0 - \Lambda_0' + b\Lambda_0')n + \sum_k \int \frac{\mathrm{d}\omega}{2\pi}\left(\frac{-\mathrm{i}me_j}{Bk_y T_\perp}\right)\left(\frac{\Omega}{k_\perp}\right)^2 \\ & \times (I_{13}+I_{23})\frac{\Omega}{\Omega_i}(\mathcal{D}-\mathcal{D}^*)b^2(\Lambda_0 - \Lambda_0' + 2b\Lambda_0')n \\ & + \sum_k \int \frac{\mathrm{d}\omega}{2\pi}\left(\frac{-\mathrm{i}me_j}{Bk_y T_\perp}\right)\left(\frac{\Omega}{k_\perp}\right)^2 (I_{13}+I_{23})\frac{\Omega}{\Omega_i}2\left(\frac{T_\perp}{T_\parallel} - 1 + \mathcal{D}\right) b^2\Lambda_0' n \\ & + \sum_k \int \frac{\mathrm{d}\omega}{2\pi}\left(\frac{-\mathrm{i}me_j}{Bk_y T_\perp}\right)\left(\frac{\Omega}{k_\perp}\right)^2 I_{23}\frac{\Omega}{\Omega_i}\left(\frac{\omega - k_\parallel V_\parallel}{\omega} - \frac{T_\perp k_y}{\Omega m\omega}\frac{\partial}{\partial x}\right) 2b^2\Lambda_0' n\end{aligned}$$

$$+\sum_k \int \frac{d\omega}{2\pi}\left(\frac{-ime_j}{Bk_yT_\perp}\right)\left(\frac{\Omega}{k_\perp}\right)^2 I_{33}\left(\frac{\Omega}{\Omega_i}\right)^2$$
$$\times(\mathcal{D}-\mathcal{D}^*)b^3(\Lambda_0-2\Lambda_0'+2b\Lambda_0')n \tag{4A.5}$$

with

$$\Upsilon=\frac{1}{2}(I_{11}+2I_{12}+I_{22})(\mathcal{D}-\mathcal{D}^*)\Lambda_0 n-(I_{13}+I_{23})\frac{\Omega}{\Omega_i}(\mathcal{D}-\mathcal{D}^*)b\Lambda_0' n$$
$$-I_{33}(\mathcal{D}-\mathcal{D}^*)\frac{\Omega^2}{\Omega_i^2}b^2\Lambda_0' n. \tag{4A.6}$$

(The suffix j, which denotes the particle species, is suppressed except for charges and ion cyclotron frequency.) In these expressions, the operator $\mathcal{D}$ is defined by the relation

$$\mathcal{D}F\equiv\left(\frac{\omega-k_\parallel V_\parallel}{\omega}\frac{T_\perp}{T_\parallel}-\frac{T_\perp k_y}{\Omega m\omega}\frac{\partial}{\partial x}\right)\left[\frac{\omega}{\sqrt{2T_\parallel/m}|k_\parallel|}Z\left(\frac{\omega-k_\parallel V_\parallel}{\sqrt{2T_\parallel/m}|k_\parallel|}\right)F\right] \tag{4A.7}$$

where

$$Z(\zeta)\equiv\frac{1}{\sqrt{\pi}}\int_{-\infty}^{\infty}dx\,\frac{\exp(-x^2)}{x-\zeta-i(+0)} \tag{4A.8}$$

is the plasma dispersion function [4.19], and

$$\Lambda_0\equiv I_0(b)\,e^{-b}\qquad \Lambda_0'=\partial\Lambda_0/\partial b \tag{4A.9}$$

$$b\equiv\frac{T_\perp}{m}\frac{k_\perp^2}{\Omega^2}=k_\perp^2\rho^2 \tag{4A.10}$$

represents the finite-gyro-radius effects. ($I_0(b)$ being the zeroth order modified Bessel function of the first kind). The quadratic form of the electromagnetic field is expressed by symbols ($I_{11}, I_{12}, I_{22}, I_{13}, I_{23}, I_{33}$) as

$$\langle\tilde{E}_y(k,\omega)\tilde{E}_y(k',\omega')\rangle=2\pi\delta_{k,-k'}\delta(\omega+\omega')I_{11} \tag{4A.11}$$

$$\frac{\omega}{k_\parallel c}\langle\tilde{E}_y(k,\omega)\tilde{B}_x(k',\omega')\rangle=2\pi\delta_{k,-k'}\delta(\omega+\omega')I_{12} \tag{4A.12}$$

$$\left(\frac{\omega}{k_\parallel c}\right)^2\langle\tilde{B}_x(k,\omega)\tilde{B}_x(k',\omega')\rangle=2\pi\delta_{k,-k'}\delta(\omega+\omega')I_{22} \tag{4A.13}$$

$$i\frac{\Omega_i k_y}{k_\perp^2 c}\langle\tilde{E}_y(k,\omega)\tilde{B}_z(k',\omega')\rangle=2\pi\delta_{k,-k'}\delta(\omega+\omega')I_{13} \tag{4A.14}$$

$$i\frac{\Omega_i k_y}{k_\perp^2 c}\frac{\omega}{k_\parallel c}\langle\tilde{B}_x(k,\omega)\tilde{B}_z(k',\omega')\rangle=2\pi\delta_{k,-k'}\delta(\omega+\omega')I_{23} \tag{4A.15}$$

$$\left(\frac{\Omega_i k_y}{k_\perp^2 c}\right)^2\langle\tilde{B}_z(k,\omega)\tilde{B}_z(k',\omega')\rangle=2\pi\delta_{k,-k'}\delta(\omega+\omega')I_{33}. \tag{4A.16}$$

REFERENCES

[4.1] Chirikov B V 1979 *Phys. Rep.* **52** 263

[4.2] Lichtenberg A J and Liebermann M A 1984 *Regular and Stochastic Motion* (New York: Springer)

[4.3] Galeev A A 1984 *Basic Plasma Physics II* ed A A Galeev and R N Sudan (Amsterdam: North-Holland) ch 6.2

Krommes J, Oberman C and Kleva R G 1983 *J. Plasma Phys.* **30** 11

Balescu R 1975 *Equilibrium and Nonequilibrium Statistical Mechanics* (New York: Wiley)

[4.4] Horton C W 1984 *Basic Plasma Physics II* ed A A Galeev and R N Sudan (Amsterdam: North-Holland) ch 6.4

[4.5] Rosenbluth M N, Hinton F L and Hazeltine R D 1972 *Phys. Fluids* **15** 116

Balescu R 1988 *Transport Processes in Plasmas 2. Neoclassical Transport Theory* (Amsterdam: North-Holland)

[4.6] Ware A A 1970 *Phys. Rev. Lett.* **25** 15

[4.7] Bickerton R D, Connor J W and Taylor J B 1971 *Nature Phys. Sci.* **229** 110

[4.8] Kennel C F and Engelman F A O 1966 *Phys. Fluids* **9** 2377

[4.9] Sagdeev R Z and Galeev A A 1969 *Nonlinear Plasma Theory* ed T M O'Neil and D L Book (New York: Benjamin) ch 1

[4.10] Yoshikawa S 1970 *Methods of Experimental Physics* vol 9 ed H R Griem and R H Loveberg (New York: Academic Press) ch 8

Liewer P 1985 *Nucl. Fusion* **25** 543

[4.11] Inoue S, Tange T, Itoh K and Tuda T 1979 *Nucl. Fusion* **19** 1252

[4.12] Dupree T H 1967 *Phys. Fluids* **10** 1049

Dupree T H 1966 *Phys. Fluids* **9** 1773

Dupree T H 1968 *Phys. Fluids* **11** 2680

Dupree T H 1972 *Phys. Fluids* **15** 334

Dupree T H 1974 *Phys. Fluids* **17** 100

See also Nishikawa K and Osaka Y 1965 *Prog. Theor. Phys.* **33** 402

[4.13] The relation of the kinetic coefficients for the macro and micro variables in the Onsager theory is explained by Landau L D and Lifshitz E M 1980 *Statistical Physics, Part 1* 3rd edn, transl. J B Sykes *et al* (Oxford: Pergamon) section 120

[4.14] Landau L D and Lifshitz E M 1987 *Fluid Mechanics* 2nd edn, transl. J B Sykes *et al* (Oxford: Pergamon) section 36

[4.15] Connor J W, Maddison G P, Wilson H R, Corrigan G, Stringer T E and Tibone F 1993 *Plasma Phys. Control. Fusion* **35** 319

[4.16] Rebut P H and Hugon M 1985 *Plasma Physics and Controlled Nuclear Fusion Research 1984* vol 2 (Vienna: IAEA) p 197

Hirshman S P and Molvig K 1979 *Phys. Rev. Lett.* **42** 648

Biskamp D and Walter M 1985 *Phys. Lett.* **109A** 34

Sydora R D, Leboeuf J N and Tajima T 1985 *Phys. Fluids* **28** 528

Lichtenberg A J, Itoh K, Itoh S-I and Fukuyama A 1992 *Nucl. Fusion* **32** 495

Fukuyama A, Itoh K, Itoh S-I, Tsuji S and Lichtenburg A J 1993 *Plasma Physics and Controlled Nuclear Fusion Research 1992* vol 2 (Vienna: IAEA) p 363

Scott B D 1992 *Phys. Fluids* B **4** 2468

Carreras B A, Sidikman K, Diamond P H, Terry P W and Garcia L 1992 *Phys. Fluids* B **4** 3115

Nordman H, Pavlenko V P and Weiland J 1993 *Phys. Plasmas* B **5** 402
Knobloch E and Weiss N O J 1983 *Physica* D **9** 379 (see also Bekki N and Karakisawa T 1995 *Phys. Plasmas* **2** 2945)
[4.17] Itoh K, Itoh S-I and Fukuyama A 1992 *Phys. Rev. Lett.* **69** 1050
Itoh K, Itoh S-I, Fukuyama A, Yagi M and Azumi M 1994 *Plasma Phys. Control. Fusion* **36** 279
[4.18] Tange T, Inoue S, Itoh K and Nishikawa K 1979 *J. Phys. Soc. Japan* **46** 266
[4.19] Fried B D and Conte S D 1961 *The Plasma Dispersion Function* (New York: Academic)

Chapter 5

Low Frequency Modes in Confined Plasmas

In this chapter, typical examples of low frequency modes in plasmas are explained. The mechanisms to determine the dispersion relation are described.

5.1 Modes and Dispersion Relations

Fluctuations in plasmas do not form a white noise, but oscillatory patterns are often excited. The spatio-temporal patterns that appear (either stationary or propagating) are called *modes*. The spectrum of mode is a characterizing information in the continuous media. If one imposes an external perturbation of the form

$$\tilde{\boldsymbol{E}}_{ext} \exp(\mathrm{i}\boldsymbol{k} \cdot \boldsymbol{x} - \mathrm{i}\omega t)$$

(where $\boldsymbol{k}$, ω can be complex), then the charged elements of plasma are displaced so as to generate a response field

$$\tilde{\boldsymbol{E}}_{induced} \exp(\mathrm{i}\boldsymbol{k} \cdot \boldsymbol{x} - \mathrm{i}\omega t).$$

If there is a relation of $(\boldsymbol{k}, \omega)$

$$\omega = \omega(\boldsymbol{k}) \tag{5.1}$$

for which the ratio $\tilde{E}_{induced}/\tilde{E}_{ext}$ becomes very large, then a pattern $\exp\{\mathrm{i}\boldsymbol{k}\cdot\boldsymbol{x} - \mathrm{i}\omega(\boldsymbol{k})t\}$ is expected to have a large amplitude and to be selectively observed. Such a pattern is called a *mode*, and the relation (5.1) is called the dispersion relation.

The search of various modes in confined plasmas is a fundamental task in investigating the fluctuations and transport in plasmas. The parameters $(\boldsymbol{k}, \omega)$ are related to the spatial-temporal scales that play roles in the transport processes.

The dispersion relation is determined by the electric conductivity tensor in equation (4.39), $\tilde{\boldsymbol{J}}_{k\omega} = (\sum_j \sigma_{j.k\omega})\tilde{\boldsymbol{E}}_{k\omega}$. The dielectric tensor is introduced (according to the convention) as

$$\varepsilon(k,\omega) = \boldsymbol{I} - \frac{\mathrm{i}c^2\mu_0}{\omega}\left(\sum_j \sigma_{j.k\omega}\right) \tag{5.2}$$

where $\boldsymbol{I}$ is the unit tensor, μ_0 is the magnetic permeability of vacuum and c is the velocity of light. Substituting equation (4.39) into Maxwell's equation one has the relation

$$\left[\varepsilon(k,\omega) - \left(\frac{kc}{\omega}\right)^2\left(\boldsymbol{I} - \frac{\boldsymbol{kk}}{k^2}\right)\right]\boldsymbol{E}_{k.\omega} = \frac{c\mu_0}{\mathrm{i}\omega}\boldsymbol{j}_{ext:k.\omega} \tag{5.3}$$

where $\boldsymbol{j}_{ext:k.\omega}$ is an externally imposed current perturbation of the Fourier component $(\boldsymbol{k},\ \omega)$. Equation (5.3) predicts that the perturbed field $\boldsymbol{E}_{k.\omega}$ can take a finite amplitude even without the external perturbation, if the condition

$$\det\left|\varepsilon(\boldsymbol{k},\omega) - \left(\frac{kc}{\omega}\right)^2\left(\boldsymbol{I} - \frac{\boldsymbol{kk}}{k^2}\right)\right| = 0 \tag{5.4}$$

is satisfied. Equation (5.4) is the condition that $\boldsymbol{E}_{k.\omega}$ has a nontrivial solution for $\boldsymbol{j}_{ext:k.\omega} = 0$. The dispersion relation (5.1) is given as a solution of equation (5.4).

The relation (5.1) could depend on the amplitude of perturbations: if it is dependent, it is a *nonlinear dispersion relation*. In the zero-amplitude limit of perturbations, equation (5.1) is the *linear* dispersion relation. The literature explains the various linear modes in plasmas in detail (e.g. [2.1–2.6]). Among various modes in plasmas, those of low frequencies have particular importance in the study of plasma transport. This is because the radial excursion of plasma elements is larger if the frequency is lower, for given amplitude of perturbations, as is shown in equation (4.7). In addition to this, fluctuations which are relevant to the transport are often excited by the plasma inhomogeneities.

5.1.1 Fluid Equations

The response of the plasma to a perturbed field is calculated based on the continuity equation, equation of motion and energy balance equation (equation of state). In the fluid description of the plasma, they are given as [2.3, 2.4]

$$\frac{\partial}{\partial t}n_j + \nabla\cdot(n_j\boldsymbol{V}_j) = 0 \tag{5.5}$$

$$m_j n_j \frac{\mathrm{d}}{\mathrm{d}t}\boldsymbol{V}_j = e_j n_j(\boldsymbol{E} + \boldsymbol{V}_j\times\boldsymbol{B}) - \nabla p_j - \nabla\cdot\boldsymbol{\Pi}_j \tag{5.6}$$

and

$$\frac{3}{2}n_j\frac{\partial}{\partial t}T_j + p_j\nabla\cdot\boldsymbol{V}_j = -\nabla\cdot\boldsymbol{q}_j - \boldsymbol{\Pi}_j\nabla\boldsymbol{V}_j + P_j \tag{5.7}$$

where

$$\frac{\mathrm{d}}{\mathrm{d}t} = \frac{\partial}{\partial t} + \boldsymbol{V} \cdot \boldsymbol{\nabla} \tag{5.8}$$

is the total time derivative, $\boldsymbol{q}_j$ is the heat flux, $\boldsymbol{\Pi}_j$ is the viscous tensor and P_j is the energy input. The off-diagonal tensor $\boldsymbol{\Pi}$ vanishes for the case of an isotropic velocity distribution. The energy input includes the energy exchange between plasma species.

The set of equations requires the knowledge on the energy transport and deformation of the distribution function; it is *not* closed in the realm of a fluid description, but needs the kinetic analysis of the plasma response. In some cases, however, a relevant and simplified expression is obtained, and is called the *equation of state*. If the temporal change is fast in comparison with the flow rate of the energy, then the adiabatic condition

$$\frac{\mathrm{d}}{\mathrm{d}t}(p_j n_j^{-\gamma}) = 0 \tag{5.9}$$

is a good approximate relation to close the set of equations. The parameter γ in this equation is the specific heat ratio. In contrast, if the energy equilibration is faster, then the simplification

$$T_j = \text{ constant} \tag{5.10}$$

is a good approximation.

These equations (5.5)–(5.7) and (5.9) (or (5.10)) describe the plasma dynamics combined with Maxwell's equation below

$$\boldsymbol{\nabla} \times \boldsymbol{B} = \mu_0 \Big(\sum_j \boldsymbol{J}_j + \varepsilon_0 \frac{\partial}{\partial t} \boldsymbol{E} \Big) \tag{5.11}$$

$$\boldsymbol{\nabla} \times \boldsymbol{E} = -\frac{\partial}{\partial t} \boldsymbol{B}. \tag{5.12}$$

5.1.2 Linearization

The evolution of perturbations is studied by the expansion of plasma parameters as

$$n = n_0 + \tilde{n} \qquad \boldsymbol{V} = \boldsymbol{V}_0 + \tilde{\boldsymbol{V}} \qquad p = p_0 + \tilde{p} \tag{5.13}$$

in the vicinity of equilibrium parameters. (Here 'equilibrium' refers to the stationary mechanical equilibrium, and is denoted by the suffix 0. The suffix j, distinguishing plasma species, is suppressed if not confused.) Terms are retained up to the first order of perturbation. The equation of motion and the continuity equation, for perturbed components, are given as

$$mn \left(\frac{\partial}{\partial t} \tilde{\boldsymbol{V}} + \boldsymbol{V}_0 \cdot \boldsymbol{\nabla} \tilde{\boldsymbol{V}} + \tilde{\boldsymbol{V}} \cdot \boldsymbol{\nabla} \boldsymbol{V}_0 \right) = en(\tilde{\boldsymbol{E}} + \boldsymbol{V}_0 \times \tilde{\boldsymbol{B}} + \tilde{\boldsymbol{V}} \times \boldsymbol{B}_0) - \boldsymbol{\nabla} \tilde{p} \tag{5.14}$$

$$\frac{\partial}{\partial t}\tilde{n} + \nabla \cdot (\tilde{n}V_0 + n_0\tilde{V}) = 0. \tag{5.15}$$

With the help of the equation of state, one obtains either

$$\frac{\partial}{\partial t}\tilde{p}n_0^{-\gamma} - \gamma p_0 n_0^{-\gamma-1}\frac{\partial}{\partial t}\tilde{n} + \tilde{V} \cdot \nabla(p_0 n_0^{-\gamma}) = 0 \qquad \text{(adiabatic limit)} \tag{5.16}$$

or

$$\tilde{T} = 0 \qquad \text{(iso-thermal limit)}. \tag{5.17}$$

The perturbed density, velocity and temperature (or pressure) ($\tilde{n}_j$, $\tilde{V}_j$, $\tilde{T}_j$) are calculated in terms of the perturbed electromagnetic fields ($\tilde{\boldsymbol{E}}$, $\tilde{\boldsymbol{B}}$).

5.1.3 Model Picture of Plasmas

A basic approximation in the plasma response to a low frequency perturbation is as follows. Namely,

(i) electrons can almost freely move along the field line, or
(ii) the parallel electric conductivity is high,

characterizing the *high-temperature* plasmas. The magnetic field is strong, i.e.,

(iii) the magnetic pressure $B^2/2\mu_0$ is much higher than the plasma pressure p.

The ratio of the plasma pressure to the magnetic pressure is called the 'plasma beta-value'. It is conventionally expressed as

$$\beta \equiv \frac{2\mu_0 p}{B^2} \tag{5.18}$$

and is considered to be much smaller than unity here. The condition

(iv) the scale-length is much longer than the Debye length, $k\lambda_D \ll 1$,

is often employed.

5.2 Sound Wave and Shear Alfvén Wave

When a perturbation amplitude is small, and a linear response of a plasma is of interest, then a calculation of the dielectric tensor follows a well defined procedure. We leave the detailed analysis of linear response functions in literature (e.g., [2.1–2.8]) and here explain two examples of the low frequency modes in magnetized plasmas, the *sound mode* and the *shear Alfvén mode*. These two branches of waves are defined in a uniform plasma and are modified in various ways in non-uniform plasmas. The sound mode and the shear Alfvén mode constitute the basis for understanding the fluctuations in confined plasmas.

A uniform, stationary and slab (planar) plasma is considered (figure 5.1). Influences of nonuniformity are discussed in the following sections.

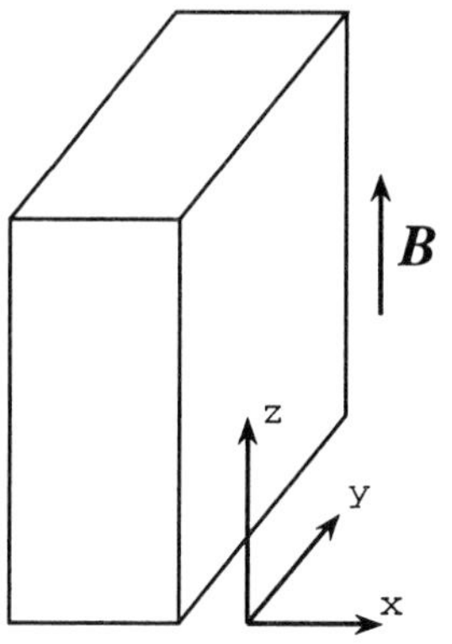

Figure 5.1. Slab plasma model with strong magnetic field.

5.2.1 Ion Sound Wave

Let us consider a perturbation that is propagating in the direction of magnetic field,

$$\boldsymbol{k} = (0, 0, k_\parallel). \tag{5.19}$$

(The propagation in the direction of the magnetic field is not an essential feature of the ion sound wave. This choice is made just for simplicity.) In terms of the fluid description, equilibrium plasma parameters (n_0, T_0) are constant and the plasma is staying still, $V_0 = 0$. In this circumstance, a response to an electrostatic perturbation,

$$\tilde{\boldsymbol{E}} = -\nabla\tilde{\phi}$$

is studied. The relevant mode is in the range

$$|k_\parallel v_{thi}| \ll \omega \ll |k_\parallel v_{the}| \tag{5.20}$$

where v_{thi} and v_{the} are thermal velocities of ions and electrons, respectively. The ion response is approximated as adiabatic. The electron response is iso-thermal, because the phase velocity $\omega/k_\parallel$ is slower than the thermal velocity v_{the}.

The equation of motion for ions is written as

$$-\mathrm{i}m_i n_i \omega \tilde{V}_{i.z} = -\mathrm{i}k_\parallel e_i n_i \tilde{\phi} - \mathrm{i}k_\parallel \tilde{p}_i. \tag{5.21}$$

The continuity equation yields the relation

$$-\mathrm{i}\omega\tilde{n}_i + \mathrm{i}k_\parallel n_0 \tilde{V}_{i.z} = 0. \tag{5.22}$$

If the adiabatic relation is used for ions, the equation of state (5.9) yields $\tilde{p}_i = \gamma_i T_i \tilde{n}_i$. Eliminating $\tilde{V}_i$ and $\tilde{p}_i$ from equations (5.21) and (5.22) with the help of the adiabatic condition, one has

$$\frac{\tilde{n}_i}{n_i} = \frac{Z_i T_e k_\parallel^2}{m_i \omega^2} \frac{1}{1 - \gamma_i k_\parallel^2 v_{thi}^2 \omega^{-2}} \frac{e\tilde{\phi}}{T_e} \tag{5.23}$$

where Z_i is the charge number of ions, $e_i = Z_i e$.

Because the electron mass is small, the equation of motion for electrons gives an approximate relation as

$$-en_e \tilde{E}_z - \mathrm{i}k_\parallel \tilde{p}_e \cong 0.$$

The fast motion of electrons tends to equilibrate the temperature along the field line, so that an iso-thermal approximation, $\tilde{T}_e \simeq 0$, is used. With the help of equation (5.17), the electron density is given as

$$\frac{\tilde{n}_e}{n_e} = \frac{e\tilde{\phi}}{T_e}. \tag{5.24}$$

The Poisson equation is written in the form of a charge neutrality condition, i.e.,

$$\tilde{n}_e \cong Z_i \tilde{n}_i. \tag{5.25}$$

Substituting the responses (5.23) and (5.24) into equation (5.25), the dispersion relation is finally given as

$$\omega^2 = c_s^2 k_\parallel^2 \tag{5.26-1}$$

with

$$c_s^2 = \frac{(Z_i T_{e0} + \gamma_i T_{0i})}{m_i}. \tag{5.26-2}$$

The dispersion relation (5.26), shown in figure 5.2, indicates that this perturbation propagates with a constant phase velocity,

$$|\omega / k_\parallel| = c_s$$

which is called the ion sound velocity.

This ion sound mode is similar to sound waves in a neutral gas. Note that the ion sound speed is finite in the limit of

$$T_{i.0} \rightarrow 0.$$

An expansion of ions in a condensed layer to a rarefied layer is governed by the electric field, not only by the pressure of ions. This is in contrast to a sound wave in neutral gas, where the expansion is governed by the pressure. The ion sound mode in plasmas has a collective nature, where plasma particles in different locations are interacting with each other through the long range electric field. Free electron motions along a field line, given in equation (5.24), play an essential role in determining the pattern of compression or decompression of ions.

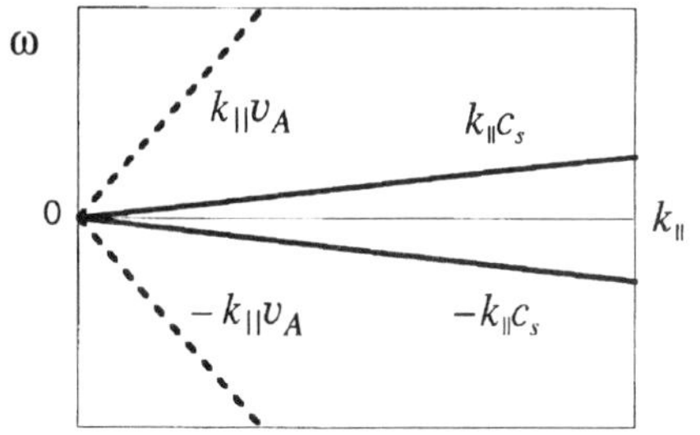

Figure 5.2. Dispersion relation of the ion sound wave (solid line) and the shear Alfvén wave (dashed line).

5.2.2 Shear Alfvén Wave

A nearly free electron motion along the field line gives another type of mode in a magnetized plasma. The magnetic perturbation can also propagate in a plasma as a mode. The *shear Alfvén mode* is accompanied by the perturbed magnetic field in the direction perpendicular to a main magnetic field, $\tilde{B}_{\perp}$. (There is another type of perturbation, which is associated with the magnetic perturbation in the direction of the main magnetic field, $\tilde{B}_{\|}$. For such a mode, a perturbation of magnetic energy associated with the wave is large. The excitation of such a mode requires the larger free energy.)

For an electromagnetic perturbation, the response in a plasma to induce a perturbed current is a key. We consider a perturbation which propagates in the z-direction, equation (5.19), with the component

$$\tilde{\boldsymbol{B}} = (\tilde{B}_x, 0, 0) \tag{5.27}$$

$$\tilde{\boldsymbol{E}} = \left(0, -\frac{\omega}{k_{\|}}\tilde{B}_x, 0\right). \tag{5.28}$$

In expressing the electric field perturbation $\tilde{E}_y$ in terms of the magnetic perturbation, $\tilde{B}_x$, we use equation (5.12).

Summing up equation (5.14) for electrons and ions, and subtracting one from the other, one has the MHD equations. Components in the direction perpendicular to the magnetic field are of interest,

$$m_i n_i \frac{\partial}{\partial t}\tilde{\boldsymbol{V}} = \tilde{\boldsymbol{J}} \times \boldsymbol{B}_0 - \boldsymbol{\nabla}\tilde{p} \tag{5.29}$$

$$\tilde{E}_y + (\tilde{\boldsymbol{V}} \times \boldsymbol{B}_0)_y - \frac{1}{en}\boldsymbol{\nabla}_y \tilde{p}_i = 0. \tag{5.30}$$

In the calculation of equation (5.30), the ion inertia effect is neglected since the change rate is slow in time compared to the ion cyclotron frequency. The

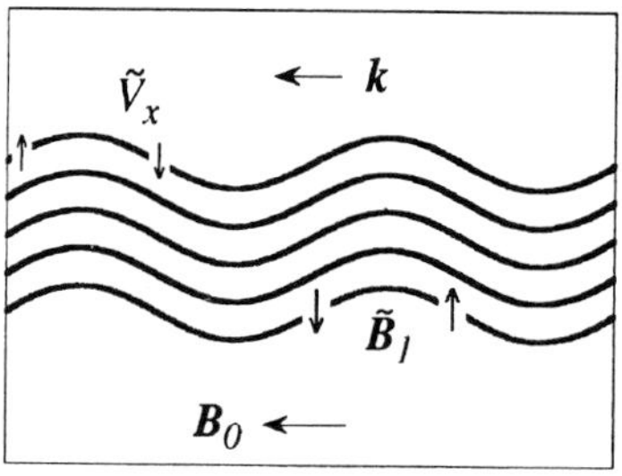

Figure 5.3. Flow and distorted magnetic field line for the propagation of a shear Alfvén wave.

electron inertia effect is also neglected. In this simplified approximation, a friction on electrons by ions is also neglected. These equations are often called the *ideal MHD equation*. For perturbations of the form of equation (5.19), equations (5.29) and (5.30) provide the relation

$$\tilde{V}_x = B_0^{-1}\tilde{E}_y \tag{5.31}$$

and

$$\tilde{J}_y = -\frac{\mathrm{i}m_i n_i \omega}{B_0^2}\tilde{E}_y = \frac{\mathrm{i}m_i n_i \omega^2}{B_0^2 k_\parallel}\tilde{B}_x. \tag{5.32}$$

Substituting equation (5.32) into Maxwell's equation,

$$\tilde{B}_x = -\frac{\mathrm{i}\mu_0}{k_\parallel}\tilde{J}_y \tag{5.33}$$

one has the dispersion relation

$$\omega^2 = k_\parallel^2 v_A^2 \tag{5.34}$$

with

$$v_A^2 = \frac{B_0^2}{m_i n_i \mu_0} \tag{5.35}$$

where v_A is called the Alfvén velocity. The dispersion relation is plotted in figure 5.2.

The patterns of current, displacement and perturbed field are shown in figure 5.3. The plasma motion, equation (5.31), is expressed as

$$\tilde{V}_x = (\tilde{B}_x/B_0)v_A.$$

This shows that the plasma motion satisfies the frozen-in condition, i.e., the distortion of the magnetic field lines coincides with the displacement of the plasma. The frozen-in condition follows from Ohm's law equation (5.30), and is satisfied generally so long as the parallel resistance is not important.

An approximately free electron motion along the field line is essential in determining the mode structures of both the ion sound mode and the Alfvén mode. The 'free' parallel motion of electrons is an idealization. In reality, however, the free electron motion is just an approximation, and a small but finite impedance affects an idealized 'free' electron motion. This impedance governs the dynamics of such modes as is explained in the next chapter.

5.3 Drift Wave and Drift-Alfvén Wave

5.3.1 Diamagnetic Drift

When a plasma is inhomogeneous, as is illustrated in figure 5.4, there appears a preferential direction for a wave. The force balance of a plasma implies that there is a current, i.e., the difference of flow velocities of ions and electrons, on the magnetic surface. A current which flows perpendicular to a magnetic field line is called a diamagnetic current. (A current in the direction of the magnetic field is called a force-free current, because it does not give rise to the Lorentz force.) The flow velocity of a plasma in the $\boldsymbol{\nabla} p \times \boldsymbol{B}$-direction is named a diamagnetic velocity. It is given in the presence of the density gradient as

$$\boldsymbol{V}_d = \frac{T_j}{e_j B} \frac{1}{n_j} \frac{\mathrm{d}n_j}{\mathrm{d}x} \hat{y} \tag{5.36}$$

in the configuration of figure 5.4. It is seen that the diamagnetic current, $\boldsymbol{J}_d = (e_i n_i V_{di} - e n_e V_{de})\hat{\boldsymbol{y}}$, satisfies the force balance equation, $\boldsymbol{J}_d \times \boldsymbol{B} = \boldsymbol{\nabla} p$, equation (3.1). A similar diamagnetic flow is generated by the temperature gradient as well.

In a microscopic picture, the flow equation (5.36) is generated by the difference of the number density of gyrating (i.e., cyclotron motion) particles (figure 5.5(a)). In a small volume of interest at location x, ions which move in the y-direction have the gyro-centre at the position with larger x, $x' = x + \rho_i$. Ions moving in the $-y$-direction have the centre at the position with smaller x, $x' = x - \rho_i$. Because of the density gradient in the x-direction, the number of particles moving in the $-y$-direction is larger. As a result of this difference, the fluid velocity results in the $-y$-direction.

In contrast, the drift due to an inhomogeneity of magnetic field causes motion of the gyro-centre of the particles in the drift direction (in the direction of $\boldsymbol{\nabla} B \times \boldsymbol{B}$). Figure 5.5(b) illustrates the case of inhomogeneous magnetic field; a field is mainly directed in the z-direction, but its magnitude gradually increases in the x-direction. Due to the small change of magnetic field across the cyclotron radius, a cyclotron orbit does not close itself on the x–y-plane, and a resultant drift in the y-direction appears. The drift velocity due to the

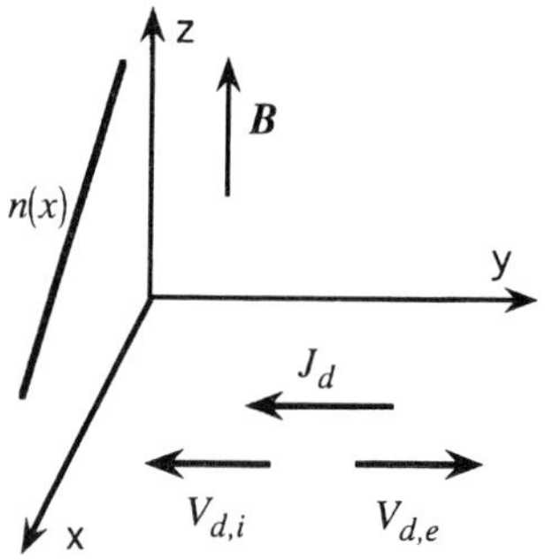

Figure 5.4. Diamagnetic flow V_d and diamagnetic current J_d in an inhomogeneous plasma.

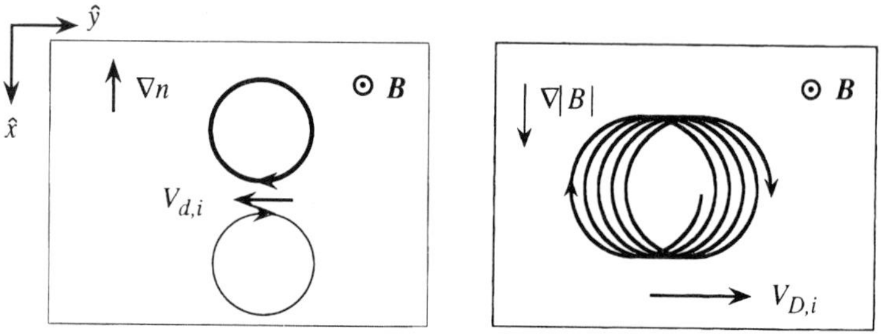

Figure 5.5. Gyro-motion and diamagnetic flow of ions in inhomogeneous plasma (a). Configuration of the plasma is the same as figure 5.4. If the magnetic field strength is varying in the x-direction, flow in the y-direction V_D also appears (b).

magnetic field inhomogeneity, $V_{D.j}$, is given as

$$V_{D.j} = \frac{W_j}{e_j B}\left(\frac{1}{|B|}\frac{\mathrm{d}|B|}{\mathrm{d}x}\right)\hat{y} \tag{5.37}$$

where W_j is the energy of particles. The diamagnetic drift and the magnetic curvature drift are in the same direction (additive) if the condition

$$\nabla p \cdot \nabla|B| > 0 \tag{5.38}$$

holds.

As is shown in the next chapters, the sign of the product $\nabla p \cdot \nabla|B|$ is important for the developments of fluctuations in inhomogeneous plasmas. The case of equation (5.38) is called *bad curvature*. (Good curvature if

$\nabla p \cdot \nabla|B| < 0$.) The name 'curvature' is used because the magnetic field strength becomes inhomogeneous when the magnetic field line is slightly curved. For a vacuum magnetic field, the scale-length of curvature and the inhomogeneity scale-length of field strength are equal (see also subsection 6.2.1).

5.3.2 Drift Wave

When the drift velocity is not negligible in comparison with the phase velocity, a preferential direction appears in a wave propagation. Consider an electrostatic perturbation, which is obliquely propagating,

$$\boldsymbol{k} = (0, k_y, k_\parallel). \tag{5.39}$$

The electric field is expressed in terms of the perturbed potential as

$$\tilde{\boldsymbol{E}} = -\mathrm{i}\boldsymbol{k}\tilde{\phi}. \tag{5.40}$$

In an inhomogeneous plasma, a new term appears in the continuity equation, equation (5.22), as

$$-\mathrm{i}\omega\tilde{n}_i + \mathrm{i}k_\parallel n_{i0}\tilde{V}_{iz} + \tilde{V}_{i.x}\frac{\mathrm{d}n_{i0}}{\mathrm{d}x} = 0 \tag{5.41}$$

with

$$\tilde{V}_{i.x} = -\mathrm{i}k_y\tilde{\phi}/B.$$

(One may expect that a Doppler shift term like $\mathrm{i}k_y V_{di}$ would appear in the time derivative, e.g., $\mathrm{i}k_y V_{di}\tilde{n}_i$ in the left-hand side of equation (5.41). However, this is not the case. Recall that particles do not move in the direction of the diamagnetic drift, as is explained in relation with figure 5.5(a) [1.6].) The equation of ion motion in the z-direction is unchanged and equation (5.21) gives a relation

$$\tilde{V}_{i.z} = k_\parallel m_i^{-1}\omega^{-1}(e_i\tilde{\phi} + n_{i0}^{-1}\gamma_i T_{i0}\tilde{n}_i) \tag{5.42}$$

with the equation of state, equation (5.9). Substituting perturbed velocities $\tilde{V}_{i.x}$ and $\tilde{V}_{i.z}$ into equation (5.41), one finds the perturbed ion density as

$$\frac{\tilde{n}_i}{n_{i0}} = \left(\omega - \frac{k_\parallel^2}{\omega m_i}\gamma_i T_{i0}\right)^{-1}\left(\frac{k_\parallel^2 T_i}{\omega m_i} - \frac{k_y T_i}{e_i B n_{i0}}\frac{\mathrm{d}n_{i0}}{\mathrm{d}x}\right)\frac{e_i\tilde{\phi}}{T_i}. \tag{5.43}$$

The free motion of electrons gives the Boltzmann relation for the electron response as equation (5.24), $\tilde{n}_e/n_e = e\tilde{\phi}/T_e$. Substitution of equations (5.24) and (5.43) into the charge neutrality condition $e_i\tilde{n}_i = e\tilde{n}_e$ gives the dispersion relation as

$$\omega^2 - \omega\omega_* - k_\parallel^2 c_s^2 = 0 \tag{5.44}$$

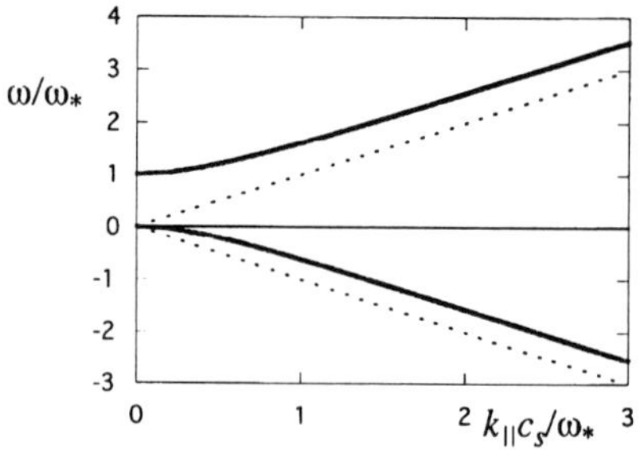

Figure 5.6. Dispersion relation for the electrostatic drift wave.

where ω_* is the drift frequency defined by

$$\omega_* = k_y V_{de} = -\frac{k_y T_e}{eBn_0}\frac{\mathrm{d}n_0}{\mathrm{d}x}. \tag{5.45}$$

The dispersion relation is shown in figure 5.6.

When the mode propagates nearly perpendicular to the magnetic field line, $k_y^2 V_{de}^2 \gg k_\parallel^2 c_s^2$, i.e.,

$$|k_y|\frac{\rho_s}{L_n} > |k_\parallel| \tag{5.46}$$

the frequency of the mode is approximately given as

$$\omega = \omega_*. \tag{5.47}$$

In equation (5.46) L_n is the scale-length of the density gradient,

$$-\nabla n/n = (1/L_n)\hat{x}$$

and ρ_s is the ion cyclotron radius evaluated by the electron temperature.

If the wave propagates mainly in the z-direction, $k_y \to 0$, the dispersion relation reduces to that of an ion sound wave.

5.3.3 Drift-Alfvén Wave

The electromagnetic mode is also influenced by an inhomogeneity. A preferential direction of propagation is chosen as equation (5.39), $\boldsymbol{k} = (0, k_y, k_\parallel)$. The perturbed magnetic field is in the x-direction, like equation (5.28), but couples with E_z due to the oblique propagation. The perturbed field is expressed as

$$\tilde{\boldsymbol{B}} = (\tilde{B}_x, 0, 0) \tag{5.48}$$

$$\tilde{\boldsymbol{E}} = (0, \tilde{E}_y, \tilde{E}_z). \tag{5.49}$$

Perturbations of magnetic and electric fields are related to each other, and are expressed as

$$\tilde{B}_x = \frac{1}{\omega}(k_y \tilde{E}_z - k_\parallel \tilde{E}_y) \tag{5.50}$$

by use of Maxwell's equation.

Plasma responses are calculated as in the preceding section. The calculation is straightforward, but a little lengthy. The explicit derivation is shown in appendix 5A. The perturbed current in the z-direction is obtained as

$$\tilde{J}_z = \frac{im_i n_{i0}}{B^2} \frac{(\omega + |\omega_{*i}|)k_y}{k_\parallel} \tilde{E}_y \tag{5.51}$$

where ω_{*i} is the ion-drift frequency,

$$\omega_{*i} = -\frac{T_i}{T_e}\omega_*. \tag{5.52}$$

Ampère's law, $\partial \tilde{B}_x / \partial y = \mu_0 \tilde{J}_z$, combines the magnetic and electric perturbations as

$$\tilde{B}_x = -\frac{(\omega + |\omega_{*i}|)}{k_\parallel v_A^2} \tilde{E}_y. \tag{5.53}$$

This relation shows that the magnetic component becomes noticeable if the phase velocity is in the range of Alfvén velocity, $|\omega / k_\parallel v_A| > 1$. The dispersion relation is given as

$$\omega^2 - \omega\omega_* - k_\parallel^2 c_s^2 = \frac{\omega^2(\omega - \omega_*)(\omega + |\omega_{*i}|)}{k_\parallel^2 v_A^2}. \tag{5.54}$$

Equation (5.54) describes drift branches (5.43) and the electromagnetic branches. In the electrostatic limit, $|\omega / k_\parallel v_A| \ll 1$, equation (5.54) reproduces the dispersion relation of the electrostatic drift waves. In an opposite limit, $|\omega / k_\parallel v_A| > 1$, it provides the drift-Alfvén mode. In the limit of $|\omega / k_\parallel v_A| > 1$, the relation $|\omega / k_\parallel c_s| \gg 1$ holds, because the sound speed is much smaller than the Alfvén velocity for the parameter of our interest, $\beta \ll 1$. Neglecting the $k_\parallel^2 c_s^2$ term in the left-hand side, one simplifies equation (5.54) as

$$(\omega - \omega_*)(\omega^2 + \omega|\omega_{*i}| - k^2 v_A^2) = 0. \tag{5.55}$$

In addition to the drift wave, $\omega \simeq \omega_*$, two other modes are obtained. The electromagnetic branch, in the small $k_\parallel$ limit, becomes the mode propagating in the direction of the ion diamagnetic drift,

$$\omega = -\frac{T_i}{T_e}\omega_*. \tag{5.56}$$

In a large $k_\parallel$ limit, $|k_\parallel| v_A \gg \omega_*$, the shear Alfvén wave is recovered. The dispersion relation, which is obtained from equation (5.54), is illustrated in figure 5.7.

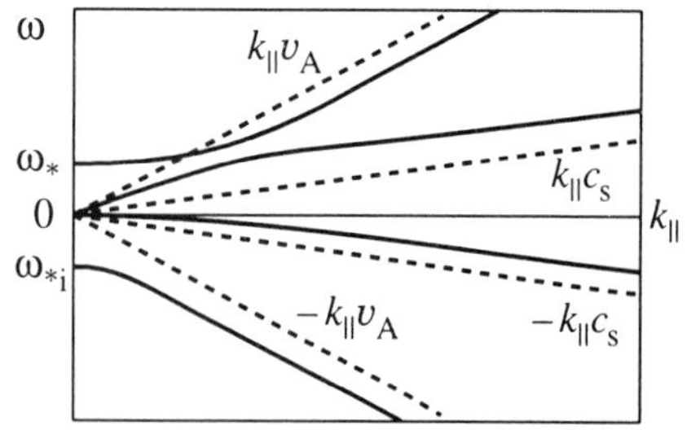

Figure 5.7. Dispersion relation for the drift wave and drift-Alfvén wave.

Appendix 5A Drift-Alfvén Wave

Equation of motion for electrons in the z-direction is given as

$$-\mathrm{i}\omega m_e n_e \tilde{V}_{e.z} = -en_e\tilde{E}_{\parallel} - \mathrm{i}k_{\parallel}T_e\tilde{n}_e + en_e V_{de}\tilde{B}_x. \tag{5A.1}$$

Neglecting the electron mass, we have

$$\frac{\tilde{n}_e}{n_e} = \frac{\mathrm{i}e\tilde{E}_{\parallel}}{k_{\parallel}T_e} - \mathrm{i}V_{de}\frac{e\tilde{B}_x}{k_{\parallel}T_e}. \tag{5A.2}$$

The magnetic perturbation appears through the Lorentz force.

The equation of motion for ions in the z-direction is expressed as

$$-\mathrm{i}\omega m_i n_i \tilde{V}_{i.z} = e_i n_i\tilde{E}_{\parallel} - \mathrm{i}k_{\parallel}\gamma_i T_i\tilde{n}_i + e_i n_i V_{di}\tilde{B}_x. \tag{5A.3}$$

The perturbed perpendicular velocity of ions is given as

$$\tilde{V}_{i.x} = \frac{\bar{\tilde{E}}_y}{B} - \frac{\mathrm{i}k_y\gamma_i T_i}{eB}\frac{\tilde{n}_i}{n_i} \tag{5A.4}$$

$$\tilde{V}_{i.y} = -\frac{\mathrm{i}m_i\omega}{eB^2}\bar{\tilde{E}}_y - \frac{m_i\omega k_y\gamma_i T_i}{e^2B^2}\frac{\tilde{n}_i}{n_i}. \tag{5A.5}$$

In this expression $\bar{\tilde{E}}_y$ indicates the electric field, which is averaged over the gyro-motion of ions. (See figure 5A.1.) It is given as

$$\bar{\tilde{E}}_y \simeq (1 - k_y^2\rho_i^2)\tilde{E}_y. \tag{5A.6}$$

The $\boldsymbol{E} \times \boldsymbol{B}$ velocity is nearly equal for ions and electrons. Owing to the larger gyro-orbit for ions, the effective electric field acting on ions is smaller than that on electrons.

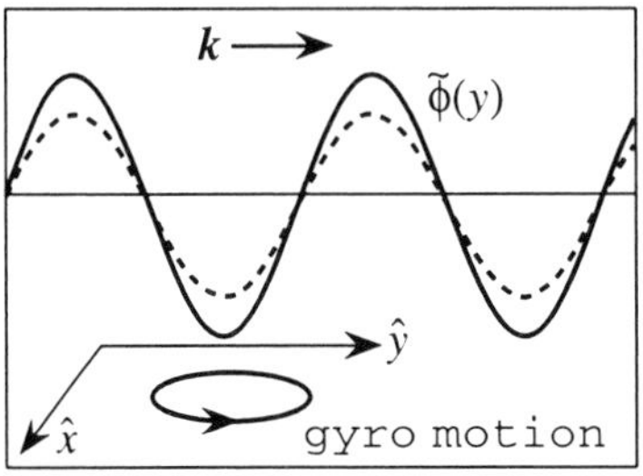

Figure 5A.1. Ions feel an averaged field (dashed line) owing to the gyro-motion. The effective field is weaker than the real field (solid line).

The ion response is calculated by substituting equations (5A.3), (5A.4) and (5A.5) into the continuity equation as

$$\frac{\tilde{n}_i}{n_{i0}} = \left(\omega - \frac{k_\parallel^2}{\omega m_i}\gamma_i T_{i0}\right)^{-1}\left[(\omega_* - \omega k_y^2\rho_i^2)\frac{\mathrm{i}e\tilde{E}_y}{k_y T_e} + \frac{k_\parallel^2 T_e}{\omega m_i}\frac{\mathrm{i}e\tilde{E}_z}{k_\parallel T_e} - \frac{\mathrm{i}ek_\parallel V_{di}}{\omega m_i}\tilde{B}_x\right]. \tag{5A.7}$$

Keeping the first order correction of the finite-gyro-radius effect, $k_y^2\rho_i^2 \ll 1$, we obtain the ion density perturbation as

$$\begin{aligned}\frac{\tilde{n}_i}{n_{i0}} &= \left(\omega - \frac{k_\parallel^2}{\omega m_i}\gamma_i T_{i0}\right)^{-1} \\ &\times \left[(\omega_* - (\omega + \omega_*)k_y^2\rho_i^2)\frac{\mathrm{i}e\tilde{E}_y}{k_y T_e} + \frac{k_\parallel^2 T_e}{\omega m_i}\frac{\mathrm{i}e\tilde{E}_z}{k_\parallel T_e} - \frac{\mathrm{i}ek_\parallel V_{di}}{\omega m_i}\tilde{B}_x\right].\end{aligned} \tag{5A.8}$$

In the electrostatic limit, this is reduced to equation (5.41) if the finite-gyro-radius effect is neglected.

The charge neutrality condition, $\tilde{n}_e/n_e = \tilde{n}_i/n_i$, with equations (5A.2) and (5A.3), gives the relation as

$$\begin{aligned}&\left\{\omega - \omega_* - \frac{k_\parallel^2 c_s^2}{\omega} + (\omega + \omega_*)k_\perp^2\rho_i^2\right\}\frac{\mathrm{i}e\tilde{E}_y}{k_y T_e} \\ &\quad = -\omega\left\{\omega - \omega_* + (\omega_* - |\omega_{*i}| - \omega)\frac{k_\parallel^2 c_s^2}{\omega^2}\right\}\frac{\mathrm{i}e\omega\tilde{B}_x}{T_e k_y k_\parallel}.\end{aligned} \tag{5A.9}$$

In a derivation of equation (5A.9), $\tilde{E}_z$ is expressed in terms of $\tilde{E}_y$ and $\tilde{B}_x$ by use of equation (5.50).

The ratio of the magnetic perturbation to the electric perturbation is given from Ampère's law. The perturbed current is calculated as

$$\nabla_\parallel \cdot \tilde{J}_z = -\nabla_\perp \cdot \{en_0(V_{i.x} - V_{e.x})\hat{\boldsymbol{x}} + en_0(V_{i.y} - V_{e.y})\hat{\boldsymbol{y}}\}. \tag{5A.10}$$

The divergence of net current vanishes, because the charge neutrality condition holds. Owing to the finite-gyro-radius effect, the divergence of the current in the x-direction remains as

$$\nabla_\perp \cdot \{en_0(V_{i.x} - V_{e.x})\hat{\boldsymbol{x}}\} = -e\frac{\mathrm{d}n_0}{\mathrm{d}x}k_y^2\rho_i^2\frac{\tilde{E}_y}{B}. \tag{5A.11}$$

The divergence of the perturbed diamagnetic current is given as

$$\nabla_\perp \cdot \{en_0(V_{i.y} - V_{e.y})\hat{\boldsymbol{y}}\} = \frac{m_i n_i k_y \omega}{B^2}\tilde{E}_y. \tag{5A.12}$$

Substitution of equations (5A.11) and (5A.12) into equation (5A.10) gives the parallel current as

$$\tilde{J}_z = \frac{m_i n_i}{B^2}\frac{\mathrm{i}(\omega + |\omega_{*i}|)k_y}{k_\parallel}\tilde{E}_y. \tag{5A.13}$$

Substituting equation (5A.13) into Ampère's law, $\partial\tilde{B}_x/\partial y = \mu_0\tilde{J}_z$, we have

$$\tilde{B}_x = \frac{\mu_0 m_i n_i}{B^2}\frac{(\omega + |\omega_{*i}|)}{k_\parallel}\tilde{E}_y = \frac{(\omega + |\omega_{*i}|)}{k_\parallel v_A^2}\tilde{E}_y. \tag{5A.14}$$

Combining equation (5A.14) and equation (5A.9), we have the dispersion relation as

$$\left\{\omega - \omega_* - \frac{k_\parallel^2 c_s^2}{\omega} + (\omega + \omega_*)k_\perp^2\rho_i^2\right\}$$
$$= \omega\left\{\omega - \omega_* + (\omega_* - |\omega_{*i}| - \omega)\frac{k_\parallel^2 c_s^2}{\omega^2}\right\}\frac{(\omega + |\omega_{*i}|)}{k_\parallel^2 v_A^2}. \tag{5A.15}$$

Noting the relation $c_s^2 \ll v_A^2$ for the parameters of our interest, $\beta \ll 1$, the right-hand side of equation (5A.15) is simplified as

$$\frac{\omega(\omega - \omega_*)(\omega + |\omega_{*i}|)}{k_\parallel^2 v_A^2}.$$

If the finite-gyro-radius effect in the left-hand side of equation (5A.15) is neglected, $k_y^2\rho_i^2 \ll 1$, equation (5A.15) becomes

$$\left(\omega - \omega_* - \frac{k_\parallel^2 c_s^2}{\omega}\right) = \frac{\omega(\omega - \omega_*)(\omega + |\omega_{*i}|)}{k_\parallel^2 v_A^2}. \tag{5A.16}$$

Equation (5A.16) provides the dispersion relation for the low frequency electromagnetic modes in inhomogeneous plasmas.

The dispersion relation (5A.16) provides two approximate relations. They are expressed for the slow branch, $\omega^2 \ll k_\parallel^2 v_A^2$, as

$$(\omega^2 - \omega_* \omega - k_\parallel^2 c_s^2) \simeq 0 \tag{5A.17}$$

and for the fast branch, $\omega^2 \gg k_\parallel^2 c_s^2$, as

$$(\omega^2 + |\omega_{*i}|\omega - k_\parallel^2 v_A^2) \simeq 0. \tag{5A.18}$$

Equation (5A.17) corresponds to the drift waves, and the latter equation (5A.18) represents the drift-Alfvén waves.

Chapter 6

Low Frequency Instabilities in Confined Plasmas

In chapter 5, dispersion relations of modes are discussed. Various plasma modes can be excited by a small amount of external perturbation, if the spatio-temporal structure satisfies the dispersion relation. In inhomogeneous and nonequilibrium plasmas, modes can be spontaneously excited, and violate the symmetry of the mechanical equilibrium state. Such a mode is called instability. The propagating pattern $\exp(\mathrm{i}\boldsymbol{k}\cdot\boldsymbol{x}-\mathrm{i}\omega t)$ grows exponentially in time if the imaginary part of the frequency, $\omega=\omega_r+\mathrm{i}\gamma$, is positive. Unstable modes are expected to appear at a high level in fluctuations. Owing to this reason, much work has been devoted to the linear stability analysis of various plasma modes. If one tries to account for all of the geometrical effects in an experimental setup, a variety in theoretical analyses on plasma instabilities would be seen (as was the case in the literature). Since inhomogeneity is one of the origins of plasma instability, an analysis in a complex geometry would be inevitable. However, we here try to reduce the geometrical complexity as far as possible, and choose typical examples which are relevant in confined plasmas.

6.1 Reactive Instability and Dissipative Instability

There are two typical mechanisms in inducing instability. They cause reactive instability and dissipative instability.

The equation of motion in the presence of a small amplitude perturbation could be rewritten in a study of the ion sound wave as

$$m_i n_i \frac{\partial^2}{\partial t^2}\tilde{V}_{iz} = T_e n_i \nabla^2 \tilde{V}_{iz} \tag{6.1}$$

which yields the dispersion relation $\omega^2 = c_s^2 k_\parallel^2$. By the introduction of a displacement $\tilde{V}_{iz} = \partial\xi_z/\partial t$, the relation is rewritten as

$$m_i \frac{\partial^2}{\partial t^2}\xi_z = -T_e k_\parallel^2 \xi_z. \tag{6.2}$$

This relation means that the restoring force, $-T_e k_\parallel^2 \xi_z$, works in the opposite direction to the displacement ξ_z, like a spring. A small perturbation does not grow in time but oscillates. In contrast, if the sign of the restoring force is reversed, i.e., the force is in the direction to increase the displacement, then a small perturbation grows exponentially. When there is a mechanism giving rise to a force in the direction of displacement, it is called *reactive instability*. Another type is *dissipative instability*.

As an analogy to plasma instabilities, let us study the motion of a point mass under gravity shown in figure 6.1. The surface to constrain the motion is given as

$$z = -\frac{1}{L}(x^2 + y^2). \tag{6.3}$$

It is well known that a point $(x, y) = (0, 0)$ is an *unstable* equilibrium point. If the position is perturbed by the amount of ξ_x in the x-direction, the force in the x-direction is given as $F_x = 2mg\xi_x/L$, where m is the mass of the particle and g is the gravity. This force causes an exponential growth of deviation ξ_x with the growth rate

$$\gamma_0 = \sqrt{\frac{2g}{L}}. \tag{6.4}$$

This belongs to a family of reactive instability.

Consider the case where the point mass in figure 6.1 is charged (charge Q) and the system is subject to a uniform vertical magnetic field B (in the

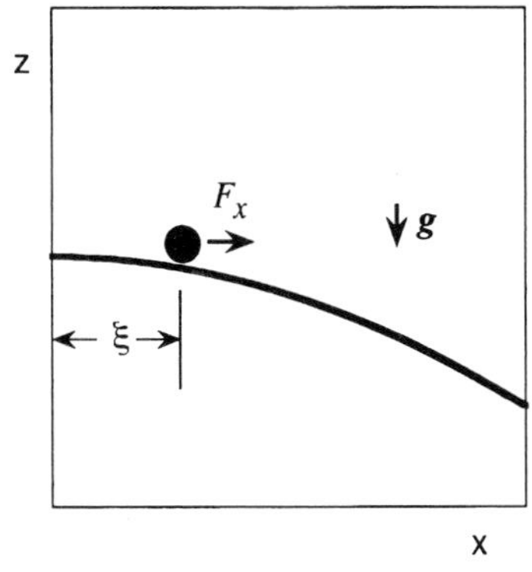

Figure 6.1. Mass point on a sliding surface.

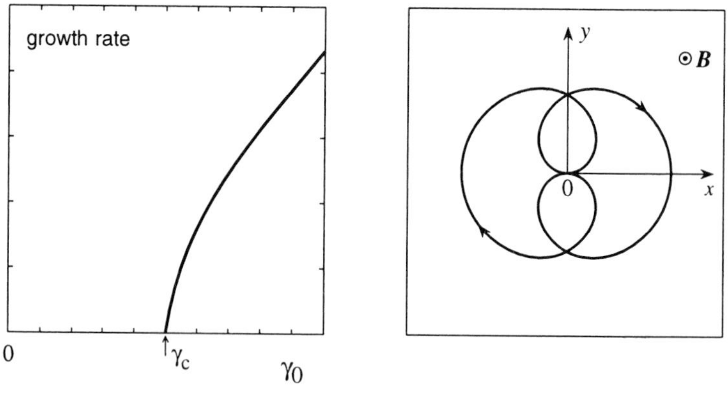

(a) (b)

Figure 6.2. Growth rate as a function of the driving parameter (a). Example of the stable orbit ($\gamma_0 < \gamma_c$) is shown in (b).

z-direction). In this situation, the Lorentz force modifies the growth rate. For temporal evolution of the type $\exp\{-\mathrm{i}\omega t\}$, the eigenvalue ω is given as

$$\omega = \frac{\pm\left(\Omega \pm \sqrt{\Omega^2 - 4\gamma_0^2}\right)}{2} \tag{6.5}$$

where $\Omega = QB/m$. The growth rate is shown in figure 6.2(a). An example of the trajectory is presented in figure 6.2(b). There is a critical value for the driving term,

$$(g/L)_c = \Omega^2/8$$

and the system is stable in the region of

$$(g/L) < (g/L)_c \qquad \text{i.e., } \gamma_0 \leq \gamma_c = \Omega/2. \tag{6.6}$$

This system shows a feature of *dissipative* instability. If there is a friction between the point mass and the surface, the equation of motion ($\boldsymbol{v} = \partial\boldsymbol{\xi}/\partial t$) is written as

$$\frac{\partial}{\partial t}\boldsymbol{v} = \gamma_0^2\boldsymbol{\xi} + \Omega\boldsymbol{v} \times \hat{\boldsymbol{z}} - \nu\boldsymbol{v}. \tag{6.7}$$

For the temporal evolution in the form of $\exp(-\mathrm{i}\omega t)$, the eigenvalue ω is given as

$$\omega = \frac{-\mathrm{i}\nu + \Omega \pm \mathrm{i}\sqrt{4\gamma_0^2 - \Omega^2 + 2\mathrm{i}\nu\Omega + \nu^2}}{2}. \tag{6.8}$$

Now the system turns out to be always unstable. In particular, near the marginal stability condition, $(g/L) \sim (g/L)_c$, the growth rate of perturbation is approximately given as

$$\mathrm{Im}(\omega) \cong \frac{1}{2}\sqrt{\nu\Omega}. \tag{6.9}$$

The growth rate depends on the fractional power of the dissipation rate. Even if the friction is small, $\nu/\Omega \ll 1$, its impact on the stability is prominent. This is also characteristic of the dissipative instability.

In the zero-driving limit, $\gamma_0 \to 0$, the growth rate of dissipative mode behaves as

$$\mathrm{Im}(\omega) \cong \frac{\nu\gamma_0^2}{\Omega^2}. \tag{6.10}$$

The growth rate is linearly proportional to the frictional coefficient.

In the presence of a dissipation, the gravitational energy is slightly dissipated by the friction. Hence, the velocity is reduced. The Lorentz force becomes less and is not enough to restore the orbit to the original point even in the region $(g/L) < (g/L)_c$. The point mass gradually slides down the slope, releasing the gravitational energy. This is an example of a mechanism where a dissipation can cause instability, and cause an exponential growth of the perturbation with a symmetry breaking [2.1]. The growth rate and typical trajectory are illustrated in figure 6.3.

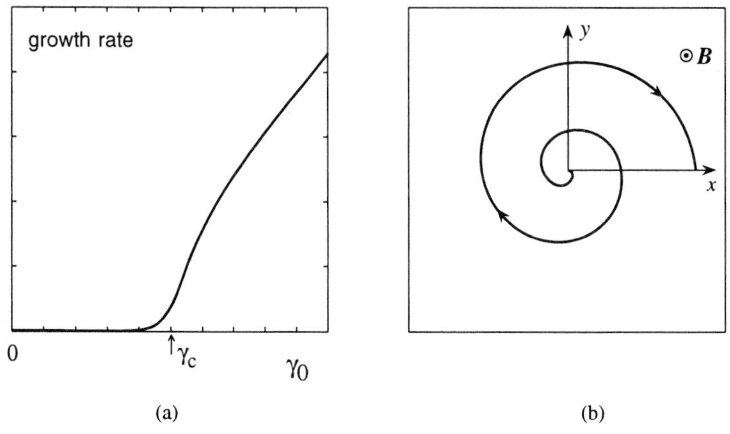

Figure 6.3. Dissipative instability exists in the region of weak gradient, $\gamma_0 < \gamma_c$. Growth rate against the driving parameter (a), and a typical example of a trajectory for the case of $\gamma_0 = \gamma_c$ and $\nu \simeq \gamma_0/100$ (b).

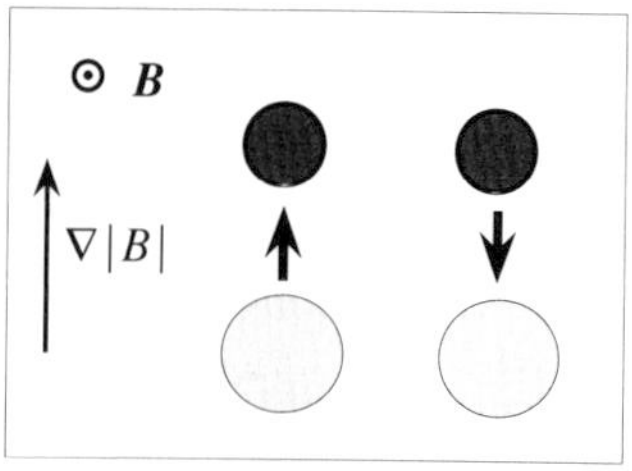

Figure 6.4. Motion of plasma in the direction of the gradient of the magnetic field. A cross-section of the plasma tube is shown. In this figure, plasma is compressed if it is moved upward, and expands when moved downward.

The dissipative instability is important from the viewpoint of nonequilibrium thermodynamics. A reactive instability usually has a larger linear growth rate; however, linear reactive instabilities could be eliminated by a proper choice of the plasma configurations [2.7–2.11]. Nonlinearities may play a dominant role in stabilizing such perturbations. In contrast, in a dissipative instability the dissipation may not be unaltered but could be *enhanced* by the development of fluctuations. If this is the case, the dissipative instability could have the nature of subcritical turbulence. This situation provides more variety in the dynamics in comparison with supercritical turbulence.

6.2 Magnetic Curvature and Pressure Gradient

An inhomogeneity of the magnetic field strength plays a role like that of the gravity in the model of figure 6.1. When plasma moves across a main magnetic field, the magnetic flux in each plasma element is approximately conserved. This nature originates from the high electrical conductivity along the magnetic field line, which is due to the approximately free electron motion. The magnetic field is often described as *frozen in* plasma [2.4, 1.6]. If this is the case, then the cross-section of the plasma element changes as it moves in the direction of the gradient of magnetic field strength. It is compressed when it moves to a high field side; it is expanded when it moves to a low field side (figure 6.4). The plasma may release the free energy when it expands. If this released energy is transformed so as to enhance a plasma displacement, as is the case of a mass point on the top of a hill, then an instability occurs. The confined state of the plasma is easily destroyed.

Owing to this nature, the influence of magnetic field inhomogeneity has long been investigated in relation to the plasma confinement. A simplest example is known as *interchange instability* [6.1]. In a real system, the curvature of the magnetic field line is not constant, and a perturbation may be localized in a particular region along the magnetic field line. In the latter case, the perturbation is called a *ballooning mode* [2.1].

6.2.1 Magnetic Well and Magnetic Hill

Consider a case where a magnetic field (mainly in the z-direction) is inhomogeneous in the x-direction. The cross-section of the x–z-plane is shown in figure 6.5, where the x-axis is taken in the direction of inhomogeneity, and the z-axis is in the direction of the strong magnetic field. The magnetic field inhomogeneity is usually induced by the curvature of the field line. The limiting case is taken as an example, where a plasma current is small and does not affect the magnitude of the magnetic field. (This condition is approximately satisfied in wide circumstances.) Ampère's law $\nabla \times \boldsymbol{B} = \mu_0 \boldsymbol{J} = 0$ provides the relation

$$B_x = \left(\frac{\partial}{\partial x} B_z \right) z \tag{6.11}$$

where the origin $z = 0$ is taken at the point where $B_x = 0$ is satisfied. The magnetic field line is expressed as, with the integration of the equation of the field line $\mathrm{d}x / B_x = \mathrm{d}z / B_z$, as

$$x = \frac{1}{2} \left(\frac{1}{B_z} \frac{\partial B_z}{\partial x} \right) z^2 .$$

It is expressed by use of a curvature R_M as

$$x = \frac{1}{2R_M} z^2 . \tag{6.12}$$

The curvature R_M is related to the gradient of the strength of the magnetic field as

$$\frac{1}{R_M} = \frac{1}{B_z} \frac{\mathrm{d}}{\mathrm{d}x} |B_z| . \tag{6.13}$$

The directions of both the gradients of plasma pressure and magnetic field strength are the key. An inhomogeneous and magnetized plasma, in an inhomogeneous magnetic field, is shown in figure 6.6. The (x, z)-cross-section is as shown in figure 6.5. When the plasma pressure is higher in the region where the magnetic field is stronger, as in figure 6.6(a),

$$\nabla p \cdot \nabla |B| > 0 \tag{6.14}$$

the configuration is called a *magnetic hill*. The magnetic field becomes weaker away from the plasma. In the opposite case, figure 6.6(b),

$$\nabla p \cdot \nabla |B| < 0 \tag{6.15}$$

it is called a *magnetic well*. The field becomes stronger away from the plasma. In the case of figure 6.6(a), the system has been known to become unstable.

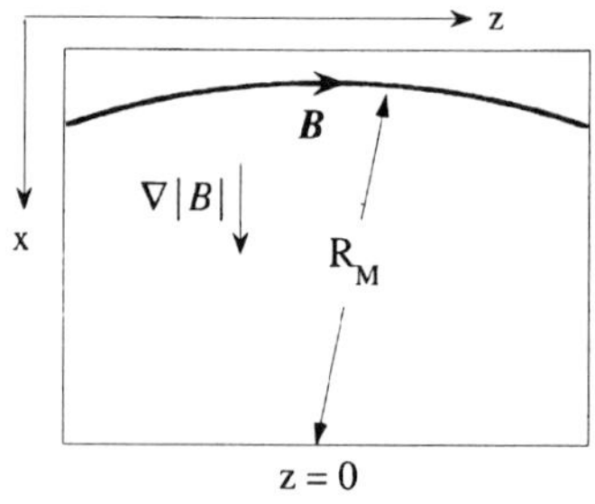

Figure 6.5. Inhomogeneity of the magnetic field. The gradient of the strength of the magnetic field is closely related to the curvature of the magnetic field line.

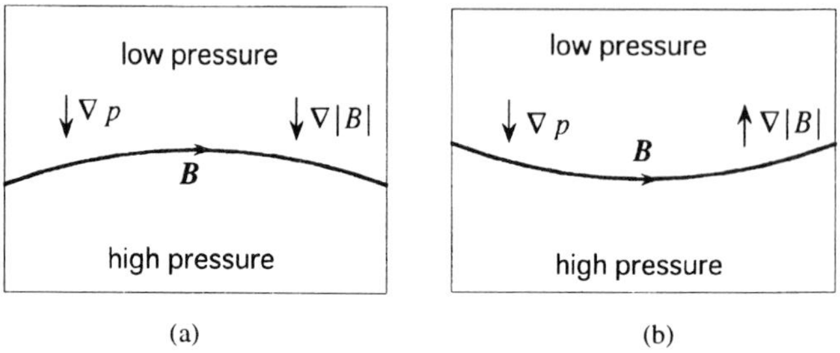

Figure 6.6. The case of a magnetic hill (unfavourable curvature, (a)) and magnetic well (favourable curvature, (b)).

6.2.2 Interchange Mode

Instabilities which appear under the configuration of a magnetic hill (figure 6.6(a)) are known by the name of interchange modes [6.1]. The growth rate is evaluated in the following.

Consider the case where a main magnetic field is in the z-direction with its gradient in the x-direction. The plasma pressure is also inhomogeneous in the x-direction (figure 6.7). This system is subject to a potential perturbation of the spatial form

$$\tilde{\phi} = \phi \exp(iky).$$

The perturbation is uniform along the magnetic field line. The consequence of this perturbation is explained in figure 6.7. The $\boldsymbol{E} \times \boldsymbol{B}$ motion appears in the x-direction. This flow causes a temporal change of density of the form

$$\tilde{n} = n_1 \exp(iky - i\pi/2).$$

A perturbed amplitude n_1 is related to the perturbed potential through the

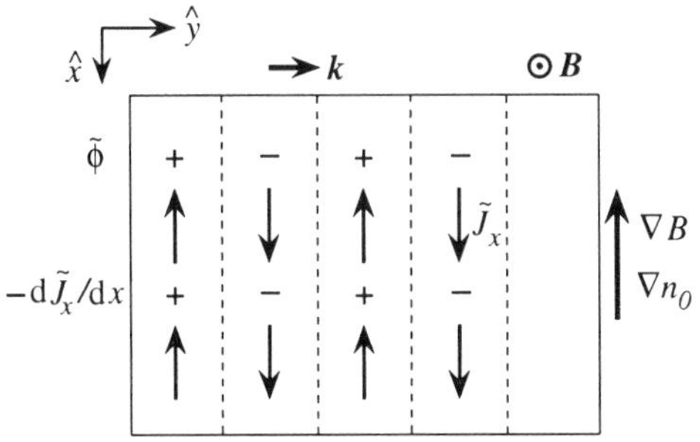

Figure 6.7. Current flow pattern (vertical arrows) associated with the potential perturbation $\tilde{\phi}$. Dotted lines denote the nodes of $\tilde{\phi}$. The magnitude of $\tilde{J}_x$ depends on x, and divergence of the perturbed diamagnetic current $\nabla \cdot \tilde{J}$ appears. (Signs of $\tilde{\phi}$ and $-\nabla \cdot \tilde{J}$ are denoted by + or −. They have the same sign in this case.)

continuity equation, which is written as

$$\partial \tilde{n}/\partial t = (\mathrm{d}n_0/\mathrm{d}x)(\mathrm{i}k\tilde{\phi}/B)$$

when the density is inhomogeneous in the x-direction. If this density perturbation appears, a perturbation in diamagnetic current is also associated with it as

$$\tilde{J}_x = -\mathrm{i}kB^{-1}(T_e + T_i)\tilde{n}.$$

A divergence of perturbed current remains, because the magnetic field B is inhomogeneous in the x-direction. The divergence of this diamagnetic current is given as

$$\partial \tilde{J}_x/\partial x = \frac{k^2}{B}\left(\frac{\mathrm{d}}{\mathrm{d}x}\frac{1}{B}\right)\frac{\mathrm{d}n_0}{\mathrm{d}x}(T_e + T_i)\int \mathrm{d}t\tilde{\phi}. \tag{6.16}$$

The growth rate is evaluated from this relation equation (6.16). The divergences of electric field and perturbed diamagnetic current are related as

$$\varepsilon_0\varepsilon_\perp \frac{\partial}{\partial t}\nabla_\perp \cdot \boldsymbol{E}_\perp = -\nabla \cdot \boldsymbol{J} \tag{6.17}$$

where

$$\varepsilon_\perp = c^2/v_A^2 \tag{6.18}$$

is the perpendicular dielectric constant in a magnetized plasma [2.3, 2.4]. (Notice that $\varepsilon_0(\varepsilon_\perp - 1)\partial E/\partial t$ is a polarization current in the magnetized plasma. Equation (6.17) is derived from Maxwell's equation and the charge conservation condition.) Substituting equation (6.16) into equation (6.17), one has the equation of motion,

$$\varepsilon_0\varepsilon_\perp \frac{\partial}{\partial t}(\mathrm{i}k_y)E_y = -\frac{k^2}{B}\left(\frac{\mathrm{d}}{\mathrm{d}x}\frac{1}{B}\right)\frac{\mathrm{d}n_0}{\mathrm{d}x}(T_e + T_i)\int \mathrm{d}t\tilde{\phi}. \tag{6.19}$$

An electric field and a potential is related as $E_y = -ik_y\tilde{\phi}$. Replacing E_y by $-ik_y\tilde{\phi}$ in equation (6.19), one has

$$\varepsilon_0\varepsilon_\perp\frac{\partial^2\tilde{\phi}}{\partial t^2} = -\frac{k^2}{B}\left(\frac{\mathrm{d}}{\mathrm{d}x}\frac{1}{B}\right)\frac{\mathrm{d}n_0}{\mathrm{d}x}(T_e+T_i)\tilde{\phi}$$

or

$$\frac{\partial^2}{\partial t^2}\tilde{\phi} = \frac{c_s^2}{L_pL_M}\tilde{\phi}. \tag{6.20}$$

Here, L_p and L_M are the gradient lengths of pressure and magnetic field, respectively,

$$\frac{1}{L_p} = -\frac{\nabla p}{p} \quad \text{and} \quad \frac{1}{L_M} = -\frac{\nabla B}{B}. \tag{6.21}$$

Equation (6.20) provides the growth rate γ as

$$\gamma^2 = \gamma_0^2 \equiv \frac{c_s^2}{L_pL_M}. \tag{6.22}$$

This result shows that a potential perturbation grows exponentially, if the sign of the coefficient of the right-hand side of equation (6.20) is positive, i.e.,

$$\frac{1}{L_pL_M} > 0. \tag{6.23}$$

The perturbation is unstable if a magnetic field becomes weaker in the direction in which the plasma pressure decreases. Under such a situation, a magnetic field is said to have a *bad curvature* (*unfavourable curvature*). In contrast, if the magnetic field becomes stronger as the pressure decreases, $L_pL_M < 0$, the system is stable. In this case the magnetic field has a *good* (*favourable*) curvature. As an example, one may imagine plasmas that are confined by the dipole field of the earth (or stars) (figure 6.8). If a ring of plasma is formed around the equator plane, the magnetic curvature is favourable for the inside part of the ring, and is unfavourable for the outside part.

In the case of interchange instability, it is possible to show that the force is in proportion to the displacement. The perturbed electric field is rewritten in terms of the $\boldsymbol{E}\times\boldsymbol{B}$ velocity, $\tilde{V}_x^{E\times B} = \tilde{E}_y/B = -ik_y\tilde{\phi}/B$. Then equation (6.19) is rewritten in the form of the equation of motion

$$m_in_i\frac{\partial}{\partial t}V_x^{E\times B} = \frac{m_in_ic_s^2}{L_pL_M}\xi_x^{E\times B} \equiv F_x \tag{6.24}$$

where $\xi_x^{E\times B}$ is the displacement by the $\boldsymbol{E}\times\boldsymbol{B}$ motion

$$\xi_x^{E\times B} = \int \mathrm{d}t\, V_x^{E\times B}. \tag{6.25}$$

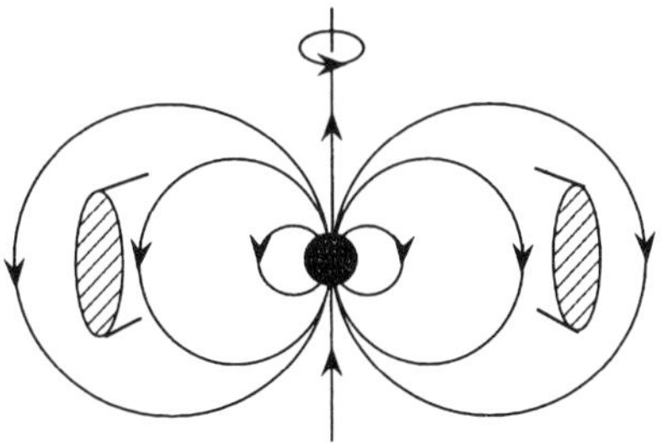

Figure 6.8. Plasma (shaded portion) confined by the dipole magnetic field of a star.

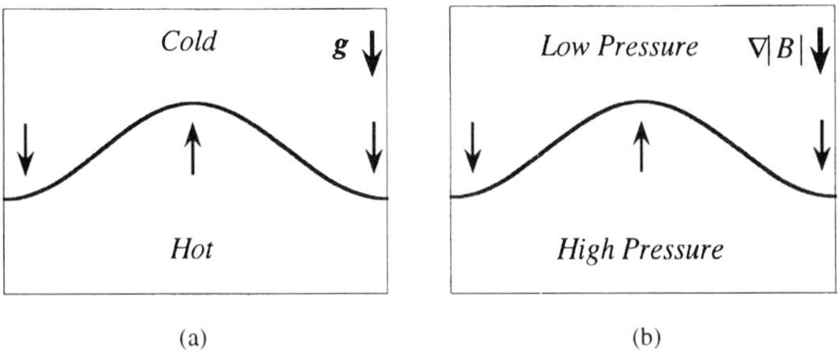

Figure 6.9. Rayleigh–Benard convection (a) and the interchange mode (b). Perturbation in the flow is shown by the arrow, and the iso-temperature (pressure) surface is also illustrated.

It is shown in equation (6.24) that the force F_x is in proportion to the displacement $\xi_x^{E\times B}$ in the x-direction. The force increases the displacement if the condition $L_p L_M > 0$ is satisfied.

This instability shows a close analogy with a Rayleigh–Taylor instability in fluid dynamics [6.2]. When water is heated from the bottom under a gravitational force, the thermal expansion causes an upward flow due to the buoyancy (figure 6.9). A convection roll is formed. In this case, the viscosity and thermal conduction tend to impede the convective motion. A roll pattern becomes unstable if the temperature gradient exceeds a threshold, which is known as a critical Rayleigh number. If water is heated from the top, the vertical temperature structure is stable. A combination of temperature gradient and gravity causes a Rayleigh–Benard convection. In a plasma, a combination of pressure gradient and effective force due to the magnetic field gradient causes an $\boldsymbol{E} \times \boldsymbol{B}$ convection.

6.2.3 Finite $k_\parallel$ Effect and Dissipative Instability

Next we consider a case where a perturbation is not uniform along a field line. In the example of section 6.2.2, no variation is introduced along the magnetic field line. In this section, it is illustrated that parallel dynamics is important for plasma instabilities.

The mode number vector has a parallel component to the magnetic field line,

$$\boldsymbol{k} = (0, k, k_\parallel)$$

and a plasma is perturbed in the x-direction as

$$\xi_x = \xi \exp(iky + ik_\parallel z - i\omega t).$$

In the presence of small but finite collisions, the equation of electron motion along the field line becomes $m_e \, \mathrm{d}\tilde{V}_{e\parallel}/\mathrm{d}t = -e\tilde{E}_\parallel - m_e \nu_{ei} \tilde{V}_{e\parallel}$. A perturbed electron current along the field line $\boldsymbol{J}_\parallel = -n_e e \boldsymbol{V}_{e\parallel}$ is given as

$$\tilde{\boldsymbol{J}}_\parallel = \frac{n_e e^2}{m_e(\nu_{ei} - i\omega)} \tilde{\boldsymbol{E}}_\parallel \tag{6.26}$$

where ν_{ei} is the electron–ion collision frequency. (If one takes a stationary limit, $\omega \to 0$, this relation reduces to the usual Ohm law, $\tilde{\boldsymbol{J}}_\parallel = (n_e e^2/m_e \nu_{ei})\tilde{\boldsymbol{E}}_\parallel$.) By use of the relation

$$\boldsymbol{E}_\parallel = -\partial \boldsymbol{A}_\parallel/\partial t - \boldsymbol{\nabla}_\parallel \phi$$

the perturbed current is expressed in terms of electrostatic potential as

$$\tilde{\boldsymbol{J}}_\parallel = \left(1 + \frac{\omega}{\omega + i\nu_{ei}} \frac{\omega_p^2}{k_\perp^2 c^2}\right)^{-1} \frac{n_e e^2}{m_e} \frac{k_\parallel}{(\omega + i\nu_{ei})} \tilde{\phi} \tag{6.27}$$

where the relation between current and vector potential,

$$\mu_0 \tilde{\boldsymbol{J}}_\parallel = k_\perp^2 \tilde{\boldsymbol{A}}_\parallel \tag{6.28}$$

is used.

In this case a bending of the magnetic field line with the nodes causes an additional restoring force and reduces the growth rate. The reduction of growth rate is understood from the observation of a pattern of perturbed current (figure 6.10). Since $\tilde{\boldsymbol{J}}_\parallel$ is not constant along the field line, but changes its direction within a wavelength $1/k_\parallel$, the divergence of $\tilde{\boldsymbol{J}}_\parallel$ is compensated by the perpendicular current $\tilde{\boldsymbol{J}}_\perp$, $\tilde{\boldsymbol{J}}_\perp \simeq (k_\parallel/k_\perp)\tilde{\boldsymbol{J}}_\parallel$. The Lorentz force $\tilde{\boldsymbol{J}}_\perp \times \boldsymbol{B}$ is directed in the direction so as to restore the deformation in the x-direction. Its magnitude is given as

$$(\tilde{\boldsymbol{J}}_\perp \times \boldsymbol{B})_\perp = \left(1 + \frac{\omega}{\omega + i\nu_{ei}} \frac{\omega_p^2}{k_\perp^2 c^2}\right)^{-1} \frac{n_e e^2}{m_e} \frac{k_\parallel^2 B}{(\omega + i\nu_{ei})k_\perp} \tilde{\phi}$$

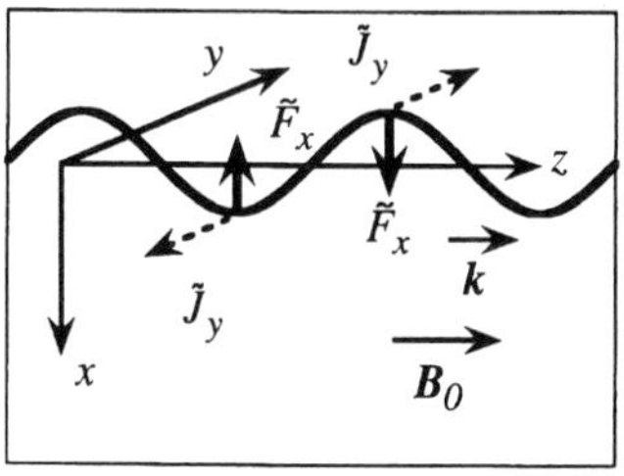

Figure 6.10. Perturbed magnetic field line and the current in the y-direction (dotted arrow). Restoring force $\tilde{F}_x$ (thick arrow) appears.

$$= \left(1 + \frac{\omega}{\omega + \mathrm{i}\nu_{ei}} \frac{\omega_p^2}{k_\perp^2 c^2}\right)^{-1} \frac{n_e e^2}{m_e} \frac{\mathrm{i}k_\parallel^2 B^2}{(\omega + \mathrm{i}\nu_{ei})k_\perp^2} \tilde{V}_x \tag{6.29}$$

where $\tilde{V}_x$ is the $\boldsymbol{E}\times\boldsymbol{B}$ velocity due to the fluctuating field, $-\mathrm{i}k\tilde{\phi}/B$. Adding this restoring force to the right-hand side of equation (6.24), the dispersion relation is now given as

$$\omega^2 + \gamma_0^2 - \left(1 + \frac{\omega}{\omega + \mathrm{i}\nu_{ei}} \frac{\omega_p^2}{k_\perp^2 c^2}\right)^{-1} \frac{\omega}{\omega + \mathrm{i}\nu_{ei}} \frac{\omega_p^2}{k_\perp^2 c^2} k_\parallel^2 v_A^2 = 0 \tag{6.30}$$

where $\gamma_0^2 = c_s^2/L_p L_M$. In the limit of $k_\parallel = 0$, it becomes equation (6.20).

We first study a limiting case of long perpendicular wavelength, i.e.,

$$\frac{\omega_p^2}{k_\perp^2 c^2} \gg 1. \tag{6.31}$$

(An opposite limit is discussed in the next subsection.) In this case, the dispersion relation equation (6.30) is simplified as

$$\omega^2 + \gamma_0^2 - \left(1 - \frac{\mathrm{i}\nu_{ei}}{\omega} \frac{k_\perp^2 c^2}{\omega_p^2}\right) k_\parallel^2 v_A^2 = 0. \tag{6.32}$$

The leading term with respect to the dissipation is retained. By use of a dc electric resistivity,

$$\eta = \mu_0 \nu_{ei} c^2 \omega_p^{-2}$$

equation (6.32) is rewritten as

$$\omega^2 + \gamma_0^2 - \left(1 - \frac{\mathrm{i}\eta k_\perp^2}{\mu_0 \omega}\right) k_\parallel^2 v_A^2 = 0. \tag{6.33}$$

Equation (6.33) predicts existences of both reactive and dissipative instabilities. A reactive instability is first explained, and a dissipative instability is discussed next.

Stabilizing effect of parallel wave number. In a collisionless limit, the dispersion relation (6.33) reduces to

$$\omega^2 + \gamma_0^2 - k_\parallel^2 v_A^2 = 0. \tag{6.34}$$

The growth rate γ is shown by the dashed line in figure 6.11. The relation (6.34) shows that there is a critical driving power γ_0^2 for a reactive instability. In the absence of electron collision, $\nu_{ei}/\omega \to 0$, equation (6.34) predicts that the mode is stable (i.e., ω is real), below a critical condition

$$\gamma_0 < |k_\parallel| v_A. \tag{6.35}$$

This result shows that the reactive instability is suppressed if the parallel mode number is large enough. Substituting the explicit form of driving term γ_0, we obtain a stability limit, equation (6.35), in the form

$$c_s/\sqrt{L_p L_M} < |k_\parallel| v_A.$$

If one introduces the ratio of sound velocity to Alfvén velocity, the stability boundary is rewritten in terms of beta value. For a plasma with singly charged ions and $T_e \simeq T_i$, the ratio of velocities is rewritten as $(c_s/v_A)^2 = \beta/2$. Equation (6.35) is read as

$$\beta < \beta_c = 2 L_p L_M k_\parallel^2. \tag{6.36}$$

This relation is often referred to as a *beta limit* in plasma confinements [2.1].

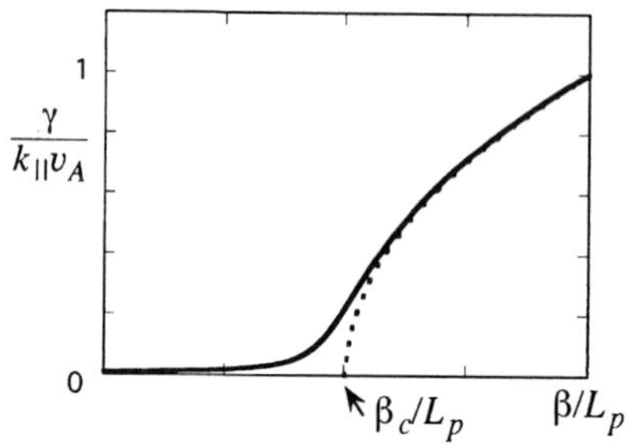

Figure 6.11. Growth rate as a function of the pressure gradient for a fixed value of $k_\parallel$. The reactive instability is shown by the dashed line. Below the critical pressure gradient, $\beta/L_p < \beta_c/L_p$, the dissipative instability remains.

The criterion (6.36) is often called the ideal MHD limit. This is based on the fact that the parallel electric field nearly vanishes, when the dispersion relation equation (6.34) is satisfied. This is seen as follows. In a limit of high density and small electron inertia, equation (6.31), the perturbed current, equation (6.27) is simplified as

$$J_{\|} \simeq \frac{ik_{\perp}^2}{\omega\mu_0}\nabla_{\|}\phi.$$

Rewriting the current, $J_{\|}$, in terms of the vector potential $A_{\|}$ by use of equation (6.28), this relation is equivalent to the equation

$$E_{\|} = -\partial A_{\|}/\partial t - \nabla\phi \simeq 0.$$

In other words, the static and inductive electric fields balance. This is because the parallel conductivity is large, so that the net parallel electric field vanishes.

Dissipative instability. A small but finite dissipation can cause an instability below the threshold condition of an ideal MHD instability, equation (6.35). A stability is obtained by a structure of perturbed current along the field line as is illustrated in figure 6.10. If this perturbed current is *impeded* by a certain mechanism, such as resistivity or current diffusion, then the resultant restoring force $\tilde{J}_{\perp} \times B$ is reduced. This becomes insufficient to provide the stability. The same situation as in figure 6.3 occurs in magnetized plasmas. When the electron collision is finite, a dissipative instability remains in the region $\gamma_0 < \gamma_c = |k_{\|} v_A|$ [6.3].

Near the marginal condition $\gamma_0 \simeq |k_{\|}|v_A$, equation (6.33) is approximated as

$$\omega^2 + \frac{i\eta k_{\perp}^2}{\mu_0\omega}\gamma_0^2 = 0. \tag{6.37}$$

The eigenfrequency ω is complex, and the relevant growth rate is estimated as

$$\mathrm{Im}(\omega) \simeq \left(\frac{\eta k_{\perp}^2}{\mu_0}\right)^{2/3} \gamma_0^{2/3} = \nu_{ei}^{1/3}\gamma_0^{2/3}\left(\frac{k_{\perp}c}{\omega_p}\right)^{2/3}. \tag{6.38}$$

The dissipative mode has a growth rate which has the fractional power of a smallness parameter ν_{ei}/γ_0. A small but finite dissipation can induce an instability with a considerably large growth rate. The growth rate of a dissipative instability is drawn by the solid line in figure 6.11.

The importance of dissipation is illustrated in equation (6.38) by taking an example of the electron–ion collisions, i.e., the resistivity. Other types of dissipation, e.g., a current diffusivity [6.4] and Landau resonance [2.4], have also been found to be influential in causing various instabilities.

6.2.4 Short Wavelength Modes

In the limit which is opposite to equation (6.31),

$$\frac{\omega_p^2}{k_{\perp}^2 c^2} \ll 1 \tag{6.39}$$

the stability of a short wavelength limit is given. This limit is also called an electrostatic approximation. (This condition is more easily satisfied in dilute (or low pressure) plasmas. Hence, it is sometimes called a limit of dilute plasmas.) In this case, the dispersion relation (6.30) is simplified as

$$\omega^2 + \gamma_0^2 - \frac{\omega}{\omega + \mathrm{i}\nu_{ei}} \frac{\omega_p^2}{k_\perp^2 c^2} k_\parallel^2 v_A^2 = 0. \tag{6.40}$$

Reactive instability. We first study the reactive instability. In the absence of electron collisions, $\nu_{ei}/\omega \to 0$, equation (6.40) predicts that the mode is stable (i.e., ω is real) below a critical condition

$$\gamma_0 < \gamma_c = \left| \frac{\omega_p}{k_\perp c} k_\parallel v_A \right| . \tag{6.41}$$

The stability boundary for the reactive instability is given in figure 6.12.

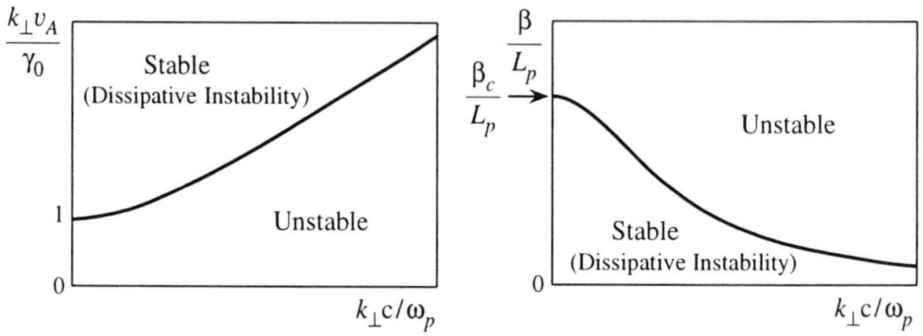

Figure 6.12. Stability boundary of the interchange mode for fixed pressure gradient (left) and for fixed $k_\parallel$ (right).

When the electron collision is finite, a dissipative instability appears in short wavelength modes. The growth rate remains in the region $\gamma_0 < \gamma_c = |k_\parallel v_A|$. Keeping the first order correction of ν_{ei}/ω, one expands the dispersion relation (6.40) as

$$\omega^2 + \left(\gamma_0^2 - \frac{\omega_p^2}{k_\perp^2 c^2} k_\parallel^2 v_A^2 \right) + \frac{\mathrm{i}\nu_{ei}}{\omega} \frac{\omega_p^2}{k_\perp^2 c^2} k_\parallel^2 v_A^2 = 0.$$

Near the marginal condition $\gamma_0 \simeq \gamma_c = |\omega_p k_\parallel v_A / k_\perp c|$, it is approximated as

$$\omega^2 + \frac{\mathrm{i}\nu_{ei}}{\omega} \frac{\omega_p^2}{k_\perp^2 c^2} k_\parallel^2 v_A^2 \simeq 0. \tag{6.42}$$

The growth rate is estimated as

$$\mathrm{Im}(\omega) = \nu_{ei}^{1/3} \gamma_0^{2/3}. \tag{6.43}$$

In the limit of an infinitesimally weak driving source, $\gamma_0 \to 0$, equation (6.30) provides the dispersion relation

$$\omega \simeq \mathrm{i}\nu_{ei}\gamma_0^2 \left(\frac{k_\perp c}{k_\parallel v_A \omega_p} \right)^2 . \tag{6.44}$$

(This approximate relation is derived by the help of the condition $|\omega| \ll \gamma_0$, i.e., $\gamma_0 \ll k_\parallel v_A \omega_p / k_\perp c$.) The dependence $\mathrm{Im}(\omega) \propto \nu_{ei}\gamma_0^2$ is seen to be the same as in equation (6.10).

In the short wavelength limit, there remains a reactive instability even if the condition (6.36) is satisfied. It has been considered that instabilities with short wavelength causes a gradual deterioration of confinement, and that they may not impose a drastic limit. Based on this conjecture, the condition (6.36) is considered to be a formula of a limiting plasma beta-value.

What is really important is the level to which these instabilities grow. The nature of turbulence, which is the evolution of these instabilities, is discussed in chapters 7–15.

6.2.5 Magnetic Shear

The results equations (6.35) and (6.41) indicate that the perturbation may become most dangerous if it is uniform along the field line, $k_\parallel = 0$. Efforts have been made in experimental research to eliminate the conditions which cause perturbations with $k_\parallel = 0$. One typical example is an introduction of magnetic shear as shown in figure 3.3. The expanded view is illustrated in figure 6.13. The direction of magnetic field lines inclines slightly from one magnetic surface to another. The magnetic field has the form

$$\boldsymbol{B} = \begin{pmatrix} 0 \\ \frac{x}{L_s} B \\ B \end{pmatrix} \tag{6.45}$$

in the vicinity of a mode rational surface, which corresponds to the surface $x = 0$ in equation (6.45).

In this magnetic configuration, the mode number in the direction of magnetic field is given as

$$k_\parallel = (\boldsymbol{k} \cdot \boldsymbol{B}) B^{-1} = k_y x / L_s \tag{6.46}$$

against a perturbation with the phase

$$\exp(\mathrm{i}k_y y).$$

The magnetic field is said to have a *shear*, and L_s is called a (magnetic) shear length. The mode number along the field line, $k_\parallel$, vanishes on the surface $x = 0$, the mode rational surface. If the magnetic field has a shear, perturbation is localized in the x-direction near the surface $\boldsymbol{k} \cdot \boldsymbol{B} = 0$.

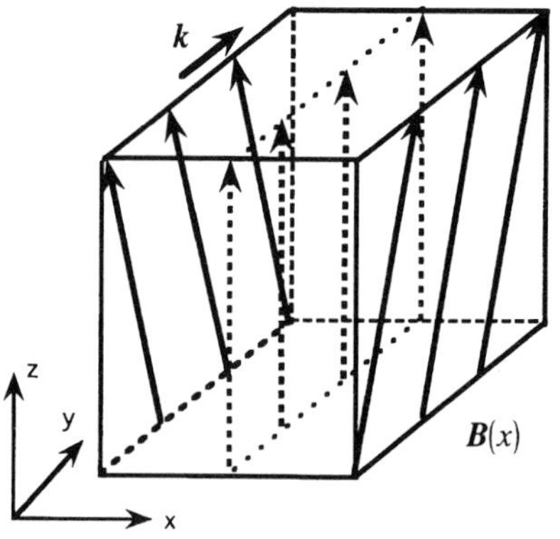

Figure 6.13. Sheared magnetic field. The dotted line indicates the mode rational surface.

In the presence of magnetic shear, the eigenvalue equation, such as equation (6.30), becomes a differential equation with respect to the variable x. This kind of procedure will be discussed at the end of this chapter.

6.2.6 Ballooning mode

The curvature of the magnetic field line is not always uniform along the field line. In the case of toroidal plasmas, the magnetic field often varies as $B \propto R^{-1}$ as is shown in figure 6.14. The curvature is bad on the outside of the torus, and is good inside. As a result of this, if one plots the driving force of the interchange mode $\nabla p \cdot \nabla B$ along the field line, it oscillates around zero with the period of $2\pi q R$. Here we simply write the variation of the curvature as a sinusoidal function as

$$c_s^2/L_p L_M = \gamma_{0M}^2\{\cos(z/2\pi q R) - h\} \qquad (6.47)$$

where γ_{0M} is the peak value of the driving term, and h is a small value. A positive value of h corresponds to an average magnetic well that makes this average negative.

We take the simplest case, where the driving term is given by a sinusoidal form equation (6.47). The mode number along the field line $k_\parallel$ is replaced by the operator $-\mathrm{i}\partial/\partial z$, and the eigenvalue equation, which is equivalent to equation (6.34), turns out to be a form of differential equation as

$$v_A^2 \frac{\mathrm{d}^2}{\mathrm{d}z^2}\psi(z) + \gamma_{0M}^2\{\cos(z/2\pi q R) - h\}\psi(z) + \omega^2\psi(z) = 0. \qquad (6.48)$$

When $c_s^2/L_p L_M$ is constant, $\psi(z)$ is given by a single Fourier component and the eigenvalue ω is given by equation (6.34). The solution of equation (6.48),

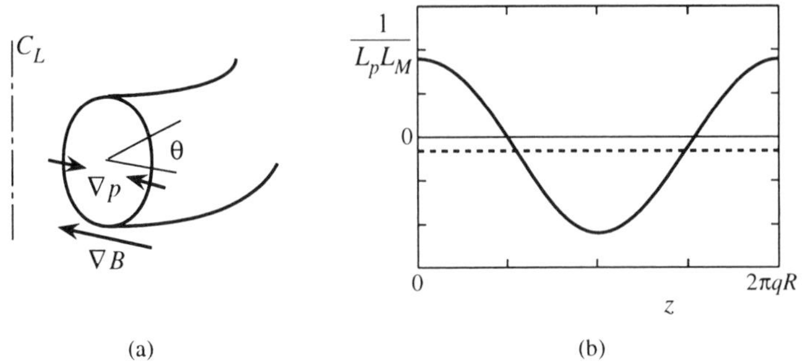

Figure 6.14. Toroidal geometry (a) and the magnetic curvature that varies along the magnetic field line (b). The average is shown by the dotted line in (b). Length z is measured along the field line in (b).

which is periodic with the period $2\pi qR$, is given by the Mathieu function

$$\psi(z) \propto ce_0(z/4\pi qR,\ 8(\pi qR/v_A)^2). \tag{6.49}$$

The eigenmode is given, in the strong drive limit, as

$$\omega^2 = -(1-h)\gamma_{0M}^2 + \frac{v_A}{2\pi qR}\gamma_{0M} \qquad \gamma_{0M} \gg v_A/2\pi qR \tag{6.50}$$

and in the weak drive limit as

$$\omega^2 = \left(h - \frac{1}{2}\left(\frac{2\pi qR}{v_A}\right)^2 \gamma_{0M}^2\right)\gamma_{0M}^2 \qquad \gamma_{0M} \ll v_A/2\pi qR. \tag{6.51}$$

The result equation (6.50) indicates that a simple estimate for the instability condition derived from equation (6.35), $\gamma_0 > k_\parallel v_A$, can be used in evaluating the region of a strong ballooning instability with the help of the estimation of $k_\parallel = 1/qR$ and the choice of the peak value $\gamma_0 \simeq \gamma_{0M}$ [2.1]. This simplification provides an estimate for the stability boundary equation (6.36) as

$$\beta < \beta_c = \frac{L_p L_M}{q^2 R^2} \simeq \frac{L_p}{q^2 R} \tag{6.36$'$}$$

where the relation $L_M \simeq R$ is used [2.1, 2.7].

The (reactive) ballooning instability disappears if

$$\gamma_{0M} \le \sqrt{2h}\frac{v_A}{2\pi qR} \tag{6.52}$$

is satisfied. This is called stabilization by use of the *average magnetic well*.

A similar argument as for the interchange mode applies to the dissipative instability of the ballooning mode. The nonlinear dissipative ballooning mode is discussed in chapter 10. The role of the magnetic shear, which also increases $k_\parallel$, is discussed in the literature [2.8, 6.5] and is not repeated here.

6.3 Drift Instabilities

When the magnetic curvature vanishes, the interchange mode becomes marginally stable; $\gamma_0 \simeq 0$. Under such circumstances, there arises a *drift instability*, which is caused by the plasma inhomogeneity. Drift waves have been subject to special attention in the research on plasma confinement. This is because this branch of oscillation appears due to an inhomogeneity of the plasma, and can become unstable in the presence of various dissipative mechanisms.

In the absence of dissipation, this mode is purely oscillating. The mode is described by an electrostatic model. The dissipation which impedes the free motion of electrons modifies the phase relation between electron density and electrostatic potential as

$$\frac{\tilde{n}_e}{n_e} = (1 - \mathrm{i}\delta_d)\frac{e\tilde{\phi}}{T_e}. \tag{6.53}$$

Here δ_d is a small but finite numerical factor, indicating a weak delay in the electron response. This is sometimes called the delay (deviation) from the adiabatic response. As was discussed in the previous section, the ion response function is given as

$$\frac{\tilde{n}_i}{n_i} = \frac{\omega_*}{\omega}\frac{e\tilde{\phi}}{T_e}. \tag{6.54}$$

The quasineutrality condition, $\tilde{n}_e = \tilde{n}_i$, provides the dispersion relation as

$$\omega = \omega_*(1 + \mathrm{i}\delta_d) \tag{6.55}$$

showing that the delay in the electron response causes an instability if $\delta_d > 0$.

In order to see the role of the inhomogeneity in the sign of phase delay, the effect of a small number of collisions is presented as an example. When there is a small but finite friction force on electrons, $-m_e n_e \nu \tilde{v}_e$, the equation of motion of electrons is given as

$$-en_e\tilde{E}_z - \mathrm{i}k_\parallel \tilde{p}_e \cong m_e n_e \nu \tilde{v}_{ez}. \tag{6.56}$$

By use of the continuity equation, $\partial\tilde{n}/\partial t + \nabla\tilde{n}\cdot V_d + n\nabla\cdot\tilde{V} = 0$, the equation of the parallel electron motion yields the response function $\tilde{n}_e$. Equation (5.24) is rewritten as

$$\frac{\tilde{n}_e}{n_e} = \left(1 + \mathrm{i}\frac{\nu(\omega_* - \omega)}{k_\parallel^2 v_{the}^2}\right)^{-1}\frac{e\tilde{\phi}}{T_e} \cong \left(1 - \mathrm{i}\frac{\nu(\omega_* - \omega)}{k_\parallel^2 v_{the}^2}\right)\frac{e\tilde{\phi}}{T_e} \tag{6.57}$$

where v_{the} is the electron thermal velocity, $v_{the}^2 = T_e m_e^{-1}$. A finite delay δ_d is explicitly given as

$$\delta_d = \frac{\nu(\omega_* - \omega)}{k_\parallel^2 v_{the}^2}. \tag{6.58}$$

The necessary condition for the instability, $\delta_d > 0$, is realized if $\omega_* > \omega$ holds. In the absence of inhomogeneity, i.e., $\omega_* = 0$, δ_d is always negative. The dissipation, like electron–ion collisions, can induce a growth of fluctuations in inhomogeneous plasmas. This is the reason why the drift wave has been thought to be particularly important.

It is noticed that this kind of phase delay is caused not only by electron collisions but also by other dissipation mechanisms (e.g., electron Landau damping).

A simple estimate of drift wave dispersion, $\omega \simeq \omega_*$, predicts that the phase difference δ_d vanishes and the stability is marginal. (That is, the determination of the stability is subtle. This is one of the reasons that there is a lot of literature on the stability analysis of drift waves.) There are, however, abundant mechanisms that make the real frequency lower than the drift frequency. If the wavelength perpendicular to the magnetic field is comparable to the ion gyro-radius, the effective potential, which the ions feel, becomes smaller as is illustrated in figure 5A.1. The effective potential $\bar{\phi}$ is given as

$$\bar{\phi} = I_0(k_\perp^2 \rho_i^2) \exp(-k_\perp^2 \rho_i^2) \tilde{\phi} \tag{6.59}$$

where I_0 is the zeroth order modified Bessel function of the first kind. It is simplified as $(1 - k_\perp^2 \rho_i^2)\tilde{\phi}$ in the limit of small finite-gyro-radius effect. The ion response is then given as

$$\frac{\tilde{n}_i}{n_i} = \frac{\omega_*}{\omega}(1 - k_\perp^2 \rho_i^2)\frac{e\tilde{\phi}}{T_e} + \frac{k_\parallel^2 c_s^2}{\omega^2}\frac{e\tilde{\phi}}{T_e}. \tag{6.60}$$

The real frequency is given as

$$\omega = \omega_* \left(1 - k_\perp^2 \rho_i^2 + \frac{k_\parallel^2 c_s^2}{\omega_*^2}\right) \tag{6.61}$$

and can become lower than ω_*. The growth rate is given as

$$\gamma = \frac{\nu \omega_*^2}{k_\parallel^2 v_{the}^2}\left(k_\perp^2 \rho_i^2 - \frac{k_\parallel^2 c_s^2}{\omega_*^2}\right). \tag{6.62}$$

The mode is unstable for the case of $k_\perp^2 \rho_i^2 > k_\parallel^2 c_s^2 \omega_*^{-2}$ or

$$k_y k_\perp \rho_i^2 > k_\parallel L_n. \tag{6.63}$$

An instability is possible for smaller parallel mode number.

This instability is also influenced by the magnetic shear. There is abundant literature.

6.4 Variations

In toroidal plasmas, the geometry gives rise to variations of mode. From the physics points of view, a similarity is observed among various instabilities. Some are explained here.

6.4.1 Trapped Particle Mode

The trapped particles influence the low frequency instability [6.6]. As is discussed in chapter 3, trapped particles are localized near the low field side of the torus. The motion of the guiding centre is illustrated schematically in figure 6.15. Trapped particles do not freely move along the magnetic field lines, and they respond to the perturbed potential in a different way compared to transit particles. Due to this difference, there arise various instabilities driven by trapped particles. The number density of transit particles is given as $n_{transit} \simeq n_e(1 - \sqrt{2\varepsilon})$ and that of trapped ones is

$$n_{trap} \simeq n_e\sqrt{2\varepsilon} \tag{6.64}$$

where $\varepsilon = r/R$ is the inverse aspect ratio.

One variation is seen in a reactive instability. The response in a collisionless limit is shown. Trapped electrons are localized to a particular poloidal angle, and the parallel motion along the field line is prohibited. When a potential perturbation such as the drift wave is considered, the transit electrons move freely so as to satisfy the Boltzmann relation. In contrast, trapped particles do not move along the field line, and the response is similar to the one in the interchange mode. The perturbed density of trapped particles is expressed as

$$\frac{\tilde{n}_{trap}}{n_e} \simeq \sqrt{2\varepsilon}\frac{\omega_*}{\omega - \omega_{Mi}}\frac{e\tilde{\phi}}{T_e}. \tag{6.65}$$

The denominator $\omega - \omega_{Me}$ includes the Doppler shift owing to the toroidal drift of trapped particles, $\omega_{Me} = k_\zeta V_{De}$. A similar relation is obtained for trapped ions.

$$\frac{\tilde{n}_{trap.i}}{n_e} \simeq \sqrt{2\varepsilon}\frac{\omega_*}{\omega - \omega_{Mi}}\frac{e\tilde{\phi}}{T_e}. \tag{6.66}$$

We take a simplified case of $T_e \simeq T_i$, and use the relation $\omega_{Mi} = -\omega_{Me}$. A simple example is an extremely low frequency mode, $|k_\parallel v_{thi}| \gg \omega$ as well as $|k_\parallel v_{the}| \gg \omega$. In this circumstance, the response of transit ions is also approximated as the Boltzmann response, i.e.,

$$\frac{\tilde{n}_{transit.e}}{n_e} \simeq (1 - \sqrt{2\varepsilon})\frac{e\tilde{\phi}}{T_e} \tag{6.67}$$

$$\frac{\tilde{n}_{transit.i}}{n_e} \simeq -(1 - \sqrt{2\varepsilon})\frac{e\tilde{\phi}}{T_e}. \tag{6.68}$$

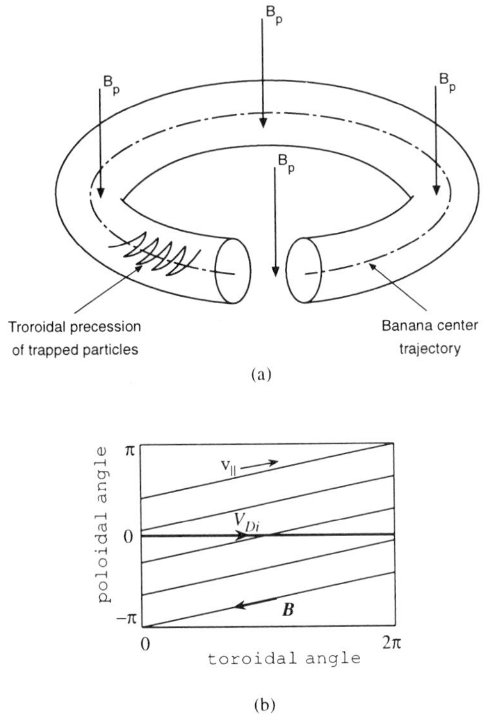

Figure 6.15. Schematic picture of the trajectory of trapped ions (a). Guiding centre motions of trapped ions (thick solid line) and untrapped ions (thin lines) are given in (b). Trapped particles move in the trough of the magnetic field, $\theta \simeq 0$, while transit particles move along the field lines.

The charge neutrality condition is rewritten as

$$2(1-\sqrt{2\varepsilon})+\sqrt{2\varepsilon}\frac{\omega_*}{\omega-\omega_{Me}}-\sqrt{2\varepsilon}\frac{\omega_*}{\omega+\omega_{Me}}\simeq 0. \tag{6.69}$$

The charge separation is not resolved, and a reactive instability appears as

$$\mathrm{Im}(\omega)\simeq\sqrt{\sqrt{2\varepsilon}\omega_{Me}\omega_*-\omega_{Me}^2}\simeq\varepsilon^{1/4}\sqrt{\omega_{Me}\omega_*}. \tag{6.70}$$

Like the interchange mode, this instability appears because the directions of the diamagnetic drift and the grad-B drift are in the same direction for trapped particles. The trapped particles are localized in the region where the magnetic curvature is bad.

This indicates that, if the drift direction of trapped particles is reversed, then this strong instability is suppressed. Drift reversal is possible, e.g., by the modification of the q-profile and the shape of the cross-section, or by strong inhomogeneity of the radial electric field [6.6–6.8]. These influences are discussed later.

Trapped particles also induce a dissipative instability. In the presence of collisions, trapped electrons have a larger collision frequency compared to the transit electrons. This is because the small pitch angle scattering leads to detrapping. The effective collision frequency of trapped particles is $1/\varepsilon$ times larger than transit particles. The ratio of the number density, $n_{trap}/n_{transit} \simeq \sqrt{2\varepsilon}$, is smaller than unity. However, the collision frequency is enhanced by the factor $1/\varepsilon$, so that the contribution from trapped particles to a mode due to collisions is larger. A small but finite number of collisions for trapped particles (but neglecting those for transit particles, for simplicity) leads to the dispersion relation (6.70) as

$$2(1-\sqrt{2\varepsilon})+\sqrt{2\varepsilon}\frac{\omega_*}{\omega-\omega_{Me}+\mathrm{i}\nu_{eff.e}}-\sqrt{2\varepsilon}\frac{\omega_*}{\omega+\omega_{Me}+\mathrm{i}\nu_{eff.i}}\simeq 0 \quad (6.71)$$

($\nu_{eff.e.i} \simeq \nu_{e.i}/\varepsilon$). Even in the absence of the effective bad magnetic curvature $\omega_{Me}\omega_* \to 0$, an instability remains. The eigenvalue is estimated as

$$\omega = -\frac{\sqrt{\varepsilon}}{2}\omega_* + \mathrm{i}\left(\frac{\varepsilon}{4}\frac{\omega_*^2}{\nu_{eff.e}} - \nu_{eff.i}\right). \quad (6.72)$$

Historically, this is called a dissipative trapped ion mode.

When the phase velocity is higher, $|k_\parallel v_{thi}| \ll \omega$, the response of transit ions deviates from the Boltzmann response. In the region of wavelength $|k_\parallel v_{thi}| \ll \omega \ll |k_\parallel v_{the}|$, the mode is the drift wave, and one must study the effect of trapped particles on the drift wave. (Figure 6.16 illustrates the regions of the trapped ion mode and trapped electron drift modes.) The drift waves $\omega \simeq \omega_*$ can be destabilized by the dissipation of trapped particles which are responding to the perturbed potential. This type of instability is called a dissipative trapped particle instability, and a lot of variations have been reported depending on the various collision frequencies.

6.4.2 Ion Temperature Gradient (ITG) Mode

The ion sound mode is destabilized by the ion temperature gradient (ITG) [6.9–6.11]. When the ion temperature is inhomogeneous in the x-direction (figure 5.1) while the density is homogeneous, the response of ions to the low frequency perturbation is calculated (figure 6.17). In this situation, the diamagnetic drift vanishes, but drift heat flow in the y-direction exists. In other words, hot ions and cold ions are moving in opposite directions in a stationary state. This instability appears due to the interactions of colder ions and hotter ions.

We consider the case

$$|k_\parallel v_{thi}| \ll \omega \ll |k_\parallel v_{the}|$$

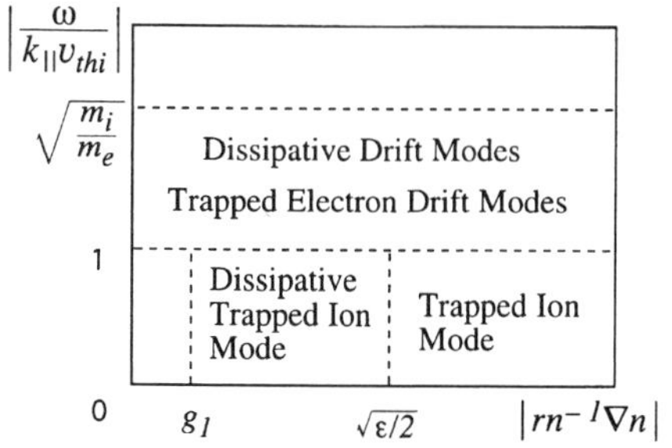

Figure 6.16. Classification of instabilities driven by density gradient. The parameter region in the density gradient and relative phase velocity is shown for the trapped ion mode and trapped electron drift modes. Criterion g_1 is given as $g_1 \sim \sqrt{\nu_{eff.e}\nu_{eff.i}\varepsilon^{-1}}rc_s^{-1}k^{-1}\rho_i^{-1}$.

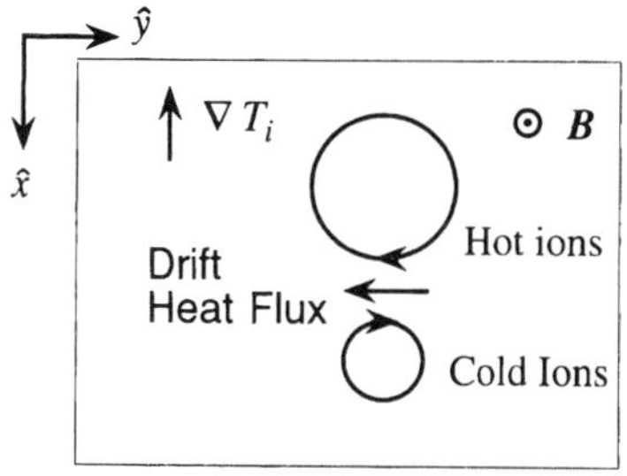

Figure 6.17. Drift heat flux in the $\nabla T \times B$ direction appears when the plasma temperature is not uniform.

which is relevant to drift and ion sound waves. The ion response function is given in chapter 5 as $\tilde{n}_i/n_0 \cong (k_{\|}^2 c_s^2/\omega^2)(e\tilde{\phi}/T_e)$ in a homogeneous plasma. In the presence of the ITG, the ion pressure perturbation due to drift heat flow appears, and the ion response is given as

$$\frac{\tilde{n}_i}{n_0} \cong \frac{k_{\|}^2 c_s^2}{\omega^2}\left(1 - \frac{\omega_{*Ti}}{\omega}\right)\frac{e\tilde{\phi}}{T_e} \tag{6.73}$$

where

$$\omega_{*Ti} = \frac{k_y}{eB}\frac{\mathrm{d}T_i}{\mathrm{d}x}. \tag{6.74}$$

The influence of the temperature gradient of ions is coupled with the parallel ion motion. The ion sound term $k_{\|}^2 v_{thi}^2/\omega^2$ is modified in the ion response. The electrons have the Boltzmann response, $\tilde{n}_e/n_0 \cong e\tilde{\phi}/T_e$, and the dispersion relation of the ion sound wave is modified as

$$1 = \frac{k_\parallel^2 c_s^2}{\omega^2}\left(1 - \frac{\omega_{*Ti}}{\omega}\right). \tag{6.75}$$

This cubic equation in ω of equation (6.75) predicts a reactive instability for

$$\left|\frac{\omega_{*Ti}}{k_\parallel c_s}\right| \geq \frac{2}{3\sqrt{3}}. \tag{6.76}$$

The growth rate is estimated as

$$\mathrm{Im}(\omega) \simeq \omega_{*Ti}^{1/3}(k_\parallel c_s)^{2/3} \qquad \text{with } \mathrm{Re}(\omega) \simeq \omega_{*Ti}^{1/3}(k_\parallel c_s)^{2/3} \tag{6.77}$$

(for $|\omega_{*Ti}/k_\parallel c_s| \gg 1$). This instability naturally causes the mixing of hotter ions and colder ions, so that it is also called the *ion-mixing mode*. Note that (6.76) and (6.77) are valid so long as the condition $|\omega/k_\parallel v_{thi}| \gg 1$ is satisfied. The numerical factor $2/3\sqrt{3}$ in equation (6.76) should be taken as an order of magnitude estimate.

Figure 6.18 shows the growth rate and real frequency as a function of the parallel mode number. This mode becomes unstable in the long wavelength region. In the limit of large parallel mode number, $|k_\parallel| c_s \gg |\omega_{*Ti}|$, the dispersion relation equation (6.75) reduces to that of the ion sound mode.

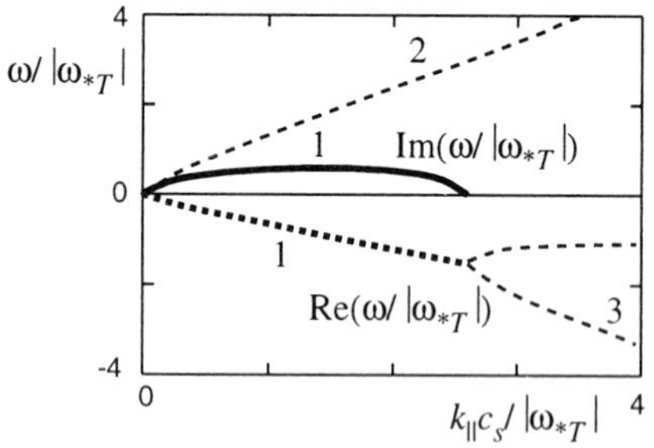

Figure 6.18. ITG mode (branch 1). Growth rate (solid line) and real frequency (dashed line) are shown. Branches 2 and 3 indicate the ion sound wave. The stable branch is shown by the thin dashed lines.

The density gradient is known to stabilize the ITG mode. The stability boundary, which is determined by the competition between density gradient and ion temperature gradient, is explained. In order to avoid the limitation $|\omega/k_\parallel v_{thi}| \gg 1$, a result based on the kinetic theory is used for the dispersion relation as [6.9]

$$1 = -\frac{T_e}{T_i}(1 + y_i Z(y_i)) - y_i Z(y_i)\frac{\omega_*}{\omega} + \frac{T_e}{T_i}\left\{y_i^2 + \left(y_i^3 - \frac{y_i}{2}\right)Z(y_i)\right\}\frac{\omega_{*Ti}}{\omega} \tag{6.78}$$

where $y_i = \omega/\sqrt{2}|k_\parallel|v_{thi}$ and $Z(y_i)$ is the plasma dispersion function. In deriving equation (6.78), the electron response is assumed to be the Boltzmann response. In the limit of $|\omega/k_\parallel v_{thi}| \gg 1$, an asymptotic relation $y_i Z(y_i) \simeq -1 - y_i^{-2}/2 - 3y_i^{-4}/4$ holds, and equation (6.78) is simplified as

$$1 = \frac{k_\parallel^2 T_e}{\omega^2 m_i}\left(1 - \frac{\omega_{*Ti}}{\omega}\right) + \frac{\omega_*}{\omega}. \tag{6.79}$$

In the absence of the temperature inhomogeneity, $\omega_{*Ti} \to 0$, this relation reproduces the one for drift waves. If one takes $\omega_* \to 0$, equation (6.75) is reproduced.

The marginal stability condition is derived from equation (6.78). When the mode is marginally stable, ω is real. The plasma dispersion function Z has a imaginary part. Therefore, at the marginal stability condition, the coefficient of the Z function and other terms must vanish, i.e., $T_e/T_i + \omega_*/\omega - (T_e/T_i)(y_i^2 - \frac{1}{2})\omega_{*Ti}/\omega = 0$ and $1 + T_e/T_i - (T_e/T_i)y_i^2\omega_{*Ti}/\omega = 0$. The real frequency and parallel wave number at the stability criterion are given as

$$\omega = \omega_* + \frac{T_e}{2T_i}\omega_{*Ti} \tag{6.80}$$

$$\frac{k_\parallel^2 v_{thi}^2}{\omega_{*Ti}^2} = \frac{T_e^2}{2T_i(T_i + T_e)}\left(\frac{1}{2} - \frac{1}{\eta_i}\right) \tag{6.81}$$

where

$$\eta_i \equiv \left(T_i^{-1}\frac{\mathrm{d}T_i}{\mathrm{d}x}\right)\left(n_i^{-1}\frac{\mathrm{d}n_i}{\mathrm{d}x}\right)^{-1}. \tag{6.82}$$

The critical condition (6.81) is shown in figure 6.19.

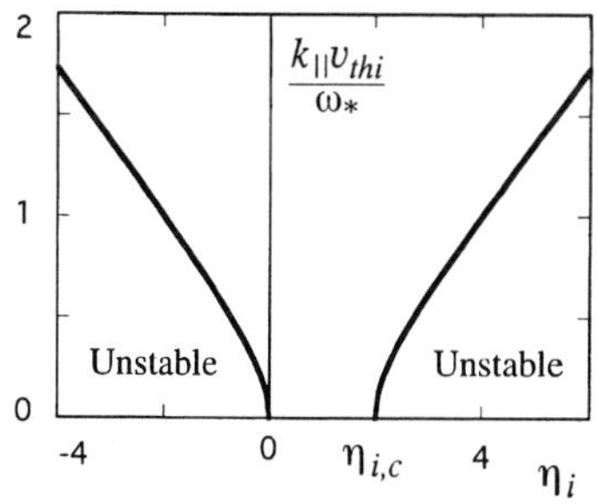

Figure 6.19. Stability/instability region for the ITG mode in the presence of the density gradient (case of $T_e = T_i$). Above the boundaries, instability is possible due to dissipation of electrons.

When the density and temperature have the gradient in the same direction, i.e., $\omega_* \omega_{*Ti} < 0$, the density gradient tends to stabilize this mode. When the ratio between the two gradients exceeds the criterion

$$\eta_i > \eta_{i.c} \tag{6.83}$$

the mode becomes unstable. The critical value $\eta_{i.c}$ satisfies $\eta_{i.c} = 2$ in the case of equation (6.78) and is of the order of unity. It has been calculated under various circumstances. (In the absence of density gradient, $\eta \to \infty$, equation (6.81) gives the stability boundary $|k_\parallel c_s / \omega_{*Ti}| = 1/2$. Comparing with this result, the analysis of the simplified fluid approximation, equation (6.76), shows a qualitative agreement.)

Even when the dispersion relation (6.78) predicts stability, there can remain a residual instability if one takes additional effects such as the finite-ion-gyro-radius effect or dissipation of electrons.

6.4.3 Toroidal ITG Mode

A cooperative effect of toroidicity and ITG causes a strong instability, which is called the toroidal ITG mode (or toroidal η_i mode) [6.11]. A simplified dispersion relation is given as

$$\frac{k_\parallel^2 c_s^2}{\omega^2} - \frac{\omega}{\omega - \omega_{*Ti}} + \frac{\bar{\omega}_{Mi}}{\omega} = 0. \tag{6.84}$$

(See [6.12] for details of derivation.) In equation (6.84), $\bar{\omega}_{Mi}$ is the Doppler shift due to the ∇B-drift of ions, $\bar{\omega}_{Mi} = k_\zeta \overline{V_{Di}}$, and the over-bar indicates an average of ∇B-drift over the region where the perturbation extends in the poloidal direction. In the absence of this Doppler shift, $\bar{\omega}_{Mi} \to 0$, equation (6.84) reduces to equation (6.75). In order to see the combined effect of the ion temperature gradient and ∇B-drift, let us take the limit of $k_\parallel \to 0$. In this limit, where equation (6.75) provides a zero growth rate, equation (6.84) yields an instability with the growth rate

$$\gamma \sim \sqrt{\omega_{*Ti} \bar{\omega}_{Mi}}. \tag{6.85}$$

The growth rate is positive if the direction of the ∇B drift of ions is the same as the ion drift heat flux (figure 6.17). This unfavourable magnetic drift appears when the magnetic curvature is bad. Stressing the importance of the bad toroidal curvature, the instability in this limit is called the toroidal ITG mode (toroidal η_i mode). The case in section 6.4.2 is called the slab ITG mode (slab η_i mode) for distinction.

In sections 6.4.2 and 6.4.3, the instability of ITG mode is derived based on an assumption of adiabatic response of electrons, i.e., $\tilde{n}_e / n_e = e\tilde{\phi}/T$. This implies that the flow of electrons is not caused by the development of ITG modes. The ITG modes, in these simple limits, do not cause particle flux, but induce selective loss of ion energy. Aiming at the explanation of anomalous ion energy flux, a lot of work has been done on the ITG modes [4.15, 6.11].

It should also be noted that the naming of the instability mode might not be the one and only possibility. In reality, several destabilizing mechanisms coexist simultaneously; one can study, e.g., destabilization of the ITG mode by trapped electrons. In such situations, the labelling of an instability by a particular name may not be unique.

6.5 Magnetic Shear and Nonlocal Mode

An introduction of the magnetic shear increases the parallel mode number, and is useful in eliminating the instability. In the case of a simplified geometry such as equation (6.45), the parallel mode number varies in the x-direction as equation (6.46),

$$k_{\parallel} = (\boldsymbol{k} \cdot \boldsymbol{B})B^{-1} = k_y x / L_s$$

where $x = 0$ corresponds to the mode rational surface, $k_{\parallel} = 0$. In a sheared magnetic field, it is necessary to study the eigenmode structure.

An example is chosen from drift waves [6.13]. Equation (6.61) is written as

$$\left(1 - \frac{\omega}{\omega_*} + \rho_i^2 \nabla_{\perp}^2 + \frac{k_{\parallel}^2 c_s^2}{\omega_*^2}\right) \tilde{\phi} = 0 \tag{6.86}$$

where $\tilde{\phi}$ is a static potential and $\mathrm{i}k_{\perp}$ is replaced by $\nabla_{\perp}$. It is noted that $k_z = 0$ is chosen, i.e., the mode rational surface is placed at $x = 0$. The perturbation is Fourier decomposed in the y-direction. We write

$$\tilde{\phi}(x, y, t) = \phi(x) \exp(\mathrm{i}k_y y - \mathrm{i}\omega t). \tag{6.87}$$

The eigenmode equation (6.86) is rewritten, with the help of equation (6.46) $k_{\parallel} = k_y x / L_s$, as a differential equation

$$\left(\rho_s^2 \frac{\mathrm{d}^2}{\mathrm{d}x^2} + 1 - \frac{\omega}{\omega_*} - k_y^2 \rho_s^2 + \frac{k_y^2 c_s^2}{\omega^2 L_s^2} x^2\right) \phi(x) = 0. \tag{6.88}$$

Owing to the dependence of $k_{\parallel}$ on x, the nabla-operator is no longer commutable with $k_{\parallel}$. The sinusoidal function $\exp(\mathrm{i}k_x x)$ no longer satisfies the dispersion relation.

Equation (6.88) must be solved as an eigenvalue equation. In a dimensionless form, $x/\rho_s \to \hat{x}$, equation (6.88) is written in the form of a Weber type equation

$$\left(\frac{\mathrm{d}^2}{\mathrm{d}\hat{x}^2} + H + \sigma^2 \hat{x}^2\right) \phi(\hat{x}) = 0 \tag{6.89}$$

with the eigenvalue

$$H = 1 - \frac{\omega}{\omega_*} - k_y^2 \rho_s^2. \tag{6.90}$$

The magnitude of the magnetic shear is represented by the parameter

$$\sigma^2 = \frac{k_y^2 c_s^2}{\omega^2} \frac{\rho_s^2}{L_s^2}. \tag{6.91}$$

Equation (6.89) is solved with a proper boundary condition. The boundary condition is chosen such that the amplitude vanishes at $|\hat{x}| \to \infty$ when the mode is growing in time. (For the damped mode, the analytic continuation is used.) The eigenfunction of equation (6.89) is given as

$$\phi(\hat{x}) = \exp\left(-\frac{\mathrm{i}\sigma^2}{4}\hat{x}^2\right) \tag{6.92}$$

with the eigenvalue

$$H = i\sigma. \tag{6.93}$$

Solving equations (6.90) and (6.93), one has the expression for the frequency.

$$\omega = (1 - k_y^2 \rho_s^2)\omega_* - \mathrm{i}\frac{k_y c_s \rho_s}{L_s} = \left(1 - k_y^2 \rho_s^2 - \mathrm{i}\frac{L_n}{L_s}\right)\omega_*. \tag{6.94}$$

In deriving the imaginary part of equation (6.94), σ is evaluated by use of the relation $\omega \simeq \omega_*$.

The dispersion relation (6.94) now shows a weak but finite damping rate associated with the magnetic shear. This damping rate has a relation to the sign of group velocity in the x-direction, equation (6.92). The eigenmode structure, illustrated in figure 6.20, shows that the wave energy is convected to the region $|\hat{x}| \to \infty$. The rate of energy propagating out is in proportion to the shear parameter $1/L_s$. (In the far distance, where the condition $|k_\parallel| v_{thi} \simeq \omega$ holds,

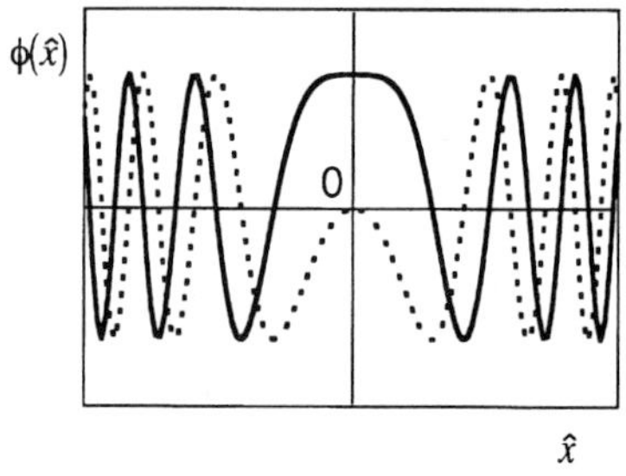

Figure 6.20. Eigenmode structure of the drift wave in a sheared magnetic field.

i.e., $|x| \simeq \rho_s L_s / L_n$, the ion Landau damping works and absorbs the wave energy [2.4].)

Comparing the eigenvalue equation (6.94) with that of a local estimate, equation (6.61), one sees that the influence of magnetic shear on the growth rate is very subtle, and careful analysis is required. The magnetic shear can also influence the phase difference of electron response [6.14]. There is plenty of literature on the linear drift instability in a sheared magnetic field [4.10, 6.15].

The propagation of the wave energy across the magnetic surface, i.e., the convective damping, is one of the principal effects of the magnetic shear. The exchange of the wave energy between particles takes place in different locations. The communication between different magnetic surfaces occurs. The nonlocal transport of the wave energy is important in analysing the turbulence-driven viscosity and related phenomena.

REFERENCES

[6.1] Rosenbluth M N and Longmire C L 1957 *Ann. Phys., NY* **1** 120

[6.2] Landau L D and Lifshitz E M 1987 *Fluid Mechanics* 2nd edn, transl. J B Sykes *et al* (Oxford: Pergamon) section 56

[6.3] Johnson J L, Greene J M and Coppi B 1963 *Phys. Fluids* **6** 1169
Johnson J L and Greene J M 1967 *Plasma Phys.* **9** 611

[6.4] Schmidt J and Yoshikawa S 1971 *Phys. Rev. Lett.* **26** 753
Kaw P J, Valeo E J and Rutherford P H 1979 *Phys. Rev. Lett.* **43** 1398

[6.5] Connor J W, Hastie R J and Taylor J B 1979 *Proc. R. Soc.* A **365** 1

[6.6] Kadomtsev B B and Pogutse O P 1971 *Nucl. Fusion* **11** 67

[6.7] Glasser A, Frieman E A and Yoshikawa S 1974 *Phys. Fluids* **17** 181

[6.8] Itoh S-I and Itoh K 1990 *J. Phys. Soc. Japan* **59** 3815

[6.9] Galeev A A, Oraevskii V N and Sagdeev R Z 1963 *Zh. Eksp. Teor. Fiz.* **44** 903 (Engl. transl. *Sov. Phys.–JETP* **17** 615)

[6.10] Coppi B, Furth H P, Rosenbluth M N and Sagdeev R Z 1966 *Phys. Rev. Lett.* **17** 377

[6.11] Further progress of analyses on the ion temperature gradient mode is surveyed in [4.15] and is seen in, e.g.,
Horton W, Estes R D and Biskamp D 1980 *Plasma Phys.* **22** 663
Waltz R E 1986 *Phys. Fluids* **29** 3684
Rewoldt G, Tang W M and Hastie R J 1987 *Phys. Fluids* **30** 807
Romaneli F 1989 *Phys. Fluids* B **1** 1018
Kotschenreuther M, Berk H L, Lebrun M J *et al* 1993 *Plasma Physics and Controlled Nuclear Fusion Research 1992* vol 2 (Vienna: IAEA) p 11
Waltz R E, Kerbel G D and Milovich J 1994 *Phys. Plasmas* **1** 2229
Kishimoto Y, Tajima T, Horton W, Lebrun M J and Kim J Y 1996 *Phys. Plasmas* **3** 1289

[6.12] Horton W Jr, Choi D-I and Tang W M 1981 *Phys. Fluids* **24** 1077

[6.13] Pearlstein L D and Berk H L 1969 *Phys. Rev. Lett.* **23** 220

[6.14] Ross D W and Mahajan S M 1978 *Phys. Rev. Lett.* **40** 324
Tsang K T, Catto P J, Whitson J C and Smith J 1978 *Phys. Rev. Lett.* **40** 327

[6.15] Tang W M 1978 *Nucl. Fusion* **18** 1089

Chapter 7

Reduced Set of Equations

Fluctuations can develop in inhomogeneous plasmas owing to various instabilities. Essential elements in the instability mechanism are explained in the preceding chapter. The main issue is the determination of characteristic parameters of turbulence (e.g., the saturation level, correlation length and correlation time). Instability modes are associated with perturbations, such as perturbations of the density, pressure and velocity (of ions and electrons), magnetic field and electric field, $\{\tilde{n}_{e.i}, \tilde{\boldsymbol{V}}_{e.i}, \tilde{p}_{e.i}, \tilde{\boldsymbol{E}}, \tilde{\boldsymbol{B}}\}$. An existence of many internal freedoms is one of the distinctive feature of plasma instabilities and is an obstacle to understand the dynamics of nonlinear phenomena. (The above perturbations have 16 elements. The magnetic perturbation and electric perturbation are connected directly, so that 13 out of 16 elements could be independent. A set of 13 nonlinearly coupled equations is still too much to be analysed.) It is a relevant starting point to choose a system of reduced number of freedoms. Such a model set of equations is called a *reduced set of equations*. In this chapter, we explain some examples of the reduced set of equations.

Basic ordering is that the typical length associated with the fluctuations is much smaller than the gradient scale length, i.e.,

$$\frac{1}{k_\perp L_p} \ll 1 \tag{7.1-1}$$

and the temporal evolution is slower than the ion cyclotron frequency

$$\Omega_i^{-1}(\mathrm{d}/\mathrm{d}t) \ll 1. \tag{7.1-2}$$

The parameter $1/k_\perp L_p$ works as the expanding parameter. This leads to the spatial *scale separations* in the modelling. The fluctuating fields evolve in much finer scale-length than the change of global structure. Therefore, the dynamics of fluctuations are analysed for a given inhomogeneity of global structure.

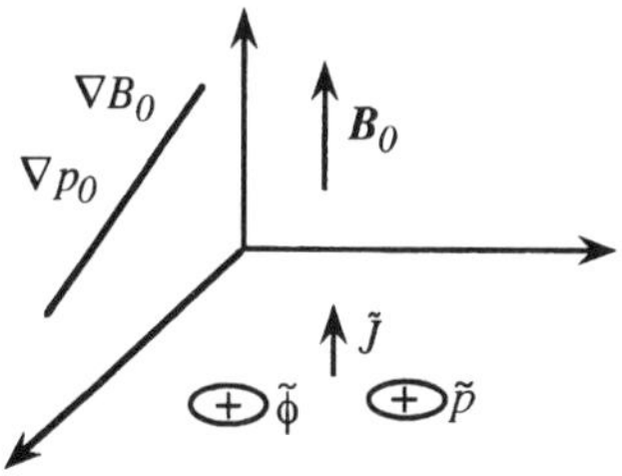

Figure 7.1. Perturbations $\{\tilde{\phi}, \tilde{J}_{\parallel}, \tilde{p}\}$ in an inhomogeneous and magnetized plasma.

7.1 Reduced Set of Equations

The fundamental mechanism of instabilities which is treated in this monograph is the interchange mode (and its variations). This mode belongs to the family of the shear Alfvén mode, and is modified by the gradients of the pressure and magnetic field as is explained in section 6.2. Variables to describe this mode are reduced to a set of three variables, $\{\tilde{\phi}, \tilde{J}_{\parallel}, \tilde{p}\}$, i.e., perturbations of a static potential, current which is parallel to a main magnetic field and pressure. Figure 7.1 schematically shows the perturbations in inhomogeneous and magnetized plasmas. The reduction of variables, through a decoupling of the compressional Alfvén mode, was first motivated by the study of the nonlinear MHD instabilities [7.1]. It is widely applied to the study of low frequency instabilities in various plasma configurations [7.2–7.6]. A systematic explanation and numerical application could be seen in the text book [2.8, 2.13]. In studying electrostatic and microscopic fluctuations, further reduction of variable is also performed. Examples include the Hasegawa–Mima equation, or Hasegawa–Wakatani equation, which describes the dynamics of two variables $\{\tilde{\phi}, \tilde{n}\}$ [7.7].

Basis of the choice of variables $\{\tilde{\phi}, \tilde{J}_{\parallel}, \tilde{p}\}$ is as follows. The perturbation is characterized by the magnetic component which is perpendicular to the main magnetic field. The perpendicular magnetic perturbation is described by the current perturbation parallel to the magnetic field, $\tilde{J}_{\parallel}$. The magnetic perturbation parallel to the main magnetic field is small,

$$\tilde{B}_{\parallel} = 0.$$

The parallel perturbation, which appears in the compressional Alfvén mode [2.4], is mixed, but the mixing occurs to the order of the β-value (i.e., the ratio of the plasma pressure to the magnetic pressure). The ion motion is dominated by the $\boldsymbol{E} \times \boldsymbol{B}$ velocity; therefore, the static potential provides the information of electric field as well as of plasma fluid velocity. The plasma pressure gradient is the driving source. The simple argument chooses a total plasma pressure for the representative variable.

The systematic deduction from equations of $\{\tilde{n}_{e.i}, \tilde{V}_{e.i}, \tilde{p}_{e.i}, \tilde{E}, \tilde{B}\}$ to equations of $\{\tilde{\phi}, \tilde{J}_\parallel, \tilde{p}\}$ uses the expansion with respect to the beta value. Details are explained, e.g., in [2.8]. The equation of motion:

$$\frac{\partial}{\partial t}\nabla_\perp^2\tilde{\phi} + \frac{1}{B}[\tilde{\phi}, \nabla_\perp^2\tilde{\phi}] = \frac{B^2}{m_i n_i}\nabla_\parallel \tilde{J}_\parallel + \frac{B}{m_i n_i}(\boldsymbol{b}\times\boldsymbol{\kappa})\cdot\boldsymbol{\nabla}\tilde{p} + \mu_c\nabla_\perp^4\tilde{\phi} \tag{7.2}$$

the Ohm law:

$$E_\parallel = -\frac{1}{en_e}\nabla_\parallel p_e + \eta_\parallel J_\parallel + \mu_0\frac{c^2}{\omega_p^2}\left(\frac{\partial}{\partial t}J_\parallel + \frac{1}{B}[\tilde{\phi}, \tilde{J}_\parallel] - \mu_{e.c}\nabla_\perp^2 J_\parallel\right) \tag{7.3}$$

and the energy balance equation

$$\frac{\partial}{\partial t}\tilde{p} + \frac{1}{B}[\tilde{\phi}, \tilde{p}] + \frac{1}{B}[\tilde{\phi}, p_0] = \chi_c\nabla_\perp^2\tilde{p} \tag{7.4}$$

constitute the set of equations for reduced variables. A sketch for the derivation of the equations in terms of $\{\tilde{\phi}, \tilde{J}_\parallel, \tilde{p}\}$ is given in appendix 7A. μ_c, $\mu_{e.c}$, and χ_c represent the collisional diffusivity for the ion momentum, electron momentum and energy, respectively. $\eta_\parallel$ is the collisional parallel resistivity. In equations (7.2)–(7.4), the Poisson bracket is defined as

$$[\phi, f] = (\boldsymbol{\nabla}_\perp\phi\times\boldsymbol{\nabla}_\perp f)\cdot\boldsymbol{b} \tag{7.5}$$

and $\boldsymbol{b} = \boldsymbol{B}_0/B_0$. In the equation of motion, $\boldsymbol{\kappa}$ indicates the inhomogeneity (i.e., curvature) of the magnetic field,

$$\boldsymbol{\kappa} = \frac{\boldsymbol{\nabla}_\perp B}{B}. \tag{7.6}$$

(In the case of a simple geometry, the gradient of the magnetic field is represented by one scale length, $L_M^{-1} = |\boldsymbol{\kappa}|$ in equation (6.21).)

Various fluctuating quantities, $\{\tilde{\boldsymbol{E}}, \tilde{\boldsymbol{B}}, \tilde{\boldsymbol{V}}, \tilde{\boldsymbol{J}}_\perp\}$, are given in terms of $\{\tilde{\phi}, \tilde{J}_\parallel, \tilde{p}\}$ as

$$\tilde{\boldsymbol{E}}_\perp = -\boldsymbol{\nabla}_\perp\tilde{\phi} \tag{7.7}$$

$$\tilde{E}_\parallel = -\nabla_\parallel\tilde{\phi} - \frac{\partial}{\partial t}\tilde{A}_\parallel \tag{7.8}$$

$$\nabla_\perp^2\tilde{A}_\parallel = -\tilde{J}_\parallel \tag{7.9}$$

$$\tilde{\boldsymbol{B}}_\perp = \boldsymbol{\nabla}_\perp\times(\tilde{A}_\parallel\boldsymbol{b}) \tag{7.10}$$

$$\tilde{B}_\parallel = 0 \tag{7.11}$$

$$\boldsymbol{\nabla}_\perp\cdot\tilde{\boldsymbol{J}}_\perp = -\nabla_\parallel\tilde{J}_\parallel \tag{7.12}$$

$$\tilde{\boldsymbol{V}}_\perp = -\frac{\boldsymbol{\nabla}\tilde{\phi}\times\boldsymbol{B}}{B^2} - \frac{m_i}{eB^2}\frac{\partial}{\partial t}(\boldsymbol{\nabla}\tilde{\phi}) \tag{7.13}$$

$$\tilde{V}_{\parallel} \simeq 0. \tag{7.14}$$

For the study of nonlinear interactions, it is convenient to use the dimensionless form. Introducing the normalization as

$$t/\tau_{Ap} \to \hat{t} \qquad \boldsymbol{r}_{\perp}/a \to \boldsymbol{r}_{\perp} \qquad a\nabla_{\perp} \to \hat{\nabla}_{\perp} \qquad qR\nabla_{\parallel} \to \hat{\nabla}_{\parallel}$$

$$\frac{\tau_{Ap}}{a^2}\frac{1}{B}\phi \to \hat{\phi} \qquad \frac{\mu_0 a}{B_p}J_{\parallel} \to \hat{J}_{\parallel} \qquad \frac{\mu_0}{B_p^2}p \to \hat{p} \qquad A_{\parallel}\frac{1}{aB_p} \to \hat{A}_{\parallel} \tag{7.15}$$

$$\frac{\tau_{Ap}}{a^2}\mu_c \to \hat{\mu}_c \qquad \frac{\tau_{Ap}}{a_2}\mu_{ec} \to \hat{\mu}_{ec} \qquad \frac{\tau_{Ap}}{a^2}\chi_c \to \hat{\chi}_c \qquad \frac{\tau_{Ap}}{\mu_0 a^2}\eta_{\parallel} \to \hat{\eta}_{\parallel}$$

where $\tau_{Ap} = qR/v_A$ is the poloidal Alfvén transit time, a reduced set of equations is simply written as

$$\frac{\partial}{\partial \hat{t}}\hat{\nabla}_{\perp}^2\hat{\phi} + [\hat{\phi}, \hat{\nabla}_{\perp}^2\hat{\phi}] = \hat{\nabla}_{\parallel}\hat{J}_{\parallel} + (\boldsymbol{b} \times \hat{\boldsymbol{\kappa}}) \cdot \hat{\boldsymbol{\nabla}}\hat{p} + \hat{\mu}_c\hat{\nabla}_{\perp}^4\hat{\phi} \tag{7.16}$$

$$\frac{\partial}{\partial \hat{t}}\hat{A}_{\parallel} = -\hat{\nabla}_{\parallel}\hat{\phi} + \alpha_c\hat{\nabla}_{\parallel}\hat{p}_e - \hat{\eta}_{\parallel}\hat{J}_{\parallel} - \frac{c^2}{\omega_p^2 a^2}\left(\frac{\partial}{\partial \hat{\eta}}\hat{J}_{\parallel} + [\hat{\phi}, \hat{J}_{\parallel}]\right) + \hat{\lambda}_c\hat{\nabla}_{\perp}^2\hat{J}_{\parallel} \tag{7.17}$$

$$\frac{\partial}{\partial \hat{t}}\hat{p} + [\hat{\phi}, \hat{p}] + [\hat{\phi}, \hat{p}_0] = \hat{\chi}_c\nabla_{\perp}^2\hat{p} \tag{7.18}$$

where

$$\hat{\nabla}_{\perp}^2\hat{A}_{\parallel} = -\hat{J}_{\parallel}.$$

In the Ohm's law, the current diffusivity due to the electron viscosity μ_e is introduced as

$$\lambda = \frac{c^2}{\omega_p^2 a^2}\mu_e \tag{7.19}$$

and a coupling coefficient α_c is defined as

$$\alpha_c = \frac{v_{ap}}{a\Omega_i}. \tag{7.20}$$

When the electron pressure perturbation is considered to be small, because of a possible large thermal conductivity along the field line, one may neglect the $\nabla_{\parallel}p_e$ term in the Ohm law. In this simplified limit, (7.17) is modified as

$$\frac{\partial}{\partial \hat{t}}\hat{A}_{\parallel} + \frac{1}{\xi}\left(\frac{\partial}{\partial \hat{t}}\hat{J}_{\parallel} + [\hat{\phi}, \hat{J}_{\parallel}]\right) = -\hat{\nabla}_{\parallel}\hat{\phi} - \hat{\eta}_{\parallel}\hat{J}_{\parallel} + \hat{\lambda}_c\hat{\nabla}_{\perp}^2\hat{J}_{\parallel} \tag{7.21}$$

with

$$\frac{1}{\xi} = \frac{c^2}{\omega_p^2 a^2}. \tag{7.22}$$

Variations of the reduced set of equations are available. An example is the study on the ion motion along the magnetic field. Instead of equation (7.14), the equation for the ion parallel motion is incorporated. The reduced set of equations for four variables, $\{\phi, J_\parallel, V_\parallel, p\}$ is given in appendix 7B. By this formulation, the transport of the parallel momentum (i.e., the momentum which is parallel to the main magnetic field) is to be studied, and instabilities which are driven by the ion temperature gradient (ITG mode) can also be investigated. Another reduced set of equations, which is relevant to study the ITG mode is given in appendix 7C.

7.2 Description of Instability

7.2.1 Linear Instability

This simplified set of equations describes the instabilities which are discussed in the preceding section. The interchange mode instability is recovered by linearizing the model equations.

A linearization here means that terms $[\hat{\phi}, \nabla_\perp^2 \hat{\phi}]$, $[\hat{\phi}, \hat{J}_\parallel]$ and $[\hat{\phi}, \hat{p}]$ are neglected in equations (7.16), (7.18) and (7.21). We put the form

$$\{\hat{\phi}, \hat{J}_\parallel, \hat{p}\} \propto \exp(\mathrm{i}\boldsymbol{k}_\perp \cdot \boldsymbol{r}_\perp + \mathrm{i}k_\parallel z - \mathrm{i}\omega t) \tag{7.23}$$

and take the x-axis in the direction of inhomogeneities (figure 7.1). The vorticity equation is transformed as

$$(\omega \hat{k}_\perp^2 + \mathrm{i}\hat{\mu}_c \hat{k}_\perp^4)\hat{\phi} = \hat{k}_\parallel \hat{J}_\parallel + \hat{\kappa}_x \hat{k}_y \hat{p}. \tag{7.24}$$

Solving $\hat{J}_\parallel$ and $\hat{p}$ in terms of $\hat{\phi}$ by using equations (7.21) and (7.18), respectively, we have

$$\hat{J}_\parallel = \frac{\hat{k}_\perp^2}{\{\hat{\omega}(1+\xi^{-1}\hat{k}_\perp^2) + \mathrm{i}(\hat{\eta}_\parallel \hat{k}_\perp^2 + \hat{\lambda}_c \hat{k}_\perp^4)\}} \hat{k}_\parallel \hat{\phi} \tag{7.25}$$

$$\hat{p} = \frac{-\hat{k}_y \, \mathrm{d}\hat{p}_0/\mathrm{d}\hat{x}}{(\hat{\omega} + \mathrm{i}\hat{\chi}_c \hat{k}_\perp^2)} \hat{\phi} \tag{7.26}$$

where a hat symbol ^ indicates the normalization (7.15). Substitution of equations (7.25) and (7.26) into equation (7.24) gives

$$(\omega \hat{k}_\perp^2 + \mathrm{i}\hat{\mu}_c \hat{k}_\perp^4)\hat{\phi} + \frac{\hat{k}_y^2 \hat{\kappa}_x \, \mathrm{d}\hat{p}_0/\mathrm{d}\hat{x}}{(\hat{\omega} + \mathrm{i}\hat{\chi}_c \hat{k}_\perp^2)} \hat{\phi} - \hat{k}_\parallel \frac{\hat{k}_\perp^2}{\{\hat{\omega}(1+\xi^{-1}\hat{k}_\perp^2) + \mathrm{i}(\hat{\eta}_\parallel \hat{k}_\perp^2 + \hat{\lambda}_c \hat{k}_\perp^4)\}} \hat{k}_\parallel \hat{\phi} = 0. \tag{7.27}$$

This equation is a generalized form for the interchange mode. A further approximation to neglect the dissipation terms and the electron inertial terms, i.e., $\hat{\eta}$, $\hat{\mu}_c$, $\hat{\chi}_c$, $\hat{\lambda}_c$, $1/\xi \to 0$ reduces equation (7.27) to

$$(\hat{\omega}^2 + \hat{\gamma}_0^2 - \hat{k}_\parallel^2)\hat{\phi} = 0 \tag{7.28}$$

or

$$\hat{\omega}^2 + \hat{\gamma}_0^2 - \hat{k}_\parallel^2 = 0 \tag{7.29-1}$$

with

$$\hat{\gamma}_0^2 = \hat{k}_y^2 \hat{k}_\perp^{-2} \hat{\kappa}_x (\mathrm{d}\hat{p}_0/\mathrm{d}\hat{x}). \tag{7.29-2}$$

Equation (7.29) describes the interchange mode and stabilization by the parallel wave number, as is described in equation (6.34). When one keeps a small but finite dissipation, equation (7.27) provides the dissipative interchange mode.

7.3 Conservation Relation

The development of fluctuations described by the nonlinear equations (7.16), (7.18), (7.21) is solved. Conservation relations are derived for this set of equations. Performing the operation $\int \mathrm{d}V \hat{\phi}^*$ on equation (7.16), $\int \mathrm{d}V\, \hat{J}_\parallel^*$ on equation (7.21) and $\int \mathrm{d}V\, \hat{p}^*$ on equation (7.18), one has

$$\frac{\partial}{2\partial \hat{t}} \int \mathrm{d}V |\hat{\nabla}_\perp \hat{\phi}|^2 = -\int \mathrm{d}V \hat{\phi}^* \hat{\nabla}_\parallel \hat{J}_\parallel - \int \mathrm{d}V \hat{\phi}^* (\boldsymbol{b} \times \hat{\kappa}) \cdot \hat{\boldsymbol{\nabla}} \hat{p}$$
$$- \hat{\mu}_c \int \mathrm{d}V |\hat{\nabla}_\perp^2 \hat{\phi}|^2 \tag{7.30}$$

$$\left(\frac{1}{2} + \frac{1}{2\xi}\right) \frac{\partial}{\partial \hat{t}} \int \mathrm{d}V |\hat{J}_\parallel|^2 = -\int \mathrm{d}V (\hat{J}^* \hat{\nabla}_\parallel \hat{\phi}) - \hat{\eta}_\parallel \int \mathrm{d}V |\hat{J}_\parallel|^2$$
$$- \hat{\lambda}_c \int \mathrm{d}V |\hat{\nabla}_\perp \hat{J}_\parallel|^2 \tag{7.31}$$

$$\frac{\partial}{2\partial \hat{t}} \int \mathrm{d}V\, \hat{p}^2 = -\int \mathrm{d}V (\hat{p}^* [\hat{\phi}, \hat{p}_0]) - \int \mathrm{d}V \hat{\chi}_c |\hat{\nabla}_\perp \hat{p}|^2. \tag{7.32}$$

In deriving equations (7.30)–(7.32), we use the relation $\int \mathrm{d}V\, A[A, B] = 0$, i.e., the nonlinearity in equations (7.16), (7.18), (7.21) indicates the energy transfer in the k-space.

Summing them up, we have an integral form as [7.8]

$$\frac{\partial}{\partial \hat{t}} \frac{1}{2} \left[\int \mathrm{d}V |\hat{\nabla}_\perp \hat{\phi}|^2 + \left(1 + \frac{1}{\xi}\right) \int \mathrm{d}V |\hat{J}_\parallel|^2 + \int \mathrm{d}V\, \hat{p}^2 \right]$$
$$= \left(\hat{\kappa}_x + \frac{\mathrm{d}\hat{p}_0}{\mathrm{d}\hat{x}}\right) \int \mathrm{d}V (\hat{p}^* \hat{\nabla}_y \hat{\phi}) - \hat{\mu}_c \int \mathrm{d}V |\hat{\nabla}_\perp^2 \hat{\phi}|^2$$
$$- \hat{\eta}_\parallel \int \mathrm{d}V |\hat{J}_\parallel|^2 - \hat{\lambda}_c \int \mathrm{d}V |\hat{\nabla}_\perp \hat{J}_\parallel|^2 - \hat{\chi}_c \int \mathrm{d}V |\hat{\nabla}_\perp \hat{p}|^2. \tag{7.33}$$

Another form of integral is also obtained as

$$\frac{\partial}{\partial \hat{t}} \frac{1}{2} \left[\int \mathrm{d}V |\hat{\nabla}_\perp \hat{\phi}|^2 + \left(1 + \frac{1}{\xi}\right) \int \mathrm{d}V |\hat{J}_\parallel|^2 - \left(\frac{\hat{\kappa}_x}{\mathrm{d}\hat{p}_0/\mathrm{d}\hat{x}}\right) \int \mathrm{d}V\, \hat{p}^2 \right]$$
$$= -\hat{\mu}_c \int \mathrm{d}V |\hat{\nabla}_\perp^2 \hat{\phi}|^2 - \hat{\eta}_\parallel \int \mathrm{d}V |\hat{J}_\parallel|^2$$
$$- \hat{\lambda}_c \int \mathrm{d}V |\hat{\nabla}_\perp \hat{J}_\parallel|^2 + \hat{\chi}_c \left(\frac{\hat{\kappa}_x}{\mathrm{d}p_0/\mathrm{d}x}\right) \int \mathrm{d}V |\hat{\nabla}_\perp \hat{p}|^2. \tag{7.34}$$

It is noticed that the energy flux in the direction of the gradient, q_{0x}, is calculated as

$$\hat{q}_{0x} = -\langle \hat{p}^* \hat{\nabla}_y \hat{\phi} \rangle \tag{7.35}$$

where the bracket indicates the average over the magnetic surface. By use of the average heat flux, equation (7.33) is written as [7.8]

$$\begin{aligned} \frac{\partial}{\partial \hat{t}} \frac{1}{2}\left[\int \mathrm{d}V |\hat{\nabla}_\perp \hat{\phi}|^2 + \left(1 + \frac{1}{\xi}\right) \int \mathrm{d}V |\hat{J}_\parallel|^2 + \int \mathrm{d}V\, \hat{p}^2 \right] \\ = -\left(\hat{\kappa}_x + \frac{\mathrm{d}\hat{p}_0}{\mathrm{d}\hat{x}}\right) \hat{q}_{0x} - \hat{\mu}_c \int \mathrm{d}V |\hat{\nabla}_\perp^2 \hat{\phi}|^2 - \hat{\eta}_\parallel \int \mathrm{d}V |\hat{J}_\parallel|^2 \\ - \hat{\lambda}_c \int \mathrm{d}V |\hat{\nabla}_\perp \hat{J}_\parallel|^2 - \hat{\chi}_c \int \mathrm{d}V |\hat{\nabla}_\perp \hat{p}|^2. \end{aligned} \tag{7.36}$$

Equation (7.36) indicates that the fluctuations are driven by the release of the free energy associated with q_{0x}, and is dissipated by the collisional diffusion in the short wavelength region.

Appendix 7A Vorticity Equation

The equation of motion

$$\frac{\partial}{\partial t} \tilde{\boldsymbol{V}}_\perp + (\tilde{\boldsymbol{V}} \cdot \nabla \tilde{\boldsymbol{V}})_\perp = \frac{1}{m_i n_i} (\tilde{\boldsymbol{J}} \times \boldsymbol{B})_\perp - \frac{1}{m_i n_i} \boldsymbol{\nabla}_\perp \tilde{p} + \mu_c \nabla_\perp^2 \tilde{\boldsymbol{V}}_\perp \tag{7A.1}$$

is rewritten in terms of the static potential. The expansion is made with respect to the ordering B^{-1}, and an approximate relation

$$\tilde{\boldsymbol{V}}_\perp = -\frac{\boldsymbol{\nabla}\tilde{\phi} \times \boldsymbol{B}}{B^2} - \frac{m_i}{eB^2} \frac{\partial}{\partial t} (\boldsymbol{\nabla} \tilde{\phi}) + \ldots . \tag{7A.2}$$

is used. Substitution of equation (7A.2) into equation (7A.1), equation (7A.1) for x- and y-components turns out to be

$$\begin{aligned} \frac{1}{B^2}\left(-\frac{\partial}{\partial t} \nabla_y \tilde{\phi} - \frac{1}{\Omega_i} \frac{\partial^2}{\partial t^2} \nabla_x \tilde{\phi} + \mu_c \nabla_\perp^2 \nabla_y \tilde{\phi} + \frac{1}{\Omega_i} \mu_c \nabla_\perp^2 \frac{\partial}{\partial t} \nabla_x \tilde{\phi} \right) \\ - \frac{1}{B^3} (\nabla_x \tilde{\phi} \nabla_y^2 \tilde{\phi} - \nabla_y \tilde{\phi} \nabla_{yx}^2 \tilde{\phi}) = \frac{1}{m_i n_i} \tilde{J}_y - \frac{1}{m_i n_i B} \nabla_x \tilde{p} \end{aligned} \tag{7A.3-1}$$

$$\begin{aligned} \frac{1}{B^2}\left(\frac{\partial}{\partial t} \nabla_x \tilde{\phi} - \frac{1}{\Omega_i} \frac{\partial^2}{\partial t^2} \nabla_y \tilde{\phi} + \mu_c \nabla_\perp^2 \nabla_x \tilde{\phi} + \frac{1}{\Omega_i} \mu_c \nabla_\perp^2 \frac{\partial}{\partial t} \nabla_y \tilde{\phi} \right) \\ - \frac{1}{B^3} (\nabla_y \tilde{\phi} \nabla_x^2 \tilde{\phi} - \nabla_x \tilde{\phi} \nabla_{yx}^2 \tilde{\phi}) = \frac{-1}{m_i n_i} \tilde{J}_x - \frac{1}{m_i n_i B} \nabla_y \tilde{p} \end{aligned} \tag{7A.3-2}$$

(The z-axis is taken in the main magnetic field.) In the left-hand side of equation (7A.3), terms are kept up to the order of B^{-3}. Applying the rotation operator to equations (7A.3), (i.e., ∇_y equation (7A.3-1)$-\nabla_x$ equation (7A.3-2)) one has

$$\frac{1}{B^2}\left(\frac{\partial}{\partial t}\nabla_\perp^2\tilde{\phi} - \mu_c\nabla_\perp^4\tilde{\phi}\right) + \frac{1}{B^3}[\tilde{\phi}, \nabla_\perp^2\tilde{\phi}] = -\frac{1}{m_i n_i}\boldsymbol{\nabla}_\perp \cdot \tilde{\boldsymbol{J}}_\perp + \frac{1}{m_i n_i B^2}[B, \tilde{p}] \tag{7A.4}$$

where the Poisson bracket is defined as

$$[f, g] \equiv (\boldsymbol{\nabla}_\perp f \times \boldsymbol{\nabla}_\perp g) \cdot \boldsymbol{b}. \tag{7A.5}$$

In obtaining equation (7A.4), several approximations are made.

(i) In the derivation of equation (7A.1) to (7A.3), $|B|$ and $\partial/\partial t$ are commuted, because an approximation $\partial|B|/\partial t \simeq 0$ is used. The decoupling of the compressional Alfvén mode, $\tilde{B}_\parallel = 0$, leads that $|B|$ is approximately constant in time.

(ii) The temporal change is assumed to be slow. The cross-field velocity of ions is composed of the $\boldsymbol{E} \times \boldsymbol{B}$ velocity and the polarization drift. In an expansion of the velocity with respect to $(\mathrm{d}/\mathrm{d}t)/\Omega$, the ion velocity is given as equation (7A.2) The second term in the right-hand side is small compared to the first term, if the temporal evolution is much slower than the ion cyclotron frequency. The polarization drift is not kept in the nonlinear term.

(iii) In the derivation of equation (7A.4), it is assumed that the gradient of the magnetic field is much weaker than the fluctuation gradient, $|\nabla B/B| \ll |\nabla\tilde{\phi}/\tilde{\phi}|$, i.e., equation (7.1-1).

(iv) The nonlinear coupling with the density perturbation is neglected. This approximation is equivalent to the Bousinesque approximation in a Rayleigh instability [7.9]. The correction with respect to this Bousinesque approximation was discussed in [7.10].

Under the charge neutrality condition, i.e.,

$$\boldsymbol{\nabla} \cdot \tilde{\boldsymbol{J}} = 0 \tag{7A.6}$$

the term $\boldsymbol{\nabla}_\perp \cdot \tilde{\boldsymbol{J}}_\perp$ is rewritten as in equation (7.12). The gradient of the static magnetic field is introduced as

$$\boldsymbol{\kappa} = \frac{\boldsymbol{\nabla}_\perp B}{B}. \tag{7A.7}$$

By including these procedures, one has the relation (7.2)

$$\frac{\partial}{\partial t}\nabla_\perp^2\tilde{\phi} + \frac{1}{B}[\tilde{\phi}, \nabla_\perp^2\tilde{\phi}] = \frac{B^2}{m_i n_i}\nabla_\parallel\tilde{J}_\parallel + \frac{B}{m_i n_i}(\boldsymbol{b} \times \boldsymbol{\kappa}) \cdot \boldsymbol{\nabla}\tilde{p} + \mu_c\nabla_\perp^4\tilde{\phi}. \tag{7A.8}$$

The Ohm law, where the finite-electron-inertia effect is kept, is derived from the equation of electron motion along the field line,

$$m_e n_e \frac{\mathrm{d}}{\mathrm{d}t}V_{\parallel e} = -en_e E_\parallel - \nabla_\parallel p_e - m_e n_e \nu_{ei} V_{\parallel e} + m_e n_e \mu_{e.c}\nabla_\perp^2 V_{\parallel e} \tag{7A.9}$$

where $\mu_{e.c}$ is the collisional viscosity of electron momentum. In a framework of the simplified equation, the Ohm law is deduced as

$$E_\parallel = -\frac{1}{en_e}\nabla_\parallel p_e + \eta_\parallel J_\parallel + \frac{m_e}{e^2 n_e}\left(\frac{\mathrm{d}}{\mathrm{d}t}J_\parallel - \mu_{e.c}\nabla_\perp^2 J_\parallel\right) \tag{7A.10}$$

with the parallel resistivity

$$\eta_\parallel = \frac{m_e}{n_e e^2}\nu_{ei}. \tag{7A.11}$$

Noting the relation

$$\frac{\mathrm{d}}{\mathrm{d}t}J_\parallel = \frac{\partial}{\partial t}J_\parallel + \tilde{\boldsymbol{V}}_\perp \cdot \nabla_\perp \tilde{\boldsymbol{J}}_\parallel = \frac{\partial}{\partial t}\tilde{J}_\parallel - \frac{1}{B}\left(\frac{\partial\tilde{\phi}}{\partial y}\frac{\partial \tilde{J}_\parallel}{\partial x} - \frac{\partial\tilde{\phi}}{\partial x}\frac{\partial \tilde{J}_\parallel}{\partial y}\right) \tag{7A.12}$$

from equation (7A.10) one has

$$E_\parallel = -\frac{1}{en_e}\nabla_\parallel p_e + \eta_\parallel J_\parallel + \mu_0 \frac{c^2}{\omega_p^2}\left(\frac{\partial}{\partial t}J_\parallel + \frac{1}{B}[\tilde{\phi}, \tilde{J}_\parallel] - \mu_{e.c}\nabla_\perp^2 J_\parallel\right). \tag{7A.13}$$

The energy conservation relation is simplified as

$$\frac{\mathrm{d}}{\mathrm{d}t}(\tilde{p} + p_0) = \chi_c \nabla_\perp^2 \tilde{p}. \tag{7A.14}$$

This equation implies that the perturbed pressure diffuses by the classical thermal diffusivity. The total time derivative is expressed like equation (7A.12) as

$$\frac{\mathrm{d}}{\mathrm{d}t}(\tilde{p} + p_0) = \frac{\partial}{\partial t}\tilde{p} + \frac{1}{B}[\tilde{\phi}, \tilde{p}] + \frac{1}{B}[\tilde{\phi}, p_0]. \tag{7A.15}$$

Equation (7A.14) is reduced to

$$\frac{\partial}{\partial t}\tilde{p} + \frac{1}{B}[\tilde{\phi}, \tilde{p}] + \frac{1}{B}[\tilde{\phi}, p_0] = \chi_c \nabla_\perp^2 \tilde{p}. \tag{7A.16}$$

Normalization is taken as

$$\begin{array}{llll} t/\tau_{Ap} \to \hat{t} & r_\perp/a \to \hat{r}_\perp & a\nabla_\perp \to \hat{\nabla}_\perp & qR\nabla_\parallel \to \hat{\nabla}_\parallel \\ \frac{\tau_{Ap}}{a^2}\frac{1}{B}\phi \to \hat{\phi} & \frac{\mu_0 a}{B_p}J_\parallel \to \hat{J}_\parallel & \frac{\mu_0}{B_p^2}p \to \hat{p} & A_\parallel \frac{1}{aB_p} \to \hat{A}_\parallel \\ \frac{\tau_{Ap}}{a^2}\mu_c \to \hat{\mu}_c & \frac{\tau_{Ap}}{a^2}\mu_{ec} \to \hat{\mu}_{ec} & \frac{\tau_{Ap}}{a^2}\chi_c \to \hat{\chi}_c & \frac{\tau_{Ap}}{\mu_0 a^2}\eta_\parallel \to \hat{\eta}_\parallel. \end{array} \tag{7A.17}$$

By use of normalization, the reduced set of equations are written in a dimensionless form as

$$\frac{\partial}{\partial \hat{t}}\hat{\nabla}_\perp^2\hat{\phi} + [\hat{\phi}, \hat{\nabla}_\perp^2\hat{\phi}] = \hat{\nabla}_\parallel \hat{J}_\parallel + (\boldsymbol{b} \times \boldsymbol{\kappa}) \cdot \hat{\boldsymbol{\nabla}}\hat{p} + \hat{\mu}_c \hat{\nabla}_\perp^4 \hat{\phi} \tag{7A.18}$$

$$\hat{E}_{\parallel} = -\alpha_c \hat{\nabla}_{\parallel} \hat{p}_e + \hat{\eta}_{\parallel} \hat{J}_{\parallel} + \frac{c^2}{\omega_p^2 a^2} \left(\frac{\partial}{\partial \hat{t}} \hat{J}_{\parallel} + [\hat{\phi}, \hat{J}_{\parallel}] \right) - \hat{\lambda}_c \hat{\nabla}_{\perp}^2 \hat{J}_{\parallel} \tag{7A.19}$$

$$\frac{\partial}{\partial t} \hat{p} + [\hat{\phi}, \hat{p}] + [\hat{\phi}, \hat{p}_0] = \hat{\chi}_c \nabla_{\perp}^2 \hat{p}. \tag{7A.20}$$

The current diffusivity is introduced in the Ohm law as

$$\hat{\lambda} = \frac{c^2}{\omega_p^2 a^2} \hat{\mu}_e. \tag{7A.21}$$

In the Ohm law the coupling coefficient is defined as

$$\alpha_c = \frac{v_{Ap}}{a \Omega_i}. \tag{7A.22}$$

Appendix 7B Four-Field Reduced Set of Equations

When the ion motion parallel to the magnetic field is included, the reduced set of equations for $\{\phi, J_{\parallel}, V_{\parallel}, p\}$ is also formulated. Derivations are given in [7.4, 7.5]. The reduced set of equations consists of: the equation of the perpendicular motion

$$\frac{\partial \hat{\nabla}_{\perp}^2 \hat{\phi}}{\partial \hat{t}} + [\hat{\phi}, \hat{\nabla}_{\perp}^2 \hat{\phi}] = \hat{\nabla}_{\parallel} \hat{J}_{\parallel} + (\boldsymbol{b} \times \hat{\boldsymbol{\kappa}}) \cdot \hat{\boldsymbol{\nabla}} \hat{p} + \hat{\mu}_{\perp c} \hat{\nabla}_{\perp}^4 \hat{\phi} \tag{7B.1}$$

the Ohm law

$$\frac{\partial \hat{A}_{\parallel}}{\partial \hat{t}} = -\hat{\nabla}_{\parallel} \hat{\phi} - \frac{1}{\xi} \left(\frac{\partial \hat{J}_{\parallel}}{\partial \hat{t}} + [\hat{\phi}, \hat{J}_{\parallel}] \right) - \hat{\eta}_{\parallel} \hat{J}_{\parallel} + \hat{\lambda}_c \hat{\nabla}_{\perp}^2 \hat{J}_{\parallel} \tag{7B.2}$$

the equation of the parallel motion

$$\frac{\partial \hat{V}_{\parallel}}{\partial \hat{t}} + [\hat{\phi}, \hat{V}_{\parallel}] = -\hat{\nabla}_{\parallel} \hat{p} + \hat{\mu}_{\parallel c} \hat{\nabla}_{\perp}^2 \hat{V}_{\parallel} \tag{7B.3}$$

and the energy balance equation

$$\frac{\partial \hat{p}}{\partial \hat{t}} + [\hat{\phi}, \hat{p}] = -\beta \hat{\nabla}_{\parallel} \hat{V}_{\parallel} + \chi_c \hat{\nabla}_{\perp}^2 \hat{p}. \tag{7B.4}$$

Equation (7B.3) is used instead of equation (7.14). The influence of the parallel ion motion on instabilities could appear through the first term in the right-hand side of equation (7B.4). If this term, which is in proportion to the beta-value, is neglected, then the set of equations (7B.1), (7B.2) and (7B.4) is equal to the three-field reduced set of equations. Parallel motion is also nonlinearly influenced through the $[\phi, V_{\parallel}]$ nonlinearity. This nonlinear term causes the nonlinear decorrelation of parallel ion motion, and gives rise to the turbulent viscosity of the parallel motion. In this set of equations, two coefficients for the ion viscosity are used: that for the perpendicular motion, $\mu_{\perp}$, and that for the parallel motion, $\mu_{\parallel}$. They may take similar values, but in principle they are different. This is the reason for introducing two coefficients of ion viscosity.

Appendix 7C Reduced Set of Equations for ITG Mode

In the dynamics of the ITG mode, there are two main differences from that in the interchange mode as is discussed in chapter 6. First, the difference of the ion response from that of electrons is important. This requires us to treat the electron pressure and ion pressure separately. Second, the coupling with the drift mode must be considered. For the transparency of the analysis, an iso-thermal limit is used for electrons,

$$\tilde{p}_e = T_{e0}\tilde{n} \tag{7C.1}$$

and the electron pressure perturbation is expressed in terms of the density perturbation. The ion pressure perturbation is related the density perturbation as $\tilde{p}_i = \gamma_i T_{i0}\tilde{n}$.

The reduced set of equations for the variables of $\{\phi, A_\parallel, V_\parallel, n, p_i\}$ has been derived as follows [7.5]. The vorticity equation

$$\frac{\partial \hat{\nabla}_\perp^2 \hat{F}}{\partial \hat{t}} + [\hat{F}, \hat{\nabla}_\perp^2 \hat{F}] - \nabla_\perp \cdot [\hat{P}_i, \nabla_\perp \hat{F}] = \beta_e^{-1}\hat{\nabla}_\parallel \hat{J}_\parallel + (z \times \hat{\kappa}) \cdot \hat{\nabla}(\hat{n} + \hat{P}_i) + \hat{\mu}_{\perp c}\hat{\nabla}_\perp^4 \hat{F}. \tag{7C.2}$$

The equation of parallel motion

$$\frac{\partial \hat{V}_\parallel}{\partial \hat{t}} + [\hat{\phi}, \hat{V}_\parallel] = -\hat{\nabla}_\parallel(\hat{n} + \hat{P}_i) + \hat{\mu}_{\parallel c}\hat{\nabla}_\perp^2 \hat{V}_\parallel \tag{7C.3}$$

the Ohm law

$$\frac{\partial \hat{A}_\parallel}{\partial \hat{t}} = -\hat{\nabla}_\parallel(\hat{\phi} - \hat{n}) - \hat{\eta}_\parallel \hat{J}_\parallel \tag{7C.4}$$

the energy balance equation of electrons

$$\frac{\partial \hat{n}}{\partial \hat{t}} + [\hat{\phi}, \hat{n}] = (\hat{z} \times \hat{\kappa}) \cdot \nabla_\perp(\hat{n} - \hat{\phi}) - \nabla_\parallel \hat{n} + \beta_e^{-1}\hat{\nabla}_\parallel^2 \hat{J} \tag{7C.5}$$

and the energy balance equation of ions

$$\frac{\partial}{\partial \hat{t}}\left(\hat{P}_i - \frac{\gamma_i T_{i0}}{T_{e0}}\hat{n}\right) + \left[(\hat{\phi} + \hat{P}_i), \left(\hat{P}_i - \frac{\gamma_i T_{i0}}{T_{e0}}\hat{n}\right)\right] = 0. \tag{7C.6}$$

In these equations, the symbols denote

$$\tilde{F} = \tilde{\phi} + \frac{\tilde{P}_i}{en_0} \qquad P_i = \frac{T_{i0}}{T_{e0}}\tilde{P}_i \qquad J_\parallel = -\nabla_\perp^2 A_\parallel. \tag{7C.7}$$

The normalization is chosen for the convenience of the drift waves, i.e., $\hat{t} = (L_n/a)t$, $\hat{x} = x/\rho_s$ and so on.

In this system of equations, the electron inertia effect is neglected in the Ohm law. If one further assumes the electrostatic limit, i.e., the inductive term in the Ohm law is neglected, then the system turns to be the dynamical equations of four variables $\{\phi, V_\parallel, n, p_i\}$. In addition to this, often employed is the approximation of the Boltzmann relation for the electron response,

$$\tilde{n} = \tilde{\phi}.$$

With this simplification, the reduced set of equations is composed of three equations for the variables of $\{\phi, V_\parallel, p_i\}$.

This set of equations describes the (resistive) ITG mode in the presence of the magnetic field inhomogeneity. The coupling between the ITG mode and the resistive interchange mode could be investigated. If one neglects the magnetic field inhomogeneity,

$$|\kappa| \to 0 \tag{7C.8}$$

this system describes the ITG mode in a slab geometry. On the other hand, if one chooses the limit that the ion pressure gradient vanishes,

$$(1 + \eta_i)\frac{T_{i0}}{T_{e0}} \to 0 \qquad \eta_i \equiv \left(\frac{\nabla T_{i0}}{T_{i0}}\right)\left(\frac{\nabla n_0}{n_0}\right)^{-1} \tag{7C.9}$$

the resistive interchange mode is described. When both the conditions of (7C.8) and (7C.9) are employed, then the weak resistive drift instability remains.

REFERENCES

[7.1] Kadomtsev B B and Pogutse O P 1974 *Zh. Eksp. Teor. Fiz.* **65** 575 (Engl. transl. *Sov. Phys.–JETP* **38** 283)

[7.2] Sykes A and Wesson J A 1976 *Phys. Rev. Lett.* **37** 140
Rosenbluth M N, Monticello D A, Strauss H and White R B 1976 *Phys. Fluids* **19** 1987

[7.3] Strauss H 1976 *Phys. Fluids* **19** 134
Strauss H 1980 *Plasma Phys.* **22** 733

[7.4] Hazeltine R D 1983 *Phys. Fluids* **26** 3242

[7.5] Yagi M 1989 *PhD Thesis* Kyoto University

[7.6] Hassam A B 1980 *Phys. Fluids* **23** 38

[7.7] Hasegawa A and Mima K 1977 *Phys. Rev. Lett.* **39** 205
Hasegawa A and Wakatani M 1983 *Phys. Fluids* **26** 2770

[7.8] Yagi M, Itoh S-I, Itoh K, Fukuyama A and Azumi M 1995 *Phys. Plasmas* **2** 4140

[7.9] Haken H 1978 *Synergetics—an Introduction* 2nd edn (Berlin: Springer) § 8.10

[7.10] Yagi M and Horton C W 1994 *Phys. Plasmas* **1** 2135

Chapter 8

Renormalization and Dressed Test Mode

By use of the reduced set of equations, we study consequences of nonlinear interactions, which appear in terms of the Poisson bracket, in unstable plasmas. An essential issue is to know how the fluctuating component of interest evolves in the presence of turbulence. The response function (conductivity tensor) in chapter 4 is obtained based on the linear response of an inhomogeneous plasma. When a fluctuating test mode is imposed on the turbulent plasma, the turbulence reacts against the imposed mode, and the response of the test mode is modified by the presence of turbulence. Namely, the response is screened by the background fluctuations.

An analogy of the Debye screening would be useful. In the Debye screening, surrounding charges react against an imposed test charge. If one takes into account of the reaction of the background charged particles, the test particle behaves with the screened mass and screened charge; such a phenomenon is described by the concept of a *dressed test particle* [2.3]. In the turbulent plasma, i.e., many other fluctuating field components coexist, the imposed mode causes a modification of the background fluctuations. The test mode, combined with the modification of the background fluctuations, reacts as a dressed mode. We use the concept of a *dressed test mode*.

The screening by background fluctuations is determined by the nonlinear operator in the basic equations. The nonlinear interactions are treated by the one-point renormalization method [8.1].

8.1 Renormalization

8.1.1 Three-Field Nonlinear Equations

We study the nonlinear development of the reduced set of equations (7.16), (7.18) and (7.21). They are formally written as

$$\frac{\partial}{\partial t}\begin{pmatrix} -\nabla_\perp^2\phi \\ (1-\xi\nabla_\perp^{-2})J \\ p \end{pmatrix} + \boldsymbol{L}\begin{pmatrix} \phi \\ J \\ p \end{pmatrix} = \boldsymbol{N} \tag{8.1}$$

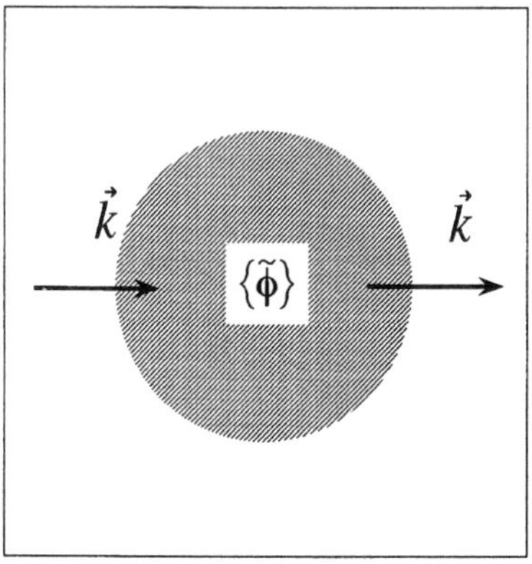

Figure 8.1. Reaction against the imposed mode (denoted by $\boldsymbol{k}$) includes the screening by the background fluctuations (symbolized by shading).

where the symbol hat ^ is suppressed for simplicity. In equation (8.1), $\boldsymbol{L}$ is the matrix of linear operators

$$\boldsymbol{L} = \begin{pmatrix} \mu_c \nabla_\perp^4 & \nabla_\parallel & (\boldsymbol{b} \times \boldsymbol{\kappa}) \cdot \boldsymbol{\nabla} \\ \xi \nabla_\parallel & \xi \eta_\parallel - \xi \lambda_c \nabla_\perp^2 & 0 \\ (\nabla p_0 \times \boldsymbol{b}) \cdot \boldsymbol{\nabla} & 0 & -\chi_c \nabla_\perp^2 \end{pmatrix} \tag{8.2}$$

and $\boldsymbol{N}$ stands for the nonlinear terms as

$$\boldsymbol{N} = \begin{pmatrix} N_u \\ N_j \\ N_p \end{pmatrix} \equiv - \begin{pmatrix} [\phi, -\nabla_\perp^2 \phi] \\ [\phi, J] \\ [\phi, p] \end{pmatrix}. \tag{8.3}$$

8.1.2 One-Point Renormalization Approximation

The nonlinear term $\boldsymbol{N}$ is treated by use of the renormalization method. A formal procedure is explained.

Nonlinear interactions. The following nonlinear process is considered. We take a test mode (denoted by $\boldsymbol{k}$) and calculate an interaction with background turbulence, which is denoted by $\boldsymbol{k}_1$. This interaction generates a driven mode (denoted by $\boldsymbol{k}_2$). Then the back-interaction of the driven mode with background fluctuation ($-\boldsymbol{k}_1$) contributes to the test mode (figure 8.2). By this procedure, the direct interaction of test mode with background fluctuations is obtained. In the calculation of a response of driven mode, the driven mode is also influenced by nonlinear interactions with background turbulence (denoted by $\boldsymbol{k}_3$). Thus the renormalized propagator is formulated as a recurrent formula of the successive nonlinear interactions and is given in a form of continued fraction.

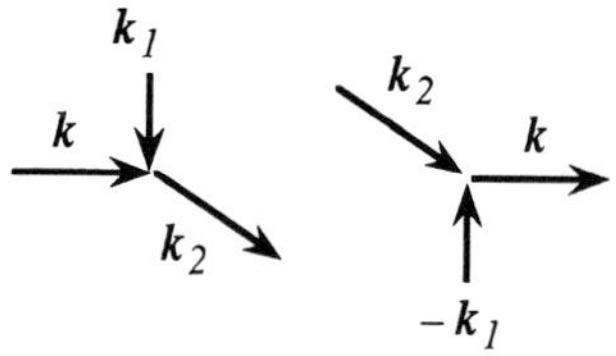

Figure 8.2. Interaction of a test mode $\boldsymbol{k}$ with a background turbulence $\boldsymbol{k}_1$ induces a driven mode $\boldsymbol{k}_2$. The back interaction drives the test mode $\boldsymbol{k}$.

Formal procedure. Equation (8.1) is symbolically written as

$$\{\boldsymbol{\partial}_T + \boldsymbol{L}\}\boldsymbol{f} = \boldsymbol{N} \tag{8.4}$$

where the linear operator related to the time derivative is introduced as

$$\boldsymbol{\partial}_T \equiv \frac{\partial}{\partial t}\begin{pmatrix} -\nabla_\perp^2 & & \\ & (1-\xi\nabla_\perp^{-2}) & \\ & & 1 \end{pmatrix} \tag{8.5}$$

and perturbed fields are abbreviated as

$$\boldsymbol{f} = \begin{pmatrix} \phi \\ J \\ p \end{pmatrix}. \tag{8.6}$$

Equation (8.4) is solved for the driven mode, $\boldsymbol{k}_2$. In the first step, the nonlinear term $\boldsymbol{N}$ for the driven mode $\boldsymbol{k}_2$ is divided into two terms:

$$\boldsymbol{N}_2 = \boldsymbol{N}_2^{drive} + \boldsymbol{N}_2'. \tag{8.7}$$

The suffix 2 denotes a driven mode $\boldsymbol{k}_2$. The term $\boldsymbol{N}_2^{drive}$ is contributed from a nonlinear coupling between a test mode $\boldsymbol{k}$ and background fluctuations $\boldsymbol{k}_1$. It represents the process $(\boldsymbol{k}+\boldsymbol{k}_1 \to \boldsymbol{k}_2)$. In calculating this interaction, background fluctuations are assumed to be *statistically independent* of the test mode $\boldsymbol{k}$. A nonlinear operator N is a quadratic form of fluctuation amplitude. Therefore the term $\boldsymbol{N}_2^{drive}$ is linear with respect to the amplitude of the test mode. The term $\boldsymbol{N}_2'$ denotes the nonlinear interaction that excludes the test mode. That is, the term $\boldsymbol{N}_2'$ indicates the process that driven mode interacts with background turbulence. Equation (8.4) is written as

$$\{\boldsymbol{\partial}_T + \boldsymbol{L}\}\boldsymbol{f}_2 - \boldsymbol{N}_2' = \boldsymbol{N}_2^{drive}. \tag{8.8}$$

The second step of the analysis is to introduce, symbolically, the nonlinear decorrelation and to write

$$\boldsymbol{N}_2' = -\boldsymbol{\Gamma}_2\boldsymbol{f}_2. \tag{8.9}$$

Elements of operator $\mathbf{\Gamma}_2$ stand for the nonlinear decorrelation due to interaction with background fluctuations. It is shown, *a posteriori*, that the operator could be given by a matrix which is expressed by the spectrum of the background turbulence. Thus equation (8.8) is written as

$$\{\partial_T + \boldsymbol{L} + \mathbf{\Gamma}_2\}\boldsymbol{f}_2 = \boldsymbol{N}_2^{drive}. \tag{8.10}$$

Although the operator in the left-hand side is screened by background turbulence, equation (8.10) is formally a linear equation with respect to a driven mode. In the next step, the driven mode is solved by an introduction of the inversion operator and is formally written as

$$\boldsymbol{f}_2 = \{\partial_T + \boldsymbol{L} + \mathbf{\Gamma}_2\}^{-1}\boldsymbol{N}_2^{drive}. \tag{8.11}$$

The amplitude of driven mode is linear with respect to test mode amplitude with the screening effect of background fluctuations.

Finally, the back-interaction to the test mode is calculated. Taking into account the back-interaction to test mode, the nonlinear term for test mode is expressed as

$$\boldsymbol{N}_k = \sum_{\boldsymbol{k}_1} \begin{pmatrix} N_u \\ N_j \\ N_p \end{pmatrix}_{\boldsymbol{k}_2 - \boldsymbol{k}_1 = \boldsymbol{k}} . \tag{8.12}$$

This term is a summation over products of driven modes and background turbulence. Substituting the form of $\boldsymbol{f}_2$ into equation (8.12), one expresses the nonlinear term $\boldsymbol{N}_k$ in terms of test mode and background fluctuations. The term yields the right-hand side of the following equation for the test mode,

$$\{\partial_T + \boldsymbol{L}\}\boldsymbol{f}_k = \boldsymbol{N}_k \tag{8.13}$$

which is then expressed in terms of test mode itself and background turbulence. The nonlinear term $\boldsymbol{N}_k$ is a nonlinear function of background turbulence, but is formally a linear function of the dressed test mode. It is shown in a symbolic manner as

$$\boldsymbol{N}_k = -\mathbf{\Gamma}_k \boldsymbol{f}_k.$$

Explicit form of the renormalized nonlinear term $\boldsymbol{N}_k$. Equation (8.10) is solved explicitly. We employ a diagonal approximation for an operator $\mathbf{\Gamma}_2$ in equation (8.10) as

$$\mathbf{\Gamma}_2 = \begin{pmatrix} \Gamma_{u2} & 0 & 0 \\ 0 & \Gamma_{j2} & 0 \\ 0 & 0 & \Gamma_{p2} \end{pmatrix} \tag{8.14}$$

where Γ_{u2}, Γ_{j2} and Γ_{p2} denote the decorrelation rate of U_2, J_2 and p_2 by the background turbulence, respectively. The explicit form of $\Gamma_{u.j.p}$ is to be determined self-consistently. This diagonal approximation equation (8.14) is introduced for analytical simplicity, but is often a good approximation. Term U is a vorticity,

$$U = -\nabla_\perp^2 \phi. \tag{8.15}$$

Then an equation for the driven mode is expressed as

$$\frac{\partial}{\partial t}\begin{pmatrix} U \\ (1-\xi\nabla_\perp^{-2})J \\ p \end{pmatrix}_{k_2} + \begin{pmatrix} -\mu_c\nabla_\perp^2 + \Gamma_u & \nabla_\parallel & (\boldsymbol{b}\times\boldsymbol{\kappa})\cdot\boldsymbol{\nabla} \\ -\nabla_\perp^{-2}\xi\nabla_\parallel & \xi\eta_\parallel - \xi\lambda_c\nabla_\perp^2 + \Gamma_j & 0 \\ -\nabla_\perp^{-2}(\nabla p_0\times\boldsymbol{b})\cdot\boldsymbol{\nabla} & 0 & -\chi_c\nabla_\perp^2 + \Gamma_p \end{pmatrix}_{k_2}\begin{pmatrix} U \\ J \\ p \end{pmatrix}_{k_2} = \begin{pmatrix} N_u \\ N_J \\ N_p \end{pmatrix}_{k_2}. \tag{8.16}$$

Equation (8.16) is solved, and the amplitude of the induced component is explicitly given as

$$U_2 = \frac{1}{K_2}\left\{N_u - \frac{\mathrm{i}k_{\parallel 2}}{\gamma_{j2}}N_j - \frac{\mathrm{i}k_{2y}\kappa_x}{\gamma_{p2}}N_p\right\} \tag{8.17}$$

$$J_2 = \frac{1}{K_2\gamma_{j2}}\left\{-\frac{\mathrm{i}k_{\parallel 2}\xi}{k_{\perp 2}^2}N_u + \left[K_2 - \frac{\xi k_{\parallel 2}^2}{\gamma_{j2}k_{\perp 2}^2}\right]N_j - \frac{\xi k_{2y}\kappa_x k_{\parallel 2}}{k_{\perp 2}^2\gamma_{p2}}N_p\right\} \tag{8.18}$$

$$p_2 = \frac{1}{K_2\gamma_{p2}}\left\{\frac{\mathrm{i}k_{2y}p_0'}{k_{\perp 2}^2}N_u + \frac{k_{2y}p_0'k_{\parallel 2}}{\gamma_{j2}k_{\perp 2}^2}N_j + \left[K_2 + \frac{k_{2y}^2 G_0}{k_{\perp 2}^2\gamma_{p2}}\right]N_p\right\} \tag{8.19}$$

and

$$K_2 = \gamma_{u2} + \frac{k_{\parallel 2}^2}{k_{\perp 2}^2}\frac{\xi}{\gamma_{j2}} - \frac{k_{2y}^2 G_0}{\gamma_{p2}k_{\perp 2}^2}. \tag{8.20}$$

In equations (8.17)–(8.20), the following notations are used.

$$\gamma_{u2} = \gamma(2) + \mu_c k_\perp^2 + \Gamma_{u2}$$

$$\gamma_{j2} = \gamma(2)(1 + \xi k_\perp^{-2}) + \xi\eta_\parallel + \mu_{e.c}k_\perp^2 + \Gamma_{j2}$$

$$\gamma_{p2} = \gamma(2) + \chi_c k_\perp^2 + \Gamma_{p2}$$

and $\gamma(2)$ is the rate of change of the k_2 mode, $\partial\{U_2, J_2, p_2\}/\partial t = \gamma(2)\{U_2, J_2, p_2\}$. The parameter that specifies an inhomogeneity is given as

$$G_0 = \kappa_x\frac{\mathrm{d}p_0}{\mathrm{d}x} \tag{8.21}$$

where p_0 is an equilibrium pressure profile. Equations (8.17), (8.18) and (8.19) are symbolically written as

$$\begin{pmatrix} U_2 \\ J_2 \\ p_2 \end{pmatrix} = \begin{pmatrix} H_{11} H_{12} H_{13} \\ H_{21} H_{22} H_{23} \\ H_{31} H_{32} H_{33} \end{pmatrix}_{k_2} \begin{pmatrix} N_u \\ N_j \\ N_p \end{pmatrix}_2 . \qquad (8.22)$$

In solving equation (8.10), a nonlinear decorrelation operator $\boldsymbol{\Gamma}$ is approximated by a diagonal matrix. The basis for this simplification is also discussed in the following subsection.

For transparency of argument, nonlinear interaction terms are taken as

$$N_{u.2} = -[\phi_1, U_k] \qquad N_{j.2} = -[\phi_1, J_k] \qquad N_{p.2} = -[\phi_1, p_k] \qquad (8.23\text{-}1)$$

for the driven mode, k_2, and

$$N_u = -[\phi_{-1}, U_2] \qquad N_j = -[\phi_{-1}, J_2] \qquad N_p = -[\phi_{-1}, p_2] \qquad (8.23\text{-}2)$$

for the test mode. (Subscripts 1 and -1 denote k_1 and $-k_1$, respectively.) This postulation implies that we take a nonlinear interaction that a mode of interest is modified by $\boldsymbol{E} \times \boldsymbol{B}$ motion of background fluctuations. A more complete form is discussed in appendix 8A. In this case, substituting equations (8.22) and (8.23-1) into equation (8.23-2), we write the nonlinear term acting on test mode as

$$\begin{pmatrix} N_u \\ N_j \\ N_p \end{pmatrix}_k = \sum_{k_1} \begin{pmatrix} H_{11} H_{12} H_{13} \\ H_{21} H_{22} H_{23} \\ H_{31} H_{32} H_{33} \end{pmatrix}_k \begin{pmatrix} [\phi_{-1}, [\phi_1, U_k]] \\ [\phi_{-1}, [\phi_1, J_k]] \\ [\phi_{-1}, [\phi_1, p_k]] \end{pmatrix} . \qquad (8.24)$$

Through this procedure, the nonlinear term for the test mode is expressed in terms of the background fluctuations.

The evolution equation of the test mode is obtained by the substitution of equation (8.24) into equation (8.13). It is given as

$$\{\partial_T + \boldsymbol{L}\} \begin{pmatrix} \phi \\ J \\ p \end{pmatrix}_k = \sum_{k_1} \begin{pmatrix} H_{11} H_{12} H_{13} \\ H_{21} H_{22} H_{23} \\ H_{31} H_{32} H_{33} \end{pmatrix}_k \begin{pmatrix} [\phi_{-1}, [\phi_1, -\nabla_\perp^2 \phi_k]] \\ [\phi_{-1}, [\phi_1, J_k]] \\ [\phi_{-1}, [\phi_1, p_k]] \end{pmatrix} . \qquad (8.25)$$

This equation describes how the test mode develops in the presence of background fluctuations.

This basic equation (8.25) has couples of characteristic features. First, it is given as a *linear* equation in terms of test mode. In the right-hand side, nonlinear interactions by background fluctuations are given as a complicated operator (matrix $\boldsymbol{H}$ and Poisson's bracket), which is nevertheless independent of other test modes. This resembles a *quasi-particle* approach in the physics of condensed matter. Second, the operator $\boldsymbol{H}$ includes nonlinear decorrelation rates Γ_{u2}, Γ_{j2} and Γ_{p2}. Third, the nonlinear decorrelation effect on k_2, $\boldsymbol{\Gamma}_2$ in equation (8.9), is physically equivalent to the right-hand side of equation (8.25).

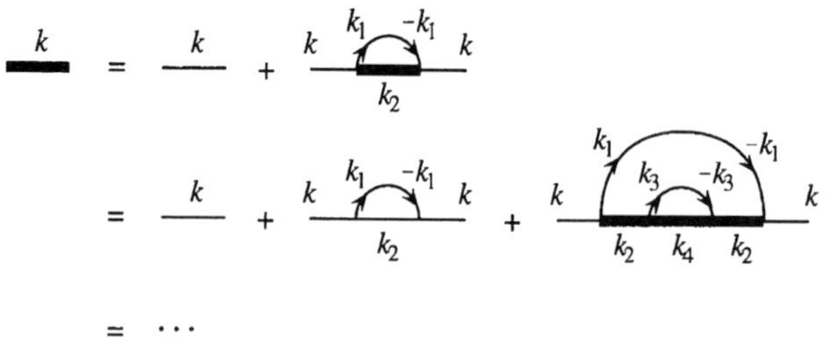

Figure 8.3. Renormalization process. Direct interaction terms are kept. The thin line indicates the linear propagator, and the thick line indicates the nonlinear propagator.

This is because a large number of fluctuation modes is assumed to be excited; the term $\mathbf{\Gamma}_2$ represents the interaction of mode $\boldsymbol{k}_2$ with background fluctuations, from which only the test mode $\boldsymbol{k}$ is separated and is not included. The $\mathbf{\Gamma}_2$ term is approximately equalized to the term from nonlinear interaction with total background fluctuations. In other words, a self-interaction is considered to be only a small portion of total nonlinear interactions.

In this sense, nonlinear effects appear both in the numerator and in the denominator of elements H_{ij}. The nonlinear operator in the right-hand side is a continued fraction with respect to fluctuation amplitudes. A symbolic diagram is illustrated in figure 8.3. In renormalized equation (8.25), the time derivative $\partial \boldsymbol{f}/\partial t$ has two terms: one is a linear response $\boldsymbol{Lf}$ and the other is a nonlinear term, which comes from an interaction of background fluctuations $\boldsymbol{k}_1$ and beat mode $\boldsymbol{k}_2$. The linear response terms are represented by thin lines. The total response including linear and nonlinear processes is represented by a thick line. The total response of mode $\boldsymbol{k}_2$ is also expressed in terms of a linear one and a nonlinear process which includes an interaction with background turbulence $\boldsymbol{k}_3$ and beat mode $\boldsymbol{k}_4 = \boldsymbol{k}_2 - \boldsymbol{k}_3$. The renormalized response is understood as a continued fraction.

8.1.3 Diffusion Approximation

Nonlinear terms in (8.25) are further simplified by a diffusion approximation. Nonlinear interactions are approximated by diffusion terms, and are expressed by diagonal elements in the matrix, as is assumed in equation (8.14). The nonlinear terms like $[\phi_{-1}, [\phi_1, U]]$ appear in equation (8.25). It is explicitly written down as

$$[\phi_{-1}, [\phi_1, \tilde{Y}]] = \left(\left| \frac{\partial \phi_1}{\partial y} \right|^2 \frac{\partial^2}{\partial x^2} + \left| \frac{\partial \phi_1}{\partial x} \right|^2 \frac{\partial^2}{\partial y^2} \right) \tilde{Y} + \left\{ \mathrm{i} k_{1y} \frac{\partial}{\partial x} \left(\phi_{-1} \frac{\partial}{\partial x} \right) \right\} \frac{\partial}{\partial y} \tilde{Y}$$

$$
- \left\{ k_{1y}^2 \frac{\partial}{\partial x} (\phi_{-1}\phi_1) \right\} \frac{\partial}{\partial x} \tilde{Y}
- \left\{ \mathrm{i}k_{1y} \left(\phi_1 \frac{\partial}{\partial x} \phi_{-1} - \phi_{-1} \frac{\partial}{\partial x} \phi_1 \right) \right\} \frac{\partial^2}{\partial x \partial y} \tilde{Y}. \tag{8.26}
$$

We postulate several *ansatz* for the features of fluctuations:

(i) The envelope of fluctuations is much smoother than the wavelength,

$$
|k_1| \gg \frac{1}{\sum_{k_1} |\phi_1|^2} \left| \nabla_x \left(\sum_{k_1} |\phi_1|^2 \right) \right|. \tag{8.27}
$$

(ii) The convective damping of fluctuations is not important,

$$
\left| \phi_1 \frac{\partial}{\partial x} \phi_{-1} - \phi_{-1} \frac{\partial}{\partial x} \phi_1 \right| \ll |k_1 \phi_{-1} \phi_1|. \tag{8.28}
$$

By the condition (i), the $\partial \tilde{Y}/\partial x$ term in equation (8.26) is neglected. The requirement (ii) leads to the fact that $\partial \tilde{Y}/\partial y$ and $\partial^2 \tilde{Y}/\partial x \partial y$ terms in equation (8.26) are unimportant. By these approximations, equation (8.26) reduces to a simplified form as

$$
[\phi_{-1}, [\phi_1, \tilde{Y}]] = \left(\left| \frac{\partial \phi_1}{\partial y} \right|^2 \frac{\partial^2}{\partial x^2} + \left| \frac{\partial \phi_1}{\partial x} \right|^2 \frac{\partial^2}{\partial y^2} \right) \tilde{Y}. \tag{8.29}
$$

For transparency of argument, we use the assumption, without lack of generality,

(iii) background fluctuations are isotropic in directions perpendicular to the magnetic field,

$$
\left| \frac{\partial \phi_1}{\partial x} \right|^2 \simeq \left| \frac{\partial \phi_1}{\partial y} \right|^2 \simeq \frac{k_{1\perp}^2}{2} |\phi_1|^2. \tag{8.30}
$$

By this simplification, equation (8.29) reduces to

$$
[\phi_{-1}, [\phi_1, \tilde{Y}]] = \frac{1}{2} k_{1\perp}^2 |\phi_1|^2 \left(\frac{\partial^2}{\partial x^2} + \frac{\partial^2}{\partial y^2} \right) \tilde{Y}. \tag{8.31}
$$

These approximations also mean that we take the contribution from the fluctuation components, which are more microscopic than the test mode of the interest, $|\boldsymbol{k}_1| > |\boldsymbol{k}|$.

Substituting equation (8.31) into the dynamical equation (8.25), we obtain the governing equation as

$$
\{\partial_T + \boldsymbol{L}\} \begin{pmatrix} \phi \\ J \\ p \end{pmatrix}_k = \sum_{k_1 >} \frac{|k_{1\perp}\phi_1|^2}{2} \boldsymbol{H}_k \nabla_\perp^2 \begin{pmatrix} -\nabla_\perp^2 \phi \\ J \\ p \end{pmatrix}_k . \tag{8.32}
$$

The symbol $k_1 >$ in the summation indicates that the contribution from the more microscopic fluctuations are taken into account.

The renormalized equation can be written for the test mode as

$$\{\partial_T + \boldsymbol{L} + \boldsymbol{\Gamma}_k\} \begin{pmatrix} \phi \\ J \\ p \end{pmatrix}_k = 0 \tag{8.33}$$

with the decorrelation operator

$$\boldsymbol{\Gamma}_k = -\sum_{k_1} \frac{|k_{1\perp}\phi_1|^2}{2} \begin{pmatrix} -H_{11}\nabla_\perp^2 & H_{12} & H_{13} \\ -H_{21}\nabla_\perp^2 & H_{22} & H_{23} \\ -H_{31}\nabla_\perp^2 & H_{32} & H_{33} \end{pmatrix}_k \nabla_\perp^2. \tag{8.34}$$

The decoction operator $\boldsymbol{\Gamma}_k$ indicates the screening effect of background fluctuations. This result confirms, *a posteriori*, the relevance of choosing the form of equation (8.9).

8.2 Dressed Test Mode

8.2.1 Screened Operator

Equations (8.33) and (8.34) describe the evolution of the test mode in the presence of background fluctuations. It is expressed as a linear response equation for the test mode,

$$\{\partial_T + \bar{\boldsymbol{L}}\} \begin{pmatrix} \phi \\ J \\ p \end{pmatrix}_k = 0 \tag{8.35}$$

with the screened operator

$$\bar{\boldsymbol{L}} = \boldsymbol{L} + \boldsymbol{\Gamma}_k. \tag{8.36}$$

The operator is modified by the background fluctuations. Therefore we call the test mode a *dressed test mode*. Conceptual model is illustrated in figure 8.4.

The form equation (8.35) could be solved by use of a mathematical method which has been used to analyse linear stability problems. The equation (8.35) includes the effects of turbulence. With the proper choice of the boundary condition, equation (8.35) is considered as an eigenvalue equation. Its eigenvalue then describes a constraint for background fluctuations and is used to solve the nature of turbulence. This procedure is called the method of the dressed test mode.

8.2.2 Diagonal Matrix Approximation

Within the matrix $\boldsymbol{\Gamma}$, the diagonal elements represent the energy transfer and effective damping. The off-diagonal elements have an importance in relation to the off-diagonal elements of the transport matrix. Keeping this in mind, we first keep the diagonal elements in the matrix $\boldsymbol{\Gamma}$. By this approximation one can learn the importance of the nonlinear instability and the nonlinear stabilization.

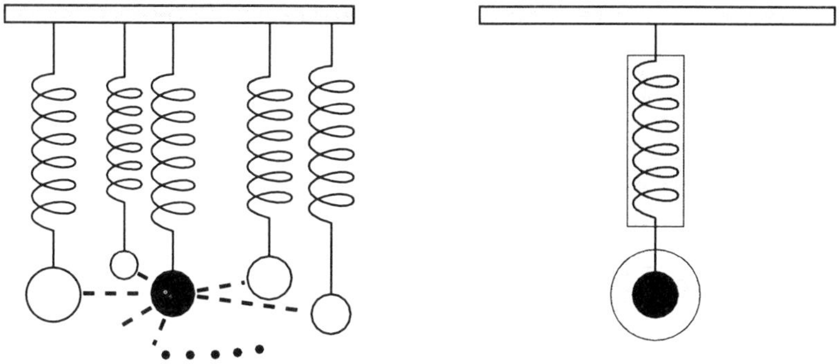

Figure 8.4. Schematic model of dressed test mode. Nonlinearly interacting oscillators (left) are represented by an independent oscillator with screened mass and screened spring constant.

The diagonal elements are explicitly written as

$$\boldsymbol{\Gamma}_k = \begin{pmatrix} \mu_N \nabla_\perp^{-4} & 0 & 0 \\ 0 & -\xi\lambda_N \nabla_\perp^2 & 0 \\ 0 & 0 & -\chi_N \nabla_\perp^2 \end{pmatrix}_k \tag{8.37}$$

where the nonlinear transfer rates of the test mode are given as

$$\mu_{N,k} = \sum_{k_1>} \frac{|k_{\perp 1}\phi_1|^2}{2} \frac{1}{K_2} \tag{8.38}$$

$$\lambda_{N,k} = \left(\frac{c}{a\omega_p}\right)^2 \mu_{eN,k} \tag{8.39}$$

$$\mu_{Ne,k} = \sum_{k_1>} \frac{|k_{\perp 1}\phi_1|^2}{2} \frac{1}{K_2} \frac{\gamma_{u2}}{\gamma_{j2}} \left[1 - \frac{G_0}{\gamma_{u2}\gamma_{p2}k_{\perp 2}^2}\right] \tag{8.40}$$

$$\chi_{N,k} = \sum_{k_1>} \frac{|k_{\perp 1}\phi_1|^2}{2} \frac{1}{K_2} \frac{\gamma_{u2}}{\gamma_{p2}} \left[1 + \frac{\xi k_{\parallel 2}^2}{k_{\perp 2}^2 \gamma_{u2}\gamma_{j2}}\right] \tag{8.41}$$

and K_2 is defined by equation (8.20). The nonlinear terms appear as the transfer rates to the higher k modes from lower k modes, and contribute to an effective viscosity in the vorticity equation, an effective current diffusivity in the Ohm law, and an effective thermal diffusion in the energy balance equation, respectively.

In this limiting case, the screened operator in equation (8.35) is given as

$$\bar{\boldsymbol{L}} = \begin{pmatrix} (\mu_N + \mu_c)\nabla_\perp^4 & \nabla_\parallel & (\boldsymbol{b} \times \boldsymbol{\kappa}) \cdot \boldsymbol{\nabla} \\ \xi \nabla_\parallel & \xi\eta_\parallel - \xi(\lambda_N + \lambda_c)\nabla_\perp^2 & 0 \\ (\nabla p_0 \times \boldsymbol{b}) \cdot \boldsymbol{\nabla} & 0 & -(\chi_N + \chi_c)\nabla_\perp^2 \end{pmatrix}_k . \tag{8.42}$$

It is noticeable that the screened operator has the same structure as the linear operator except the dissipation terms are replaced as

$$\mu_c \to \mu_{N.k} + \mu_c$$
$$\lambda_c \to \lambda_{N.k} + \lambda_c \tag{8.43}$$
$$\chi_c \to \chi_{N.k} + \chi_c.$$

8.2.3 Effective Mass Screened by Turbulence

The screening by the background turbulence is expressed by a concept of the effective mass of particles. Substitution of the screened operator $\bar{L}$ into equation (7.27) leads to

$$\left(1 + \frac{\mathrm{i}\mu_{N.k}k_\perp^2}{\omega}\right)\omega^2\phi + \frac{(k_y/k_\perp)^2 G_0}{1 + \mathrm{i}\chi_{N.k}k_\perp^2/\omega}\phi - \frac{k_\parallel^2}{1 + \xi^{-1}k_\perp^2(1 + \mathrm{i}\mu_{eN.k}k_\perp^2/\omega)}\phi = 0. \tag{8.44}$$

For clarity of argument, the strongly turbulent limit is considered and collisional dissipation is omitted. The first, second and third terms of equation (8.44) correspond to the inertia effect, the driving by the pressure gradient and the restoring force by the bending of magnetic field line, respectively. The last term is modified by the electron mass which is in proportion to the parameter $1/\xi$. This equation could be understood in such a way that the ion mass and electron mass are replaced by the effective masses as

$$m_i \to m_{i.eff} = m_i\left(1 + \frac{\mathrm{i}\mu_{N.k}k_\perp^2}{\omega}\right) \tag{8.45}$$

$$m_e \to m_{e.eff} = m_e\left(1 + \frac{\mathrm{i}\mu_{eN.k}k_\perp^2}{\omega}\right). \tag{8.46}$$

The effective mass could be heavier because of the turbulent fluctuations. The driving term, which includes G_0 in equation (8.44), is also screened by a factor $(1 + \mathrm{i}\omega^{-1}\chi_{N.k}k_\perp^2)^{-1}$.

8.2.4 Renormalization and Elimination of Divergence

The concept of *renormalization* is used in a wide area of physics in order to resolve unphysical divergence. This is also true for the renormalization process in plasma turbulence. The nonlinear response (nonlinear transfer rate) includes the denominator K_2. If one calculates the linear response, the denominator satisfies a relation

$$\gamma K_2 = \gamma^2 + \xi k_{\parallel 2}^2 k_{\perp 2}^{-2} - G_0 k_y^2 k_{\perp 2}^{-2}.$$

Notice that the relation

$$\gamma K_2 = 0$$

is the dispersion relation of the linear mode. There appears a divergence in the response function at the poles of linear dispersion relations. In the linear response method, one treats this divergence by taking the principal part in the summation. By the introduction of the renormalization, where the secular terms are summed up, the divergence is removed from the response function. The finite value $\mathbf{\Gamma}_k$, which is given from the contribution of the finite-amplitude fluctuations, remains even in the vicinity of the dispersion relation of the linear mode.

The removal of the divergence is not limited to the model discussed here. It holds under wide circumstances, including the kinetic treatment of plasma dynamics [4.4, 8.1, 8.2].

8.2.5 Turbulence-Driven Flux

The turbulence-driven heat flux is expressed by use of the formalism of the renormalization. Equation (7.35) describes the energy flux as $\hat{q}_{0x} = -\langle \hat{p}^* \hat{\nabla}_y \hat{\phi} \rangle$. By use of equations (8.35) and (8.42), the pressure fluctuation is expressed in terms of the static potential fluctuation as

$$p_k = \frac{\nabla_y \phi_k}{\gamma_k + (\chi_{N.k} + \chi_c) k_\perp^2} \frac{\mathrm{d}p_0}{\mathrm{d}x}. \tag{8.47}$$

Substituting equation (8.47) into equation (7.35), and rewriting the symbol of summation from k to k_1, one has the expression as

$$q_{0x} = -\left(\frac{1}{2} \sum_{k_1} \frac{|k_{1\perp} \phi_1|^2}{\gamma_1 + (\chi_{N.1} + \chi_c) k_{1\perp}^2} \right) \frac{\mathrm{d}p_0}{\mathrm{d}x}. \tag{8.48}$$

In the derivation of equation (8.48), the simplification of the isotropy of background fluctuations, equation (8.30), is used. If the transport coefficient is introduced as

$$q_0 = -\chi_T \frac{\mathrm{d}p_0}{\mathrm{d}x} \tag{8.49}$$

the turbulence-driven transport coefficient is expressed as

$$\chi_T = \sum_{k_1} \frac{|k_{1\perp} \phi_1|^2}{2\gamma_{p1}} \tag{8.50}$$

where $\gamma_{p1} = \gamma(1) + \Gamma_{p1}$.

Transport coefficient for macro variable and micro variable. The transport coefficient is compared to the nonlinear transfer rate of the fluctuating mode. In the long wavelength limit, the nonlinear transfer rate has a limiting value as

$$\lim_{k \to 0} \chi_k = \sum_{k_1} Q_k^2 \frac{|k_{\perp 1} \phi_1|^2}{2\gamma_{p1}} \tag{8.51}$$

with the weighting coefficient

$$Q_k^2 = \frac{1 + \xi k_{\parallel 1}^2 k_{\perp 1}^{-2} \gamma_{j1}^{-1} \gamma_{u1}^{-1}}{1 + \xi k_{\parallel 1}^2 k_{\perp 1}^{-2} \gamma_{j1}^{-1} \gamma_{u1}^{-1} - G_0 k_y^2 k_{\perp 1}^{-2} \gamma_{u1}^{-1} \gamma_{p1}^{-1}} \tag{8.52}$$

where $\gamma_{u1} = \gamma(1) + \mu_c k_{\perp 1}^2 + \Gamma_{u1}$ and $\gamma_{j1} = \gamma(1)(1 + \xi k_{\perp 1}^{-2}) + \xi \eta_\parallel + \mu_{e.c} k_{\perp 1}^2 + \Gamma_{j1}$. It is shown that the coefficient satisfies the relation $Q_k \sim \mathrm{O}(1)$; comparison between equations (8.50) and (8.51) gives the relation

$$\chi_T \sim \lim_{k \to 0} \chi_k \tag{8.53}$$

aside from a numerical factor of 1–2. The terms which are in proportion to ξ or G_0 in equation (8.52) come from the coupling of three variables $\{\phi, J_\parallel, p\}$. The off-diagonal term in the operator $\boldsymbol{L}$, which is essential in the drive of the instability, causes this coupling. If the driving mechanisms are absent, i.e.,

$$G_0 = 0$$

an equality

$$Q_k = 1 \tag{8.54}$$

holds. In this case, the relation holds as

$$\chi_T = \lim_{k \to 0} \chi_k. \tag{8.55}$$

In other words, the transport coefficient is the long wavelength limit of the nonlinear damping rate of fluctuations in the absence of driving $G_0 = 0$.

This result is compared to the characteristics near thermal equilibrium. For a linear nonequilibrium system, the Onsager *ansatz*, i.e., the decorrelation rate for the micro dynamics equals the macro variable [4.13], is applied. This *ansatz* leads to the famous Onsager relation for the symmetry of the transport matrix. In turbulent plasmas, the state is far from thermal equilibrium, and the Onsager *ansatz* may hold as an approximate relation. The deviation of χ_T from the nonlinear decorrelation rate of micro fluctuations is caused by the finite value of G_0. This term characterizes the degree of nonequilibrium of the system. Nevertheless, an approximate relation $\chi_T \sim \chi_{k \to 0}$ is shown to be satisfied.

8.3 Mean Field Approximation

The nonlinear set of equations is modified as equations (8.35) with (8.42). Equations (8.35) and (8.42) are the set of linear equations for various modes $\boldsymbol{k}$. In principle, the coefficients for the nonlinear screening, $\{\mu_{N.k}, \lambda_{N.k}, \chi_{N.k}\}$, depend on the choice of the test mode. Since the fluctuations could be composed of a nearly infinite number of modes, equations (8.35) and (8.42) constitute the almost infinite number of coupled linear equations. If this set of equations could be solved, then the approximate solution of the fluctuation spectrum is obtained. This process is briefly explained in chapter 13. In order to obtain the analytic insight, however, we first employ the approximation where the set of coefficients $\{\mu_{N,k}, \lambda_{N,k}, \chi_{N,k}\}$ is represented by a set of common coefficients, $\{\mu_N, \lambda_N, \chi_N\}$, for all $\boldsymbol{k}$. This approximation resembles to the mean field approximation in the physics of phase transition [8.3].

The mean field approximation is employed and a set of nonlinear coefficients is chosen as

$$\mu_N = \Sigma \frac{|k_{\perp 1}\phi_1|^2}{2} \frac{1}{K_1} \tag{8.56}$$

$$\mu_{e.N} = \Sigma \frac{|k_{\perp 1}\phi_1|^2}{2} \frac{1}{K_1} \frac{\gamma_{u1}}{\gamma_{j1}} \left[1 - \frac{G_0}{\gamma_{u1}\gamma_{p1}k_{\perp 1}^2} \right] \tag{8.57}$$

$$\chi_N = \Sigma \frac{|k_{\perp 1}\phi_1|^2}{2} \frac{1}{K_1} \frac{\gamma_{u1}}{\gamma_{p1}} \left[1 + \frac{\xi k_{\parallel 1}^2}{k_{\perp 1}^2 \gamma_{u1}\gamma_{j1}} \right]. \tag{8.58}$$

The nonlinear transfer rates equations (8.56)–(8.58) are determined by the fluctuation $\{\phi_1\}$, so the nonlinear coefficients μ_N, μ_{Ne} ($\mu_{Ne} \equiv \xi^{-2}\lambda_N$) and χ_N are related to each other.

The screened operator is expressed as

$$\bar{\boldsymbol{L}} = \begin{pmatrix} (\mu_N + \mu_c)\nabla_\perp^4 & \nabla_\parallel & (\boldsymbol{b} \times \boldsymbol{\kappa}) \cdot \boldsymbol{\nabla} \\ \xi \nabla_\parallel & \xi\eta_\parallel - \xi(\lambda_N + \lambda_c)\nabla_\perp^2 & 0 \\ (\nabla p_0 \times \boldsymbol{b}) \cdot \boldsymbol{\nabla} & 0 & -(\chi_N + \chi_c)\nabla_\perp^2 \end{pmatrix} \tag{8.59}$$

in which the nonlinear interaction of the background turbulence is represented by a set of three scalar quantities, $\{\mu_N, \lambda_N, \chi_N\}$. In this framework, the turbulence-driven transport coefficient is evaluated as

$$\chi_T \sim \chi_N \tag{8.60}$$

It is noted that the equation (8.59) has a similarity to that of the eddy viscosity model of nonlinear interactions in fluid dynamics [8.4]. In the analysis of nonlinear plasma turbulence, equation (8.35) is solved as an eigenvalue equation. Explicit examples are given in the following chapters.

Before closing this chapter, an additional note is made on the role of the incoherent part of the nonlinear interactions. Background fluctuations can also drive the test mode randomly through nonlinear interactions. Such contributions are not kept here, but could be treated as a random noise term in the Langevin equation [8.5]. Consideration of the spontaneous noise term shows that the result of turbulence and turbulent transport is not modified qualitatively by the noise [8.6].

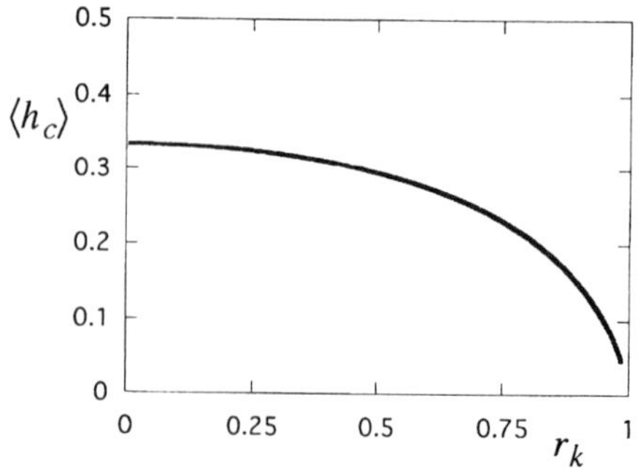

Figure 8.5. Average $\langle h_c \rangle$ as a function of $r_k = |\boldsymbol{k}|/|\boldsymbol{k}_1|$.

Appendix 8A Nonlinear Corrections

The nonlinear interaction terms

$$N_u = [\phi_1, U_k] \qquad N_j = [\phi_1, J_k] \qquad N_p = [\phi_1, p_k]$$

are kept in equation (8.23). This implies that we take the nonlinear interaction where the mode of interest is modified by the $\boldsymbol{E} \times \boldsymbol{B}$ motion of background fluctuations. This choice is made for analytic transparency of the renormalization process. There are more complicated nonlinear terms; however, the characteristic nature of the renormalized operator does not alter even if the more complete terms are kept. Let us show the terms in the vorticity more carefully. The nonlinear term of the vorticity is written as

$$N_u = [\phi_1, U_k] + [\phi_k, U_1]. \tag{8A.1}$$

The second term is neglected in the main text. Noting the relation $U = k_\perp^2 \phi$, equation (8A.1) is modified as

$$N_u = \left(1 - \frac{k_{1\perp}^2}{k_\perp^2}\right)[\phi_1, U_k]. \tag{8A.2}$$

By this correction, the term H_{11} in equation (8.25) is modified by the factor

$$\left(1 - \frac{k_{1\perp}^2}{k_{2\perp}^2}\right)\left(1 - \frac{k_{1\perp}^2}{k_\perp^2}\right) \equiv h_c. \tag{8A.3}$$

If one expresses $\cos\varphi = (\boldsymbol{k} \cdot \boldsymbol{k}_1)/|\boldsymbol{k}||\boldsymbol{k}_1|$ and $|\boldsymbol{k}| = r_k|\boldsymbol{k}_1|$ (i.e., φ is the angle between two vectors $\boldsymbol{k}$ and $\boldsymbol{k}_1$, and r_k is the ratio between the magnitudes of

$\boldsymbol{k}$ and $\boldsymbol{k}_1$), the average of the coefficient h_c is calculated. One calculates the average $\langle h_c \rangle = \frac{1}{2}\int_0^{\pi} \sin\varphi h_c \,\mathrm{d}\varphi$ and obtains

$$\langle h_c \rangle = \left\{1 - \frac{1}{2r_k}\ell n\left(\frac{1+r_k}{1-r_k}\right)\right\}\left(1 - \frac{1}{r_k^2}\right). \tag{8A.4}$$

The right-hand side is a positive numerical factor of order unity. For instance, in the limit of the short wavelength background fluctuations, $r_k \to 0$, one has

$$\lim_{r_k \to 0} \langle h_c \rangle \to \tfrac{1}{3}. \tag{8A.5}$$

Figure 8A.1 illustrates the average $\langle h_c \rangle$.

Appendix 8B Renormalization for the Four-Field Reduced Set of Equations

The dynamical equation for four-field variables $\{\phi, J_\parallel, p, V_\parallel\}$ are formally written as

$$\frac{\partial}{\partial t}\begin{pmatrix} -\nabla_\perp^2 \phi \\ (1 - \xi \nabla_\perp^{-2}) J \\ p \\ V_\parallel \end{pmatrix} + \boldsymbol{L}\begin{pmatrix} \phi \\ J \\ p \\ V_\parallel \end{pmatrix} = \boldsymbol{N} \tag{8B.1}$$

with

$$\boldsymbol{L} = \begin{pmatrix} \mu_{\perp c}\nabla_\perp^4 & \nabla_\parallel & (\boldsymbol{b} \times \boldsymbol{\kappa}) \cdot \boldsymbol{\nabla} & 0 \\ \xi \nabla_\parallel & \xi \eta_\parallel - \xi \lambda_c \nabla_\perp^2 & 0 & 0 \\ (\nabla p_0 \times \boldsymbol{b}) \cdot \boldsymbol{\nabla} & 0 & -\chi_c \nabla_\perp^2 & \beta \nabla_\parallel \\ 0 & 0 & \nabla_\parallel & -\mu_{\parallel c}\nabla_\perp^2 \end{pmatrix} \tag{8B.2}$$

and

$$\boldsymbol{N} = \begin{pmatrix} N_u \\ N_j \\ N_p \\ N_V \end{pmatrix} = -\begin{pmatrix} [\phi, -\nabla_\perp^2 \phi] \\ [\phi, J] \\ [\phi, p] \\ [\phi, V_\parallel] \end{pmatrix}. \tag{8B.3}$$

By use of the same procedure as leading to (8.22), the driven mode is calculated as

$$U_2 = \frac{1}{K_2}\left\{N_u - \frac{\mathrm{i}k_{\parallel 2}}{\gamma_{j2}}N_j - \frac{\mathrm{i}k_{2y}\kappa_x\gamma_{v2}}{\Gamma_{vp2}}N_p - \frac{k_{2y}\kappa_x\beta k_{\parallel 2}}{\Gamma_{vp2}}N_v\right\} \tag{8B.4}$$

$$\begin{aligned} J_2 = \frac{1}{K_2}\Bigg\{ & -\frac{\mathrm{i}k_{\parallel 2}\xi}{\gamma_{j2}k_{\perp 2}^2}N_u + \left[\frac{K_2}{\gamma_{j2}} - \frac{\xi k_{\parallel 2}^2}{\gamma_{j2}^2 k_{\perp 2}^2}\right]N_j \\ & - \frac{\xi k_{2y}\kappa_x\gamma_{v2}k_{\parallel 2}}{\gamma_{j2}k_{\perp 2}^2\Gamma_{vp2}}N_p + \frac{\mathrm{i}\xi k_{2y}\kappa_x\beta k_{\parallel 2}^2}{\gamma_{j2}k_{\perp 2}^2\Gamma_{vp2}}N_v\Bigg\} \end{aligned} \tag{8B.5}$$

$$p_2 = \frac{1}{K_2}\left\{\frac{\mathrm{i}\gamma_{v2}k_{2y}p_0'}{\Gamma_{vp2}k_{\perp 2}^2}N_u + \frac{k_{2y}p_0'\gamma_{v2}k_{\parallel 2}}{\Gamma_{vp2}\gamma_{j2}k_{\perp 2}^2}N_j\right\}$$
$$+\frac{1}{K_2}\left\{\frac{\gamma_{v2}}{\Gamma_{vp2}}\left[K_2 + \frac{G_0k_{2y}^2\gamma_{v2}}{k_{\perp 2}^2\Gamma_{vp2}}\right]N_p - \frac{\mathrm{i}\beta k_{\parallel 2}}{\Gamma_{vp2}}\left[K_2 + \frac{G_0k_{2y}^2\gamma_{v2}}{k_{\perp 2}^2\Gamma_{vp2}}\right]N_v\right\} \tag{8B.6}$$

and

$$V_{\parallel 2} = \frac{1}{K_2}\left\{\frac{k_{\parallel 2}k_{2y}p_0'}{\Gamma_{vp2}k_{\perp 2}^2}N_u - \frac{\mathrm{i}k_{2y}p_0'k_{\parallel 2}^2}{\Gamma_{vp2}\gamma_{j2}k_{\perp 2}^2}N_j\right\}$$
$$+\frac{1}{K_2}\left\{-\frac{\mathrm{i}k_{\parallel 2}}{\Gamma_{vp2}}\left[K_2 + \frac{G_0k_{2y}^2\gamma_{v2}}{k_{\perp 2}^2\Gamma_{vp2}}\right]N_p + \frac{\gamma_{p2}}{\Gamma_{vp2}}\left[K_2 - \frac{G_0k_{2y}^2\beta k_{\parallel 2}^2}{\gamma_{p2}k_{\perp 2}^2\Gamma_{vp2}}\right]N_v\right\} \tag{8B.7}$$

where

$$K_2 = \gamma_{u2} + \frac{k_{\parallel 2}^2}{K_{\perp 2}^2}\frac{\xi}{\gamma_{j2}} - \frac{k_{2y}^2G_0\gamma_{v2}}{\Gamma_{vp2}k_{\perp 2}^2} \tag{8B.8}$$

$\gamma_{v2} = \gamma(2) + \mu_{\parallel c}k_{\perp 2}^2 + \Gamma_{v2}$, Γ_v is the nonlinear decorrelation rate for parallel momentum and

$$\Gamma_{vp2} = \gamma_{p2}\gamma_{v2} + \beta k_{\parallel 2}^2. \tag{8B.9}$$

Equation (8.22) is reproduced, if the limit $\beta \to 0$ is taken in equations (8B.4)–(8B.9).

As in (8.32), the renormalized equation is expressed as

$$\{\partial_T + \boldsymbol{L}\}\begin{pmatrix}\phi\\ J\\ p\\ V_\parallel\end{pmatrix}_k = \sum_{k_1>}\frac{|k_{1\perp}\phi_1|^2}{2}\boldsymbol{H}_k\nabla_\perp^2\begin{pmatrix}-\nabla_\perp^2\phi\\ J\\ p\\ V_\parallel\end{pmatrix}_k. \tag{8B.10}$$

The elements of the 4×4 matrix $\boldsymbol{H}$ are given as the coefficients in equations (8B.4)–(8B.7). The nonlinear decorrelation is taken into account, and the dynamical equation for the test mode is given as

$$\frac{\partial}{\partial t}\begin{pmatrix}-\nabla_\perp^2\phi\\ (1-\xi\nabla_\perp^{-2})J\\ p\\ V_\parallel\end{pmatrix}_k + (\boldsymbol{L}+\boldsymbol{\Gamma})_k\begin{pmatrix}\phi\\ J\\ p\\ V_\parallel\end{pmatrix}_k = 0. \tag{8B.11}$$

The diagonal approximation gives the nonlinear decorrelation matrix $\boldsymbol{\Gamma}$ as

$$\boldsymbol{\Gamma}_k = \begin{pmatrix}\mu_{\perp k}\nabla_\perp^4 & 0 & 0 & 0\\ 0 & -\xi\lambda_k\nabla_\perp^2 & 0 & 0\\ 0 & 0 & -\chi_k\nabla_\perp^2 & 0\\ 0 & 0 & 0 & -\mu_{\parallel k}\nabla_\perp^2\end{pmatrix}_k \tag{8B.12}$$

where

$$\mu_{\perp k} = \sum_{k_1} \frac{|k_{\perp 1}\phi_1|^2}{2} \frac{1}{K_2} \tag{8B.13}$$

$\lambda_k = \xi^{-1}\mu_{ek}$ with

$$\mu_{ek} = \sum_{k_1} \frac{|k_{\perp 1}\phi_1|^2}{2} \frac{1}{K_2} \frac{\gamma_{u2}}{\gamma_{j2}} \left[1 - \frac{\gamma_{v2}}{\gamma_{u2}} \frac{k_{2y}^2 G_0}{\Gamma_{vp2} k_{\perp 2}^2} \right] \tag{8B.14}$$

$$\chi_k = \sum_{k_1} \frac{|k_{\perp 1}\phi_1|^2}{2} \frac{1}{K_2} \frac{\gamma_{u2}\gamma_{v2}}{\Gamma_{vp2}} \left[1 + \frac{\xi k_{\parallel 2}^2}{k_{\perp 2}^2 \gamma_{u2}\gamma_{j2}} \right] \tag{8B.15}$$

$$\mu_{\parallel k} = \sum_{k_1} \frac{|k_{\perp 1}\phi_1|^2}{2} \frac{1}{K_2} \frac{\gamma_{p2}}{\Gamma_{vp2}} \left[K_2 - \frac{k_{2y}^2 G_0 \beta k_{\parallel 2}^2}{\Gamma_{vp2}\gamma_{p2} k_{\perp 2}^2} \right] \tag{8B.16}$$

and K_2 is given by equation (8B.8). The mean field approximation can also be made for the coefficients (8B.13)–(8B.16).

REFERENCES

[8.1] Yoshizawa A 1998 *Turbulence Modelling and Statistical Theory of Hydrodynamic and Magnetohydrodynamic Flows* (Amsterdam: Kluwer)

[8.2] Cook I and Taylor J B 1973 *J. Plasma Phys.* **9** 131
Taylor J B and McNamara B 1971 *Phys. Fluids* **14** 1492
Krommes J A 1984 *Basic Plasma Physics II* ed A A Galeev and R N Sudan (Amsterdam: North-Holland) ch 5.5
Horton C W and Choi D 1979 *Phys. Rep.* **49** 273

[8.3] Kittel C 1966 *Introduction to Solid State Physics* 3rd edn (New York: Wiley) ch 15

[8.4] Yoshizawa A 1984 *Phys. Fluids* **27** 1377
Canuto V M and Goldman I 1985 *Phys. Rev. Lett.* **54** 430

[8.5] Kraichnan R H 1970 *J. Fluid Mech.* **41** 189
Krommes J A 1996 *Phys. Rev* E **53** 4865
Krommes J A 1998 *Plasma Phys. Control. Fusion* **40** at press

[8.6] Itoh S-I and Itoh K 1998 Statistical theory of subcritically-excited strong turbulence in inhomogeneous plasmas *Research Report* IPP-III/234 Max-Planck-Institut fur Plasmaphysik

Chapter 9

Self-Sustained Turbulence I—Current Diffusive Interchange Mode

In this chapter, we discuss an example of turbulence and turbulent transport in an inhomogeneous plasma. We choose the simplest situation, that a global plasma pressure and a magnetic field vary only in the $\hat{x}$-direction; global structures of plasma and magnetic field are assumed to be uniform in other directions. The interchange mode is the simplest instability being driven by a combination of pressure and magnetic field gradients. It is first shown that this instability could be excited by nonlinear mechanisms. In other words, it belongs to a class of *subcritical turbulence*. A stationary state of turbulent plasma is derived from a solution of the dispersion relation of the dressed test mode (the nonlinear dispersion relation). The gradient parameter, G_0, is shown to be an *order parameter* which characterizes nonequilibrium plasma transport.

The subject is determination of characteristic quantities of turbulence and turbulent transport coefficients: they include turbulence level, typical decorrelation lengths, decorrelation times and transport coefficients. In this chapter, a mean field approximation is employed. (An analysis of the spectrum is made in chapter 13.) In this framework, the length of micro-scale dynamics is represented by only one parameter. We take the mode number (wavelength) of the micro-mode as a representative of the characteristic scale-length of micro-fluctuations. By this simplification, characteristic quantities are expressed in terms of the order parameter G_0.

9.1 Nonlinear Instability

9.1.1 Nonlinear Dispersion Relation

The dynamical equation of the fluctuation mode in a turbulent plasma is expressed by use of the method of the dressed test mode as equation (8.35),

$$\{\partial_T + \bar{L}\} \begin{pmatrix} \phi \\ J \\ p \end{pmatrix}_k = 0. \tag{9.1}$$

The screened operator, which includes the effects of turbulent shielding, is given as

$$\bar{\boldsymbol{L}} = \begin{pmatrix} \bar{\mu}\nabla_\perp^4 & \nabla_\parallel & (\boldsymbol{b}\times\boldsymbol{\kappa})\cdot\boldsymbol{\nabla} \\ \xi\nabla_\parallel & \xi\eta_\parallel - \xi\bar{\lambda}\nabla_\perp^2 & 0 \\ (\nabla p_0 \times \boldsymbol{b})\cdot\boldsymbol{\nabla} & 0 & -\bar{\chi}\nabla_\perp^2 \end{pmatrix} \tag{9.2}$$

with

$$\bar{\mu} = \mu_N + \mu_c \qquad \bar{\lambda} = \lambda_N + \lambda_c \qquad \bar{\chi} = \chi_N + \chi_c. \tag{9.3}$$

As in the preceding chapter 8, all quantities are normalized, and the symbol ˆ is suppressed.

This equation predicts whether a test mode grows or damps under given background fluctuations. The equation is expressed as a linear equation for the test mode, and the influences of the background fluctuations are renormalized in the coefficients equation (9.3). The plasma structure is invariant in the $\hat{y}$- and $\hat{z}$-directions, so that each (k_y, k_z)-Fourier component can be solved separately. The test mode is Fourier decomposed as

$$\phi(\boldsymbol{x}, t) = \sum_{k_x, k_y} \phi(k_x, k_y)\exp(\mathrm{i}k_x x + \mathrm{i}k_y y + \gamma t). \tag{9.4}$$

We choose the magnetic configuration with a magnetic shear as equation (6.45),

$$\boldsymbol{B} = (0, x/L_s, 1)B. \tag{9.5}$$

This means that the origin $x = 0$ is chosen at a mode rational surface of the test mode. The choice is based on the consideration that the interchange mode could be excited if the parallel mode number is small enough. The parallel mode number is given as equation (6.46) and is expressed by the introduction of the shear parameter s as

$$k_\parallel = s k_y x. \tag{9.6}$$

In applying the local Cartesian coordinates to a toroidal plasma, the shear parameter is related to the pitch of the field line as

$$s = rq^{-1}(\mathrm{d}q/\mathrm{d}r)$$

where q is the safety factor.

Substituting equation (9.4) into equation (9.1) and taking the same procedure as used to derive the relation (7.27) one obtains

$$\left\{(\gamma + \bar{\mu}k_\perp^2)k_\perp^2 - \frac{k_y^2 G_0}{\gamma + \bar{\chi}k_\perp^2} + k_\parallel \frac{k_\perp^2}{\gamma(1+\xi^{-1}k_\perp^2) + (\eta_\parallel k_\perp^2 + \bar{\lambda}k_\perp^4)} k_\parallel\right\}$$
$$\times \phi(k_x, k_y) = 0. \tag{9.7}$$

It is noted that the mode number $k_\parallel$ depends on x as is given by equation (9.6); therefore variables $k_\parallel$ and $k_\perp$ are not always commutable. Noting the relation

$k_\parallel = sk_y(\mathrm{i}\partial/\partial k_x)$ from equation (9.6), equation (9.7) is written as the second-order ordinary differential equation with respect to k_x as

$$\left\{\frac{\partial}{\partial k_x}\frac{s^2k_y^2k_\perp^2}{\gamma(1+\xi^{-1}k_\perp^2)+(\eta_\parallel k_\perp^2+\bar{\lambda}k_\perp^4)}\frac{\partial}{\partial k_x}+\frac{k_y^2G_0}{\gamma+\bar{\chi}k_\perp^2}-(\gamma+\bar{\mu}k_\perp^2)k_\perp^2\right\}$$
$$\times\phi(k_x;k_y)=0. \tag{9.8}$$

(In equations (9.7) and (9.8), k_y is a parameter to indicate the Fourier component, not a variable.) The boundary condition, that the mode is localized in the vicinity of the mode rational surface, is rewritten as that an integral $\int_{-\infty}^{\infty}\mathrm{d}k_x|\phi(k_x)|^2$ exists. It requires that

$$\phi(k_x;k_y)\to 0 \qquad \text{as } |k_x|\to\infty. \tag{9.9}$$

The dynamics of the test mode, which is governed by equation (9.7), is discussed in the following. It is noted that the contribution of the perturbed parallel current (the first term of the right-hand side of equation (9.8)) includes two dissipation mechanisms, i.e., the resistivity and the current diffusivity. The resistive effect appears as $\eta_\parallel k_\perp^2$ and the current diffusive effect has a form $\lambda k_\perp^4$. For the microscopic mode, the term with the higher power to $k_\perp$ could have the larger influence. Therefore, we concentrate on the study of the current diffusivity in this chapter. The effect of the resistivity is discussed in chapter 11.

9.1.2 Nonlinear Instability

Equation (9.7) shows that the interchange mode could be excited by *nonlinear mechanisms*. It is shown that the third term of the left-hand side of equation (9.7) is the restoring force associated with the bend of the magnetic field line. The electron inertia effects, from either the electron–ion collisions $\eta_\parallel$ or the cross-field diffusion (λ), impede the free electron motion. These effects reduce the restoring force, and hence destabilize the mode. This is the mechanism of the dissipative instability. As is shown in equation (8.40), the growth of fluctuation level enhances the effective current diffusivity. (In other words, the electron effective mass becomes heavier as is shown by equation (8.46).) Then the restoring force becomes smaller. The nonlinear effect makes the growth rate larger. The schematic chain of interactions is illustrated in figure 9.1.

Current diffusive mode. The nonlinear excitation is shown quantitatively. For the transparency of the argument, we study that the electron dissipation is dominated by the current diffusive term,

$$\bar{\lambda}k_\perp^2 \gg \eta_\parallel.$$

For the demonstration of the nonlinear destabilization mechanism, the correction to the current diffusivity is kept. In this case, equation (9.8) is simplified as

$$\left\{\frac{\partial}{\partial k_x}\frac{1}{\bar{\lambda}k_\perp^2}\frac{\partial}{\partial k_x}+\gamma^{-1}s^{-2}G_0-\frac{\gamma k_\perp^2}{s^2k_y^2}\right\}\phi(k_x;k_y)=0. \tag{9.10}$$

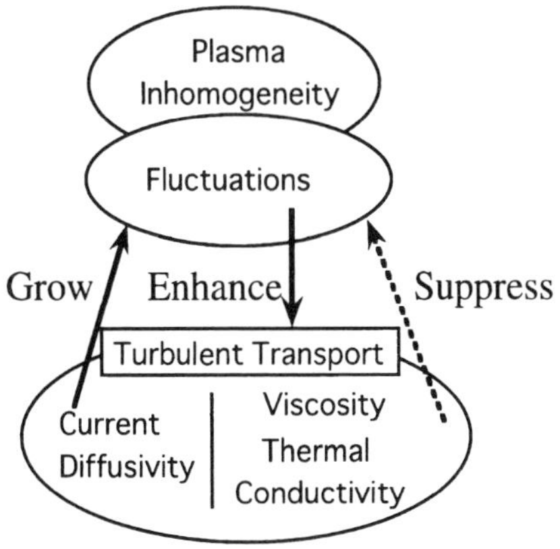

Figure 9.1. Nonlinear link between the fluctuations and turbulent transport. Through the current diffusivity, they enhance each other.

The condition $\bar{\lambda} k_\perp^4 \gg \gamma(1 + \xi^{-1} k_\perp^2)$ is also used to lead equation (9.10). The growth rate is estimated by the variational method (see appendix 9A for the details), and is given as

$$\gamma = \frac{2^{4/5}}{3^{3/5}} \frac{k_y^{4/5} G_0^{3/5}}{s^{2/5}} \bar{\lambda}^{1/5}. \tag{9.11}$$

This result shows that the growth rate becomes higher if the current diffusivity becomes larger. In addition to this, the power index to the current diffusivity is 1/5 and is very small. Even if a very small amount of current diffusivity appears under some mechanisms, the growth rate becomes considerably large. Figure 9.2 illustrates a chained link between the fluctuations, growth rate and current diffusivity. This nonlinear drive continues to work until other nonlinearities, i.e., ion viscosity and thermal conductivity, are enhanced to balance with it.

9.1.3 Reduction of Variables

Equation (9.8), which is to be solved, represents a set of an infinite number of linear equations for various values of k_y. In order to estimate a turbulent level, we employ a method of reduction of variables. Instead of treating all components $\{\phi(k_x, k_y)\}$ for each value of k_y, to which the set of eigenvalues $\{\mu_{N,k}, \lambda_{N,k}, \chi_{N,k}\}$ corresponds, we choose several representative quantities. A mean field approximation is used, and a turbulent state is represented by $\{\mu_N, \lambda_N, \chi_N\}$ with turbulent level ϕ and typical decorrelation scale length k_y^{-1}. Five variables $\{\mu_N, \lambda_N, \chi_N; \phi, k_y\}$ are used. Reduction of variables is a very crude approximation for nonlinear system of inhomogeneous plasmas.

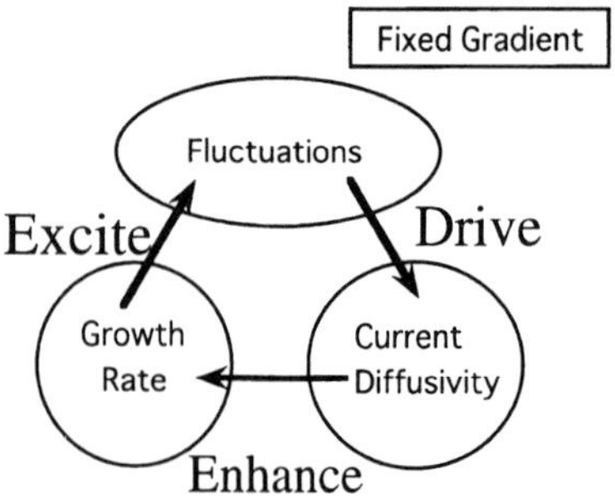

Figure 9.2. Chained link between the fluctuations, current diffusivity and growth rate.

Nevertheless, it provides a clear view on the characteristic features of the development of nonlinear nonequilibrium systems.

Reduced variables $\{\mu_N, \lambda_N, \chi_N; \phi, k_y\}$ are related to each other. Relations are explicitly given in equations (8.56)–(8.58). The denominator is rewritten by use of numerical coefficient C_1 as $K_1 = \gamma_{u1}(1 + C_1)$, where C_1 is defined as

$$C_1 = \frac{\xi k_{\parallel 1}^2}{k_{\perp 1}^2} \frac{\gamma_{u1}}{\gamma_{j1}} - \frac{k_{1y}^2 G_0}{k_{\perp 1}^2} \frac{\gamma_{u1}}{\gamma_{p1}}. \tag{9.12}$$

It will be shown that the coefficient C_1 is of order unity. Noting the relation $\gamma_{u1} = \gamma + \mu_N k_\perp^2$, nonlinear transfer rates are then expressed as

$$\mu_N = \sum_{k_1} \frac{|\phi_1|^2}{2(1+C_1)} \frac{1}{(\mu_N + \gamma k_{1\perp}^{-2})} \tag{9.13}$$

$$\mu_{e.N} = \sum_{k_1} \frac{|\phi_1|^2}{2(1+C_1)} \frac{[1 - \gamma_{u1}^{-1} \gamma_{p1}^{-1} k_{\perp 1}^{-2} k_{1y}^2 G_0]}{(\mu_{e.N} + \gamma k_{1\perp}^{-2})} \tag{9.14}$$

and

$$\chi_N = \sum_{k_1} \frac{|\phi_1|^2}{2(1+C_1)} \frac{[1 + \gamma_{u2}^{-1} \gamma_{j2}^{-1} k_{\perp 2}^{-2} \xi k_{\parallel 2}^2]}{(\chi_N + \gamma k_{1\perp}^{-2})}. \tag{9.15}$$

These relations show that an average fluctuation level, $\sqrt{\langle|\phi^2|\rangle}$, and nonlinear transfer rates $(\mu_N, \mu_{e.N}, \chi_N)$ are expressed by the moments of fluctuation spectrum. The weighting functions are different, but the coefficients (such as C_1) are of order unity. The coefficients $(\mu_N, \mu_{e.N}, \chi_N)$ take similar values.

9.1.4 Eigenvalue Relation for Stationary State

Equation (9.8) is solved for the dressed test mode which is screened by given background fluctuations. The growth rate of the test mode is given as a functional of background fluctuations equation (9.11). The marginal stability condition (i.e., nonlinear neutral condition) for the test mode, if found, is considered as the equation that specifies the stationary turbulence.

The marginal condition, $\gamma = 0$, leads equation (9.8) to be

$$\left\{ \frac{\partial}{\partial k_x} \frac{s^2 k_y^2}{(\eta_\parallel + \bar{\lambda} k_\perp^2)} \frac{\partial}{\partial k_x} + \frac{k_y^2 G_0}{\bar{\chi} k_\perp^2} - \bar{\mu} k_\perp^4 \right\} \phi(k_x; k_y) = 0 \tag{9.16}$$

where the coefficients for the dissipation are given as $\bar{\mu} = \mu_N + \mu_c$, $\bar{\lambda} = \lambda_N + \lambda_c$ and $\bar{\chi} = \chi_N + \chi_c$. The case of $\bar{\lambda} k_\perp^2 \gg \eta_\parallel$ is studied. The marginal condition for the current diffusive interchange mode (CDIM) is given as

$$\left\{ \frac{\partial}{\partial k_x} \frac{s^2 k_y^2}{\bar{\lambda} k_\perp^2} \frac{\partial}{\partial k_x} + \frac{k_y^2 G_0}{\bar{\chi} k_\perp^2} - \bar{\mu} k_\perp^4 \right\} \phi(k_x; k_y) = 0. \tag{9.17}$$

For an analytic insight, this eigenvalue equation is remodelled under the WKB approximation, and $\mathrm{d}\phi/\mathrm{d}k_x$ is neglected;

$$\left\{ \frac{\mathrm{d}^2}{\mathrm{d}k_x^2} + \frac{\bar{\lambda} G_0}{\bar{\chi} s^2} - \frac{\bar{\lambda} \bar{\mu} k_\perp^6}{s^2 k_y^2} \right\} \phi(k_x; k_y) = 0. \tag{9.18}$$

A typical mode number

$$k_0 = \left(\frac{s^2}{\bar{\lambda} \bar{\mu}} \right)^{1/6} \tag{9.19}$$

and normalized pressure gradient are introduced as

$$\Im = \frac{\bar{\lambda}^{2/3} G_0}{s^{4/3} \bar{\chi} \bar{\mu}^{1/3}}. \tag{9.20}$$

This number $\Im$ (Itoh number) is to be obtained as an eigenvalue. Then the eigenvalue equation is rewritten as

$$\left\{ \frac{\mathrm{d}^2}{\mathrm{d}u^2} + \Im - \frac{1}{b} (b + u^2)^3 \right\} \phi(u; b) = 0 \tag{9.21}$$

where $b = (k_y/k_0)^2$. The eigenvalue $\Im$ is approximately obtained from equation (9.21) by use of the WKB method as

$$(\Im - b^2)^{2/3} b^{1/6} \int_0^1 \sqrt{1 - z^6}\, \mathrm{d}z = \frac{\pi}{4}. \tag{9.22}$$

By use of the numerical constant

$$C_{CD} = \left(\frac{4}{\pi} \int_0^1 \sqrt{1 - z^6}\, \mathrm{d}z \right)^{-3/2} \tag{9.23}$$

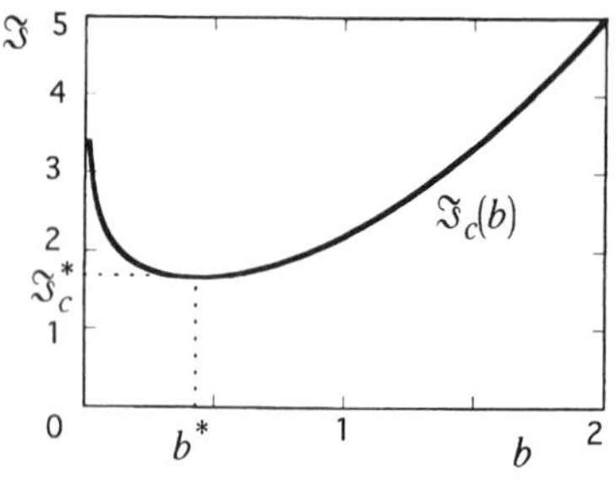

Figure 9.3. Nonlinear marginal stability line for the current diffusive interchange mode. The stability boundary is expressed in terms of parameters $\Im$ and b.

the eigenvalue $\Im$ is obtained as

$$\Im = \Im_c(b) = b^2 + C_{CD}b^{-1/4}. \tag{9.24}$$

The eigenvalue $\Im$ is now given as a function of the mode number b, and is illustrated in figure 9.3.

When $\Im$ is above a critical Itoh number $\Im_c^*$, the mode is subject to a nonlinear instability. This result depicts the balance condition between the nonlinear growth (through current diffusion) and nonlinear damping (through ion viscosity and energy diffusion). As is discussed in equation (9.11), the nonlinear instability exists if the current diffusivity is increased. However, nonlinear terms μ_N and χ_N increase together with λ_N. The stationary state is realized when the normalized pressure satisfies equation (9.24).

The normalized pressure gradient $\Im$ takes the minimum $\Im_c^*$ of

$$\Im_c^* = 9(C_{CD}/8)^{8/9} \tag{9.25}$$

at the mode number satisfying

$$b = b^* = (C_{CD}/8)^{4/9}. \tag{9.26}$$

These results are slightly modified by additional effects in plasmas. For instance, a compressibility effect is discussed in appendix 9B. Parallel ion motion is solved, and viscosity of parallel momentum is derived.

9.1.5 Normalized Number $\Im$ and Rayleigh Number

The normalized pressure gradient

$$\Im = \frac{\lambda^{2/3} G_0}{s^{4/3} \chi \mu^{1/3}}$$

plays an important role in dictating the evolution of turbulence in an inhomogeneous plasma. This particular combination of parameters describes an essential feature of the plasma turbulence. As is illustrated in figure 6.9, a certain similarity holds between this turbulence and the Rayleigh–Benard convection. It is well known that the buoyancy-driven instability in a fluid which is heated from the bottom, i.e., the Benard convection problem, is governed by the Rayleigh number

$$\Re_a \equiv \frac{d^4 \alpha g \nabla T}{\chi \mu}.$$

At the critical Rayleigh number, the Benard convection becomes unstable. In the Rayleigh number, the temperature gradient coupled with the gravity $\alpha g \nabla T$ has the role of the driving parameter. This driving parameter is divided by the viscosity and thermal diffusivity, which are stabilizing the mode. In the case of inhomogeneous and magnetized plasma, G_0, which is the combination of the pressure gradient and the magnetic field inhomogeneity, has a role of the driving parameter. The normalized number $\Im = \lambda^{2/3} s^{-4/3} \chi^{-1} \mu^{-1/3} G_0$ reflects that the ion viscosity and thermal diffusivity suppress the instability. However, the third component of the dissipation, the current diffusivity, appears as the numerator in the normalized pressure gradient $\Im$. This shows that the dissipation process can *destabilize* the mode in plasmas. This destabilization process by the dissipation is one of the characteristic features of turbulence and turbulent transport in plasmas.

9.2 Ratio of Nonlinear Transfer Rates

9.2.1 Rate of Nonlinear Transfer

In a stationary state, $\gamma \simeq 0$, the relationships between the various transfer rates are more clearly seen. Explicit forms are derived from equations (9.13)–(9.15) as

$$\mu_N^2 = \sum_{k_1} \frac{|\phi_1|^2}{2(1+C_1)} \tag{9.27}$$

$$\mu_{e.N}^2 = \sum_{k_1} \frac{|\phi_1|^2}{2(1+C_1)} \left[1 - \frac{k_{1y}^2 G_0}{\mu_N \chi_N k_{\perp 1}^6} \right] \tag{9.28}$$

and

$$\chi_N^2 = \sum_{k_1} \frac{|\phi_1|^2}{2(1+C_1)} \left[1 + \frac{\xi k_{\parallel 2}^2}{\mu_N \mu_{e.N} k_{\perp 1}^6} \right]. \tag{9.29}$$

There is a sum rule that the nonlinear transfer rates must satisfy, i.e., [9.1]

$$\mu_{e.N}^2 + \chi_N^2 = \mu_N^2 + \sum_{k_1} \frac{|\phi_1|^2}{2} \tag{9.30}$$

or

$$\mu_{e.N}^2 + \chi_N^2 = \mu_N^2 + \tfrac{1}{2} \langle |\phi|^2 \rangle. \tag{9.31}$$

9.2.2 Turbulent Prandtl Number

It is shown that all terms, μ_N, $\mu_{e.N}$, χ_N and average fluctuation level, $\sqrt{\langle|\phi|^2\rangle}$, are of the same order of magnitude. Ratios μ_N/χ_N and $\mu_{e.N}/\chi_N$ are more or less constant in comparison with the magnitude of χ_N itself. It is a characteristic feature that the turbulent Prandtl numbers are close to unity [9.2]

$$\mu_T/\chi_T \sim \mu_{e.T}/\chi_T \sim \mathrm{O}(1). \tag{9.32}$$

9.3 Stationary Turbulence and Transport Coefficient

9.3.1 Least Stable Mode

We here consider a stationary turbulence, where the averaged fluctuation level does not grow. Such a state is realized under the nonlinear marginal stability condition for the least stable mode. The fluctuation level would continue to grow if there remains a nonlinear instability.

The turbulent state here is characterized by a reduced set of variables, $\{\mu_N, \mu_{e.N}, \chi_N, \phi, k_y\}$. As is stated in the beginning of this chapter, the characteristic length of micro-scale fluctuations, e.g., the decorrelation length of micro-fluctuations, ℓ_{corr}, is represented by a 'wavelength' of the dressed test mode, $\ell_{corr} \simeq k_\perp^{-1}$. Under the condition for nonlinear marginal stability, variables $\{\mu_N, \mu_{e.N}, \chi_N, \phi, k_y\}$ turn out to depend to each other. Within the framework of the mean field approximation, we shall determine the characteristic poloidal mode number, k_y, by which the micro-correlation length is evaluated.

A characteristic number k_y is determined as that of the least stable mode, which becomes stable last at the largest value of turbulent transport coefficient. The nonlinear marginal stability condition can be formulated as equation (9.24). Both normalized parameters $b = k_y^2(\bar{\lambda}\bar{\mu}/s^2)^{1/3}$ and $\Im = \bar{\lambda}^{2/3}s^{-4/3}\bar{\chi}^{-1}\bar{\mu}^{-1/3}G_0$ include nonlinear transfer rates. Relation (9.24), $\Im = b^2 + C_{CD}b^{-1/4}$, is rewritten as an implicit function between parameter k_y and nonlinear transfer rates as

$$\frac{\bar{\lambda}^{2/3}G_0}{s^{4/3}\bar{\chi}\bar{\mu}^{1/3}} = (\bar{\lambda}\bar{\mu})^{2/3}s^{-4/3}k_y^4 + C_{CD}(\bar{\lambda}\bar{\mu})^{-1/12}s^{1/6}k_y^{-1/2}. \tag{9.33}$$

For a fixed value of the gradient parameter, χ_N could be solved as a function of the parameter k_y for the nonlinear marginal condition. The result is shown in figure 9.4. The curve takes the maximum value at value $k_y = k_*$. This determines the least stable mode.

In the large k_y limit, equation (9.33) is simplified as

$$\frac{\bar{\lambda}^{2/3}G_0}{s^{4/3}\bar{\chi}\bar{\mu}^{1/3}} \simeq (\bar{\lambda}\bar{\mu})^{2/3}s^{-4/3}k_y^4.$$

The asymptotic relation

$$\chi_N \sim G_0\sqrt{\chi_N/\mu_N}k_y^{-2} \tag{9.34}$$

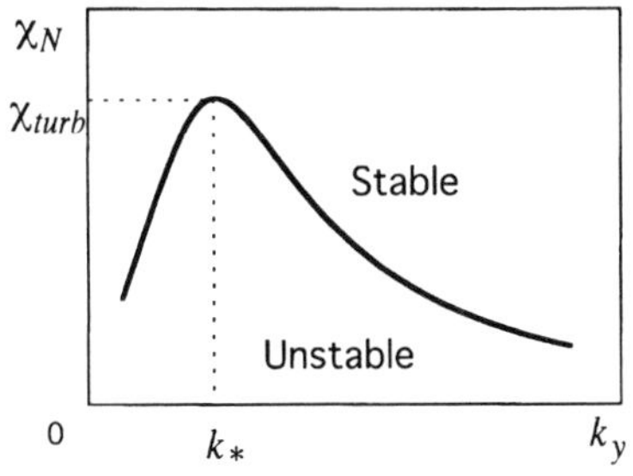

Figure 9.4. The relation $\chi_N(k_y)$ for the nonlinear marginal condition. Nonlinear transfer rate is maximum at $k_y = k_*$.

holds for the curve in figure 9.4.

9.3.2 Turbulence and Transport

The least stability condition determines the characteristic mode number as

$$k_* = \sqrt{b^*}\left(\frac{s^2}{\lambda_N \mu_N}\right)^{1/6} \tag{9.35}$$

with $b^* = 0.43$ and $\Im_c^* = 1.67$. This condition, in turn, provides the maximum value of χ_N from equation (9.20) as

$$\chi^* = \left[\Im_c^{*-3/2}\frac{\mu_{e.N}}{\sqrt{\chi_N \mu_N}}\right]\frac{1}{\xi s^2}G_0^{3/2} \simeq \frac{1}{\xi s^2}G_0^{3/2}. \tag{9.36}$$

For the marginal stability condition of the least stable mode, we have

$$\chi_N = \chi^* \tag{9.37}$$

$$\mu_N \simeq \frac{1}{\xi s^2}G_0^{3/2} \tag{9.38}$$

$$\mu_{e.N} \simeq \frac{1}{\xi s^2}G_0^{3/2} \tag{9.39}$$

and

$$\chi_{turb} \simeq \frac{1}{\xi s^2}G_0^{3/2}. \tag{9.40}$$

Turbulence-driven transport coefficients also obey the rule

$$\mu_{turb} \simeq \mu_{e.turb} \simeq \chi_{turb} \simeq \frac{1}{\xi s^2}G_0^{3/2}. \tag{9.41}$$

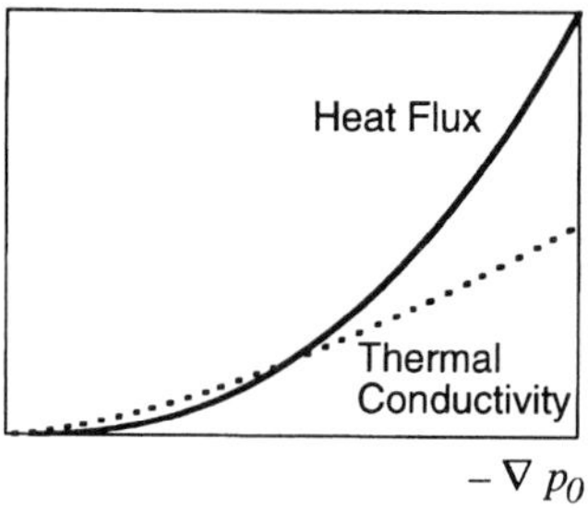

Figure 9.5. Nonlinear dependence of the cross-field heat flux on the pressure gradient. Thermal conductivity also increases as the gradient increases.

These results show the essential feature of the turbulence-driven transport. The turbulent coefficients are expressed in terms of the global structural parameters; they depend on the gradient. As the gradient becomes steeper, the turbulence becomes stronger, and the transport coefficient increases. The cross-field flux,

$$q_x = -\chi \frac{\mathrm{d}p_0}{\mathrm{d}x} \tag{9.42}$$

is a higher order function of the gradient. In the strong turbulence limit, the dependence of

$$q_x \propto \left(\frac{\mathrm{d}}{\mathrm{d}x} p_0 \right)^{5/2} \tag{9.43}$$

is obtained. Figure 9.5 schematically shows the dependences of the global heat flux and thermal conductivity on the pressure gradient.

9.3.3 Fluctuations in Strong Turbulence

Level and correlation of fluctuations are derived. Substitution of equation (9.38) into equation (9.27) gives the fluctuation level as

$$\sum_{k_1} |\phi_1|^2 \simeq 2(1 + C_1)\xi^{-2} s^{-4} G_0^3. \tag{9.44}$$

We have an amplitude as

$$\sqrt{\langle |\phi|^2 \rangle} = \sqrt{\sum_{k_1} |\phi_1|^2} \simeq \frac{1}{\xi s^2} G_0^{3/2}. \tag{9.45}$$

The characteristic mode number is given from equations (9.35), (9.38) and (9.39) as

$$k_* \simeq s\xi^{1/2} G_0^{-1/2}. \tag{9.46}$$

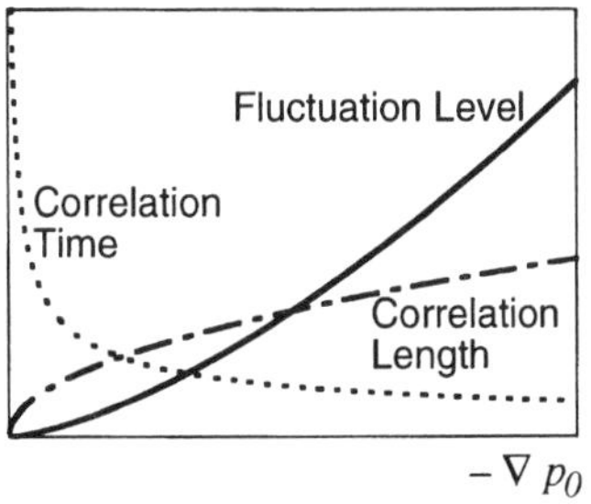

Figure 9.6. Fluctuation level, $\sqrt{\langle\phi^2\rangle}$, correlation length, ℓ_{corr}, and correlation time, τ_{corr}, are shown as a function of the pressure gradient.

The correlation length ℓ_{corr} is estimated by $\ell_{corr} \simeq k_*^{-1}$ and is expressed as

$$\ell_{corr} \simeq s^{-1}\xi^{-1/2}G_0^{1/2}. \tag{9.47}$$

The nonlinear decorrelation rate $\tau_{dec}^{-1} = \mu_N k_*^2$ is given as

$$\tau_{corr}^{-1} \simeq G_0^{1/2}. \tag{9.48}$$

9.4 Order Parameter and Critical Exponent

Equations (9.45)–(9.48) show an important role of the gradient in controlling the fluctuations. Figure 9.6 illustrates the dependence on the gradient parameter. The fluctuation level becomes higher and the correlation length becomes longer as the pressure gradient increases. At the same time, the correlation time becomes shorter. The parameter G_0, which denotes the pressure gradient coupled with magnetic field gradient, plays the role of controlling the fluctuations.

This situation is compared to thermal fluctuations near a critical point [9.3]. In the case of critical phenomena in solids and liquids, anomalies of thermal properties have been known near the critical point. For instance, the susceptibility χ_{sus} and the correlation length in the Ising spin system obey the rules

$$\chi_{sus} \simeq \left|1 - \frac{T}{T_c}\right|^{-\gamma} \qquad (\gamma \simeq 1.24) \tag{9.49}$$

and

$$\ell_{corr} \simeq \left|1 - \frac{T}{T_c}\right|^{-\nu} \qquad (\nu \simeq 0.625) \tag{9.50}$$

where T_c is the critical temperature.

The result of strong plasma turbulence shows that the turbulence level and correlation length are dependent on the gradient parameter G_0. The result suggests a similarity between the parameter G_0 and the distance to the critical point as

$$G_0 \leftrightarrow \left|1 - \frac{T}{T_c}\right|^{-1}. \tag{9.51}$$

As is shown in equations (9.45)–(9.48), characteristic quantities of fluctuations are expressed in terms of the power law of the gradient parameter.

The fundamental issue of the transport in a confined plasma is that a state is far from the thermal equilibrium. The parameter G_0, that is the pressure gradient multiplied by the gradient of the magnetic field, plays the role of an order parameter. The transport coefficient, χ, the correlation length, ℓ_{corr}, and the decorrelation rate, γ_{dec}, are given as the power law of G_0, as

$$\chi \propto G_0^{\eta} \tag{9.52}$$

$$\ell_{corr} \propto G_0^{\nu} \tag{9.53}$$

$$\gamma_{dec} \propto G_0^{\varepsilon}. \tag{9.54}$$

In the case of the current diffusive turbulence we have

$$(\eta, \nu, \varepsilon) = (\tfrac{3}{2}, \tfrac{1}{2}, \tfrac{1}{2}). \tag{9.55}$$

The indices (γ, ν) in equations (9.49) and (9.50) are called *critical exponents*. The power indices to G_0 in equations (9.52)–(9.54) are considered to be *nonequilibrium exponents*.

Appendix 9A Growth Rate of Current Diffusive Interchange Mode

The eigenvalue equation is given as equation (9.10)

$$\left\{\frac{\partial}{\partial k_x}\frac{1}{k_\perp^2}\frac{\partial}{\partial k_x} + \frac{\bar{\lambda}G_0}{\gamma s^2} - \frac{\gamma\bar{\lambda}k_\perp^2}{s^2 k_y^2}\right\}\phi(k_x; k_y) = 0. \tag{9A.1}$$

In order to obtain the eigenvalue by means of the variational principle, this equation is rewritten as

$$\left(\frac{\mathrm{d}}{du}\frac{1}{1+u^2}\frac{\mathrm{d}}{du} + g - \Omega(1+u^2)\right)\phi(u) = 0 \tag{9A.2}$$

with

$$u = k_x/k_y \qquad g = \bar{\lambda}G_0 k_y^4 \gamma^{-1} s^{-2} \qquad \Omega = \gamma\bar{\lambda}k_y^4 s^{-2}. \tag{9A.3}$$

The functional is defined as

$$\Re = \int_{-\infty}^{\infty} du\phi^*(u)\left(\frac{\mathrm{d}}{du}\frac{1}{1+u^2}\frac{\mathrm{d}}{du} + g - \Omega(1+u^2)\right)\phi(u) \tag{9A.4}$$

with a trial function (even-parity mode)

$$f_0(u) = \frac{\sigma^{1/2}}{\pi^{1/4}} \exp\left(-\frac{\sigma^2 u^2}{2}\right). \tag{9A.5}$$

The eigenvalue of equation (9A.2) is solved by

$$\Re[\sigma] = 0 \tag{9A.6}$$

$$\frac{\partial}{\partial\sigma}\Re[\sigma] = 0. \tag{9A.7}$$

The functional is calculated to be

$$\Re[\sigma] = -\sigma^4 + \frac{2}{\sqrt{\pi}}\sigma^5 \exp(\sigma^2)\mathrm{Erfc}(\sigma) + g - \left(1 + \frac{1}{2\sigma^2}\right)\Omega. \tag{9A.8}$$

It is shown, *a posteriori*, that σ is a smallness parameter. Therefore the second term is considered to be small in comparison with the first term, and the last term is replaced by $\Omega/2\sigma^2$ in equation (9A.8). Equation (9A.8) is approximated as

$$\Re[\sigma] \simeq -\sigma^4 + g - \frac{\Omega}{2\sigma^2}. \tag{9A.9}$$

Equation (9A.9) is substituted into equations (9A.6) and (9A.7). One has the eigenvalue

$$g = \frac{3}{2^{4/3}}\Omega^{2/3} \tag{9A.10}$$

with

$$\sigma = 2^{-1/3}\Omega^{1/6}. \tag{9A.11}$$

Substituting the definitions for g and Ω, equation (9A.3), into the eigenvalue equation (9A.10), one has the expression of growth rate as

$$\gamma = \frac{2^{4/5}}{3^{3/5}} \frac{k_y^{4/5} G_0^{3/5}}{s^{2/5}} \bar{\lambda}^{1/5}. \tag{9A.12}$$

It is useful to study a mode with different parity. The trial function is chosen as

$$f_1(u) = \frac{2^{1/2}\sigma}{\pi^{1/4}} u \exp\left(-\frac{\sigma^2 u^2}{2}\right). \tag{9A.13}$$

By means of the same procedure as leading to equation (9A.9), the functional is calculated as

$$\Re[\sigma] = -2\sqrt{\pi}\sigma^3 + g - \frac{3\Omega}{2\sigma^2}. \tag{9A.14}$$

Substituting equation (9A.14) into equations (9A.6) and (9A.7), one has

$$g = \frac{5\pi^{1/5}}{2^{3/5}}\Omega^{3/5}. \tag{9A.15}$$

By use of equation (9A.3), the growth rate is given as

$$\gamma = \frac{2^{3/8}}{\pi^{1/8}5^{5/8}} \frac{k_y G_0^{5/8}}{s^{1/2}} \bar{\lambda}^{1/4}. \tag{9A.16}$$

In both cases, the growth rate is given in terms of a fractional power of the current diffusivity.

Appendix 9B Finite-Beta Correction

The ion dynamics along the field line is included as a correction of the finite-beta effects, and the reduced set of equations is formulated in appendix 8B. The effect is briefly discussed here [9.4]. The renormalized equations are given as a linear form for the dressed test wave with diffusion coefficients $(\mu_\perp, \lambda, \chi, \mu_\parallel)$ in equation (8B.11). Eliminating J and $V_\parallel$ from the set of equations, we have

$$k_\parallel \frac{k_\perp^2}{\gamma(1+\xi^{-1}k_\perp^2)+\lambda k_\perp^4} k_\parallel \tilde{\phi} + (\gamma + \mu_\perp k_\perp^2) k_\perp^2 \tilde{\phi} + \mathrm{i} A_k \tilde{p} = 0 \tag{9B.1}$$

and

$$\left(\gamma + \chi k_\perp^2 + \beta k_\parallel \frac{1}{\gamma + \mu_\parallel k_\perp^2} k_\parallel\right) \tilde{p} - \mathrm{i} G_k \tilde{\phi} = 0. \tag{9B.2}$$

(The tilde denotes a dressed test wave in this set of equations. Operators are defined as $\mathrm{i} A_k p_k = (\boldsymbol{b} \times \boldsymbol{\kappa}) \cdot \nabla p_k$ and $G_k = k_{02}(\mathrm{d}p_0/\mathrm{d}r)$.) The term which is proportional to β, i.e., the last term in the bracket of the left-hand side of equation (9B.2), denotes the effect of parallel compressibility. If this term is neglected, the result reduces to that in the main text.

The effect of parallel compressibility is treated perturbatively. It is shown *a posteriori* that the correction is of a higher order in the inverse aspect ratio, and this expansion is validated. Expanding the pressure perturbation with respect to β, we obtain $\tilde{p}$ in terms of $\tilde{\phi}$ from equation (9B.2) as

$$\tilde{p} = \frac{\mathrm{i} G_k}{\gamma + \chi k_\perp^2} \left\{1 - \beta k_\parallel \frac{1}{\gamma + \mu_\parallel k_\perp^2} k_\parallel \frac{1}{\gamma + \chi k_\perp^2}\right\} \tilde{\phi}. \tag{9B.3}$$

Substituting this expression into equation (9B.1), we have

$$k_\parallel \frac{k_\perp^2}{\gamma(1+\xi^{-1}k_\perp^2)+\lambda k_\perp^4} k_\parallel \tilde{\phi} + (\gamma + \mu_\perp k_\perp^2) k_\perp^2 \tilde{\phi} - \frac{k_\theta^2 G_0}{\gamma + \chi k_\perp^2} \times \left\{1 - \beta k_\parallel \frac{1}{\gamma + \mu_\parallel k_\perp^2} k_\parallel \frac{1}{\gamma + \chi k_\perp^2}\right\} \tilde{\phi} = 0. \tag{9B.4}$$

By a similar procedure in the text, the eigenvalue equation is rewritten as

$$\frac{\mathrm{d}}{\mathrm{d}k} \frac{k_\theta^2 s^2}{\lambda k_\perp^2} \frac{\mathrm{d}}{\mathrm{d}k} \phi(k) + \mu_\perp k_\perp^4 \phi(k) - \frac{k_\theta^2 G_0}{\chi k_\perp^2} \left\{1 - \beta k_\theta^2 s^2 \frac{\mathrm{d}}{\mathrm{d}k} \frac{1}{\mu_\parallel k_\perp^2} \frac{1}{\mathrm{d}k} \frac{1}{\chi k_\perp^2}\right\} \phi(k) = 0. \tag{9B.5}$$

The approximation for the neglect of the first derivative $\mathrm{d}\phi/\mathrm{d}k$ is used, which yields

$$\left(1+\beta\frac{\lambda}{\chi}\frac{G_0k_\theta^2}{\mu_\parallel\chi k_\perp^4}\right)\frac{\mathrm{d}^2}{\mathrm{d}k^2}\phi-\frac{\lambda\mu_\perp k_\perp^6}{k_\theta^2s^2}\phi+\frac{\lambda}{\chi}\frac{G_0}{s^2}\phi=0. \tag{9B.6}$$

The correction appears in the first term and is estimated as

$$\beta\frac{\lambda}{\chi}\frac{G_0k_\theta^2}{\mu_\parallel\chi k_\perp^4}\simeq\frac{2}{G_0}s^2\beta. \tag{9B.7}$$

In this estimation, the coefficients of β in the left-hand side are evaluated by use of the eigenvalue in the absence of β-correction. By substituting equation (9B.7) into equation (9B.6), we obtain the eigenmode equation as

$$\frac{\mathrm{d}^2}{\mathrm{d}k^2}\phi-\frac{1}{(1+2G_0^{-1}s^2\beta)}\frac{\lambda\mu_\perp k_\perp^6}{k_\theta^2s^2}\phi+\frac{1}{(1+2G_0^{-1}s^2\beta)}\frac{\lambda}{\chi}\frac{G_0}{s^2}\phi=0.$$

This equation is identical to equation (9.18) with the following replacement

$$\lambda\to\frac{1}{(1+2G_0^{-1}s^2\beta)}\lambda.$$

The evaluations of turbulence quantities and of turbulent transport coefficients are given by this replacement into equations (9.35)–(9.41).

REFERENCES

[9.1] Itoh S-I, Itoh K, Fukuyama A, Yagi M, Azumi M and Nishikawa K 1994 *Phys. Plasmas* **1** 1154

[9.2] Itoh K, Itoh S-I, Fukuyama A, Yagi M and Azumi M 1993 *J. Phys. Soc. Japan* **62** 4269

[9.3] Landau L D and Lifshitz E M 1980 *Statistical Physics, Part 1* 3rd edn, transl. J B Sykes *et al* (Oxford: Pergamon) section 148

[9.4] Itoh K, Itoh S-I, Fukuyama A, Yagi M and Azumi M 1994 *Plasma Phys. Control. Fusion* **36** 1501

Chapter 10

Self-Sustained Turbulence II—Current Diffusive Ballooning Mode

A combination of plasma pressure gradient and magnetic field gradient (curvature) is shown to play an important role for turbulent transport in the study of preceding chapter. When a gradient of magnetic field is not uniform along a field line, both the average value and the local variation of gradient are important in determining fluctuation characteristics. When the average curvature of magnetic field is favourable, fluctuations of the interchange mode grow little. However, there remains a certain region where the gradient (curvature) of magnetic field is unfavourable along the magnetic field line. Fluctuations could be excited in this region. Such fluctuations, i.e., the ballooning modes, are discussed in this chapter.

Figure 10.1 illustrates a poloidal cross-section of the axisymmetric torus. Circles represent magnetic surfaces, i.e., contours of the pressure. The gradient vector of the magnetic field, $\boldsymbol{\kappa}$, changes its direction relative to ∇p, according to the poloidal angle. The gradient of the magnetic field is expressed as

$$\boldsymbol{b} \times \boldsymbol{\kappa} = (\kappa_0 \sin\theta)\hat{\boldsymbol{r}} + (\langle\kappa\rangle + \kappa_0 \cos\theta)\hat{\boldsymbol{\theta}} \tag{10.1}$$

where κ_0 is a peak of (bad) magnetic curvature, $\langle\kappa\rangle$ is the average, and $\hat{\boldsymbol{r}}$ and $\hat{\boldsymbol{\theta}}$ are unit vectors in radial and poloidal directions, respectively. (In the simple model equation of (6.47), the average is expressed by a parameter $h = \langle\kappa\rangle/\kappa_0$ and the gradient length has a relation $1/L_M = \kappa_0$.) The condition $\langle\kappa\rangle < 0$ is satisfied in the system of an average magnetic well.

The magnetic curvature is determined by the magnetic field configuration, which is shown in figure 10.1. An approximate relation is given in the limit of $R/r \gg 1$ [2.4]

$$\kappa_0 \simeq \frac{1}{R}$$

$$\langle\kappa\rangle \simeq \left[-\left(1-\frac{1}{q}\right)\frac{r}{aR} + \frac{2\Delta'}{R}\right]$$

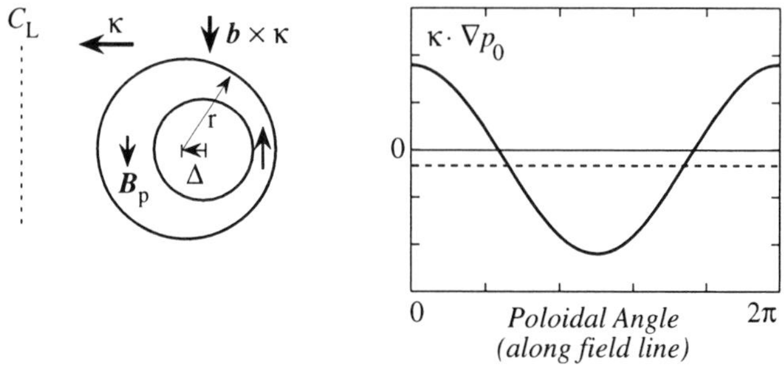

Figure 10.1. Poloidal cross-section of the magnetic surfaces, i.e., contours of the pressure p_0. (a) Δ denotes a shift of the outer magnetic surface with respect to the magnetic axis. Product $\boldsymbol{\kappa} \cdot \nabla p_0$ is shown in (b). The dotted line in (b) indicates an average (magnetic well, in this case).

where R and a are the major and minor radii of the torus, respectively, and Δ is the shift of the centre of the magnetic surface of interest from the magnetic axis which is explained in section 3.2 (figure 3.4).

10.1 Ballooning Transformation

In the analysis of interchange mode turbulence, coefficient $\boldsymbol{b}\times\boldsymbol{\kappa}$ in the eigenmode equation, equations (9.1) and (9.2), is considered to be constant. Consequently, equation (9.1) is reduced to a set of *ordinary differential equations*; so that each (m, n)-Fourier component

$$\phi(r, \theta, \zeta) = \sum_{m,n} \phi(r)_{m,n} \exp(im\theta - in\zeta)$$

is treated separately. In toroidal plasmas, the sign of coefficient $\boldsymbol{b}\times\boldsymbol{\kappa}$ has poloidal angle dependence. If the average of $\boldsymbol{\kappa}$ is in a direction of favourable curvature, an interchange mode does not grow: modes are localized in a bad curvature region. Owing to the axial symmetry of the torus, Fourier decomposition is applicable, but the (r, θ)-dependence is not necessarily separated. One writes

$$\phi(r, \theta, \zeta) = \sum_{n} \phi_n(r, \theta) \exp(-in\zeta) \tag{10.2}$$

where $\phi_n(r, \theta)$ is a periodic function with respect to poloidal angle. Therefore, the eigenvalue is obtained from a solution of *partial differential equations*. Instead of directly solving a partial differential equation, we obtain an approximate eigenvalue by the WKB method.

10.1.1 Ballooning Transformation

An eigenvalue in periodic potential has been studied associated with, e.g., solid state physics [10.1]. A fast oscillating part and a slowly varying part are separated. Similar to this, a transformation called the ballooning transformation has been used in plasma physics [6.5, 10.2]. The Fourier component with toroidal mode number n is expressed in terms of a slowly varying function $\hat{\phi}(\theta)$ and a rapidly varying Eikonal as

$$\phi_n(r,\theta,\zeta) = \sum_{\ell=-\infty}^{\infty} \hat{\phi}(\theta + 2\pi\ell)\exp\left[inq_0\frac{(r-r_{mn})}{r_{mn}}\left\{\int_0^{\theta} \hat{\iota}(\theta)\,\mathrm{d}\theta + 2\pi s\ell\right\}\right] \times \exp[in(q_0\theta - \zeta)] \tag{10.3}$$

where r_{mn} is a rational surface of the (m,n)-component, $q(r_{mn}) = q_0 = m/n$, $q' = [\mathrm{d}q/\mathrm{d}r]_{r=r_{mn}}$, s is the shear parameter $s = q' r_{mn} q_0^{-1}$ and $\hat{\iota}(\theta)$ indicates the angle of the field line with respect to an iso-phase surface of perturbation,

$$\hat{\iota}(\theta) = s + \alpha\cos\theta. \tag{10.4}$$

The term $\alpha\cos\theta$ shows a modulation of the pitch of the field line, and is caused by Shafranov shift of magnetic surfaces. The parameter that characterizes pressure gradient is given as

$$\alpha = -Rq^2\frac{\mathrm{d}}{\mathrm{d}r}\beta. \tag{10.5}$$

This WKB transformation has been applied to a ballooning mode in plasma physics, and is called the *ballooning transformation*.

10.1.2 Short Wavelength Approximation

Under this WKB approximation, a perpendicular wave number is given from an Eikonal. In a limit of $1/n \to 0$, the following replacement holds:

$$\nabla_\theta \to inq \qquad \nabla_r \to inq\int_0^{\theta}\hat{\iota}(\theta)\,\mathrm{d}\theta \qquad \nabla_\perp^2 \to -n^2q^2f^2 \tag{10.6-1}$$

$$(\boldsymbol{b}\times\boldsymbol{\kappa})\cdot\boldsymbol{\nabla} \to \kappa_0 H(\theta)(inq) \tag{10.6-2}$$

where weighting functions are given as

$$f^2 = 1+\left(\int_0^{\theta}\hat{\iota}(\theta)\,\mathrm{d}\theta\right)^2 \qquad H(\theta) = \left\{\frac{\langle\kappa\rangle}{\kappa_0}+\cos\theta+\sin\theta\int_0^{\theta}\hat{\iota}(\theta)\,\mathrm{d}\theta\right\}. \tag{10.7}$$

The parallel derivative is replaced by

$$\nabla_\parallel \to \mathrm{d}/\mathrm{d}\theta \tag{10.8}$$

where θ becomes a variable with an extended interpretation (i.e., the ballooning coordinate) of equation (10.3), and does not represent a poloidal angle. In order to avoid a confusion between a poloidal angle and a coordinate under a ballooning transformation, we introduce a variable η for the ballooning coordinate and replace $\theta \to \eta$ in equations (10.6)–(10.8).

10.1.3 Ballooning Mode Eigenvalue Equation

By setting $\partial/\partial t = \gamma$, one reduces the partial differential equation equation (9.1) to an ordinary differential equation as

$$\begin{pmatrix} (\gamma + \bar{\mu} n^2 q^2 f^2) n^2 q^2 & \mathrm{d}/\mathrm{d}\eta & -inq\kappa_0 H(\eta) \\ \mathrm{d}/\mathrm{d}\eta & (n^{-2} q^{-2} \xi \gamma + \eta_\parallel + \bar{\lambda} n^2 q^2 f^2) & 0 \\ inq\beta' R & 0 & \gamma + \bar{\chi} n^2 q^2 f^2 \end{pmatrix} \times \begin{pmatrix} \hat{\phi}(\eta) \\ \hat{J}(\eta) \\ \hat{p}(\eta) \end{pmatrix} = 0. \tag{10.9}$$

(Variables with hats, $\hat{\phi}$, $\hat{J}$ and $\hat{p}$, are functions in the ballooning coordinates in this chapter. All quantities are normalized as in (7.15).) The eigenmode equation is given as

$$\frac{\mathrm{d}}{\mathrm{d}\eta} \frac{f^2}{\gamma + \hat{\eta}_\parallel f^2 + \lambda n^4 q^4 f^4} \frac{\mathrm{d}}{\mathrm{d}\eta} \hat{\phi}(\eta) + \frac{\alpha}{\gamma + \chi n^2 q^2 f^2} H(\eta) \hat{\phi}(\eta) - (\gamma + \mu n^2 q^2 f^2) f^2 \hat{\phi}(\eta) = 0. \tag{10.10}$$

For the case of equation (10.4), i.e., the case with the high aspect ratio and circular cross-sections, explicit forms are given as

$$f^2 = 1 + (s\eta - \alpha \sin \eta)^2 \tag{10.11a}$$

$$H(\theta) = \left\{ \frac{\langle \kappa \rangle}{\kappa_0} + \cos \eta + \sin \eta (s\eta - \alpha \sin \eta) \right\}. \tag{10.11b}$$

The potential function $-H(\eta)$ is illustrated in figure 10.2.

Equation (10.10) constitutes the basic equation for the ballooning mode in the presence of background fluctuations. This could also be compared to the

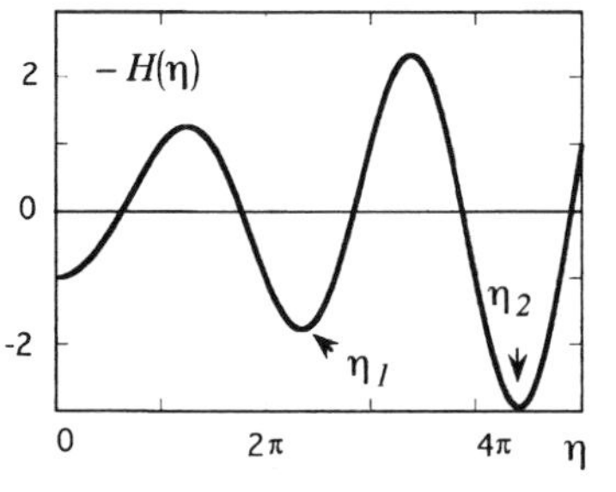

Figure 10.2. Shape of the potential $-H(\eta)$ for the ballooning mode ($s = 0.2$).

previous linear analysis [2.8]. If cross-field transport effects are neglected, i.e., μ, λ, $\chi \to 0$, equation (10.10) reduces to

$$\frac{\mathrm{d}}{\mathrm{d}\eta}\frac{\gamma f^2}{\gamma + \hat{\eta}_\parallel f^2}\frac{\mathrm{d}}{\mathrm{d}\eta}\hat{\phi}(\eta) + \alpha H(\eta)\hat{\phi}(\eta) - \gamma^2 f^2\hat{\phi}(\eta) = 0 \tag{10.12}$$

which is known as an equation for the resistive ballooning mode [2.11, 10.3]. If one further neglects the parallel resistivity, one has

$$\frac{\mathrm{d}}{\mathrm{d}\eta}f^2\frac{\mathrm{d}}{\mathrm{d}\eta}\hat{\phi}(\eta) + \alpha H(\eta)\hat{\phi}(\eta) - \gamma^2 f^2\hat{\phi}(\eta) = 0 \tag{10.13}$$

i.e., the ideal ballooning mode equation [6.5].

10.2 Nonlinear Instability and Self-Sustained Turbulence

10.2.1 Reactive Instability

Equation (10.10) describes nonlinear as well as linear instabilities. A linear instability in the absence of resistivity and cross-field transport (μ, λ, $\chi \to 0$) has been called an ideal ballooning mode. A linear reactive instability is predicted to occur if the pressure gradient exceeds a critical value

$$\alpha > \alpha_c^{MHD}(s) \tag{10.14}$$

where superscript *MHD* indicates a result from the linear and ideal MHD theory. This criterion has been often discussed in terms of a (linear) beta-limit. A linear stability boundary is plotted in figure 10.3. (When magnetic shear is weak, a stability region exists in a high pressure gradient region. This region has been called the second stability region.)

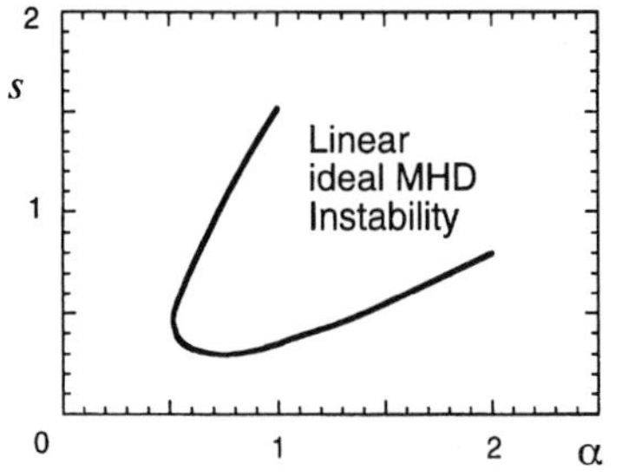

Figure 10.3. s–α diagram for ideal MHD ballooning instability.

10.2.2 Dissipative Instability and Nonlinear Instability

In the presence of dissipation, it is possible for a dissipative instability to occur. Dissipation is either classical parallel resistivity or current diffusivity. The latter provides a strong nonlinear instability. For an analytic insight, we first consider only drive by dissipation, and discard damping by viscous and thermal transport ($\mu, \chi \to 0$). It is approximated that a mode is localized near a bad curvature region, $\eta \simeq 0$, i.e., $H(\eta) \simeq 1$. Under this approximation, equation (10.10) is simplified as

$$\frac{\mathrm{d}^2}{\mathrm{d}\eta^2}\hat{\phi}(\eta) + \left(\frac{\eta_\parallel}{\gamma} + \frac{\lambda n^4 q^4 s^2}{\gamma}\eta^2\right)(\alpha - s^2\gamma^2\eta^2)\hat{\phi}(\eta) \simeq 0. \tag{10.15}$$

Using the WKB approximation, we obtain a dispersion relation as

$$\frac{2}{3}\sqrt{\frac{\eta_\parallel \alpha^2}{\gamma^3 s^2}}\sqrt{1+a_r}\left(\frac{K(k_c)-E(k_c)}{k_c^2} + 2E(k_c) - K(k_c)\right) = \pi(\ell + \tfrac{1}{2}) \tag{10.16}$$

for the ℓth eigenmode, where $a_r = \lambda\alpha/(\eta_\parallel \gamma^2)$ denotes the ratio of two dissipation processes, $k_c^2 = a_r/(1+a_r)$, $K(k_c)$ and $E(k_c)$ are the first and second kind of elliptic function, respectively. Solutions of two limiting cases are obtained as

$$\gamma \simeq \frac{1}{(2\ell+3)^{2/3}}\eta_\parallel^{1/3} s^{-2/3}\alpha^{2/3} \qquad \text{(resistive mode)} \tag{10.17}$$

for $a_r \ll 1$, and

$$\gamma \simeq \left(\frac{4}{\pi}\right)^{2/5} \frac{n^{4/5}q^{4/5}}{(2\ell+3)^{2/5}}\lambda^{1/5} s^{-2/5}\alpha^{3/5} \qquad \text{(current diffusive mode)} \tag{10.18}$$

for $a_r \gg 1$ [10.4].

The growth rate of a current diffusive ballooning mode critically depends on current diffusivity with a small power,

$$\gamma \propto \lambda^{1/5}. \tag{10.19}$$

Consequently a small but finite value of current diffusivity is enough to cause an instability, the growth rate of which is in a range of Alfvén transit time. The condition for a current diffusive mode to occur is given as $a_r \geq 1$ or

$$\lambda > \eta_\parallel^{5/3} n^{-2/3} q^{-2/3} s^{-4/3} \alpha^{1/3}. \tag{10.20}$$

An example of the solution of equation (10.10) was calculated in [10.4], and the growth rate is shown in figure 10.4.

The dependence of growth rate on the dissipation, equation (10.20), shows that nonlinear interactions can enhance the growth rate. As in the case of the current diffusive interchange mode, figure 9.2, a nonlinear link between fluctuation level, turbulent transport and growth rate exists. Owing to this nonlinear link, a self-sustained turbulence develops independent of the linear stability condition.

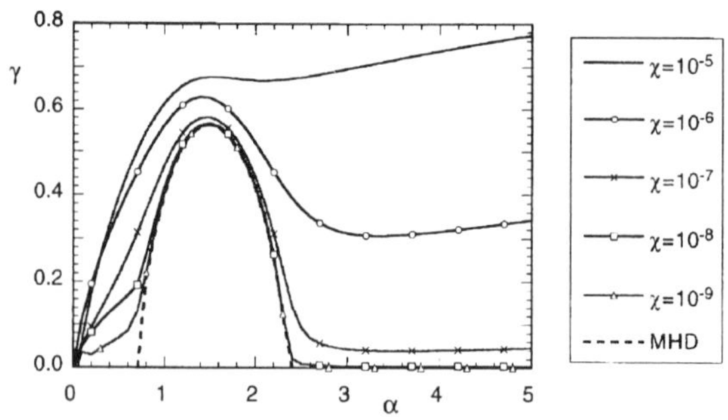

Figure 10.4. Growth rate of the current diffusive ballooning mode as a function of the pressure gradient. The dashed line indicates the result of the linear ideal MHD calculation. Even in the stability region for the ideal mode, strong instability is predicted to occur. (γ is normalized to τ_{Ap}^{-1}. Parameters: $\lambda/\chi = 10^{-3}$, $\mu = \chi$, $q = 3$, $n = 30$, $s = 1$, $\langle\kappa\rangle = -1/9$. $\chi\tau_{Ap}a^{-2}$ is chosen as 10^{-9}, 10^{-8}, 10^{-7}, 10^{-6}, 10^{-5}, respectively.)

10.3 Nonlinear Marginal Stability Condition

10.3.1 Marginal Condition

Nonlinear instability, which is driven by enhanced current diffusivity, reaches a stationary state owing to the damping effects from the enhanced viscosity and thermal conductivity. The nonlinear marginal stability condition is derived. The subject of study is an estimation of turbulent quantities such as $\hat{\phi}$, τ_{corr}, ℓ_{corr}. As in the case of chapter 9, a mode number of the dressed test mode is taken as a representative value for micro-scale fluctuations. In this chapter, the current diffusive mode is studied, and the resistivity effect is neglected. In the next chapter, a transition to a resistive instability is discussed.

In a stationary state, equation (10.10) is reduced to the form

$$\frac{\mathrm{d}}{\mathrm{d}\eta}\frac{f^2}{\lambda_N n^4 q^4 f^4}\frac{\mathrm{d}}{\mathrm{d}\eta}\hat{\phi}(\eta) + \frac{\alpha}{\chi_N n^2 q^2 f^2}H(\eta)\hat{\phi}(\eta) - \mu_N n^2 q^2 f^4 \hat{\phi}(\eta) = 0 \quad (10.21)$$

where collisional diffusion coefficients are neglected in comparison with nonlinear terms. For an analytic insight, this equation is approximated by a Weber type equation as

$$\frac{\mathrm{d}^2}{\mathrm{d}\eta^2}\hat{\phi}(\eta) + (\sigma_1 - \sigma_2\eta^2)\hat{\phi}(\eta) = 0 \quad (10.22)$$

with

$$\sigma_1 = \frac{\alpha\lambda_N n^2 q^2}{\chi_N} - \mu_N\lambda_N n^6 q^6 \quad (10.23\text{-}1)$$

$$\sigma_2 = \frac{\alpha\lambda_N n^2 q^2}{\chi_N}\left(\tfrac{1}{2} + \alpha - s\right) + 3\mu_N\lambda_N n^6 q^6 (s-\alpha)^2 \tag{10.23-2}$$

which is valid in the limit of $s < \alpha + 1/2$.

The approximation equation (10.22) is used for perturbations which are strongly localized at a minimum of potential $-H(\eta)$ near the origin, $\eta \sim 0$. The potential $-H(\eta)$ has many minima other than the one at $\eta \simeq 0$ as is shown in figure 10.2. All modes are not necessarily localized at $\eta \simeq 0$. Perturbations which are not localized near the origin are discussed in appendix 10A.

For convenience, two normalized quantities are introduced. The combination of the gradient and transport coefficients

$$\Im_b \equiv \frac{\alpha\lambda_N^{2/3}}{\chi_N \mu_N^{1/3}} \tag{10.24}$$

and the normalized mode number

$$N \equiv nq\left(\frac{\mu_N \chi_N}{\alpha}\right)^{1/4} \tag{10.25}$$

is used. It is noted that normalized parameter $\Im_b$ has the same property as in the case of interchange modes, equation (9.20). (The suffix b in equation (10.24) indicates the ballooning mode.) The difference is that G_0 in (9.20) is replaced by α. Both G_0 and α are in proportion to the pressure gradient. A difference of geometry gives rise to a modification from G_0 to α. By use of normalized variables $\Im_b$ and N, the coefficients σ_1 and σ_2 are expressed as $\sigma_1 = \Im_b^{3/2}(N^2 - N^6)$ and $\sigma_2 = \Im_b^{3/2}\{(1/2 + \alpha - s)N^2 + 3(s-\alpha)^2 N^6\}$.

The eigenvalue is obtained by the relation

$$\sigma_2 = \sigma_1^2 \tag{10.26}$$

which is rewritten as

$$\Im_b = f_1^{2/3}(N). \tag{10.27}$$

The dependence on the normalized wave number, $f_1(N)$, is given as

$$f_1(N) = \frac{\left(\frac{1}{2} + \alpha - s\right) + 3(s-\alpha)^2 N^4}{N^2(1-N^4)^2}. \tag{10.28}$$

The eigenfunction of the fundamental mode is considered, and is estimated to be

$$\hat{\phi}(\eta) = \exp(-(\eta/\eta_0)^2) \qquad 2\eta_0^{-2} = \frac{\alpha^{3/2}\lambda_N}{\chi_N^{3/2}\mu_N^{1/2}} N^2(1-N^4). \tag{10.29}$$

10.3.2 Least Stable Mode

Eigenvalue equation (10.27) gives the condition for the stability boundary of a test mode. The least stable mode is obtained from the optimum solution with respect to mode number N. A minimum value of $f_1(N)$ is estimated as

$$f_1(N) \geq F(s,\alpha)^{-1} \equiv (1+2\alpha-2s)\sqrt{2+\frac{6(s-\alpha)^2}{(1+2\alpha-2s)}} \tag{10.30}$$

at

$$N = N_* \simeq \left\{2+\frac{6(s-\alpha)^2}{(1+2\alpha-2s)}\right\}^{-1/4}. \tag{10.31}$$

For the least stable mode, the marginal condition is given as

$$\Im_b = F(s,\alpha)^{-2/3}. \tag{10.32}$$

This relation describes the nonlinear eigenvalue which must be satisfied in order to establish a stationary state.

10.4 Turbulent Transport Coefficient and Fluctuations

10.4.1 Transport Coefficients

The nonlinearly least stable mode determines the stationary state. From equations (10.24) and (10.32), we have

$$\chi_N = F(s,\alpha)\frac{\lambda_N}{\sqrt{\chi_N \mu_N}}\alpha^{3/2}. \tag{10.33}$$

We here again note the fact that the ratios of nonlinear transfer rates, μ_N/χ_N and $\mu_{e.N}/\chi_N$, are insensitive to the turbulence level, compared to the magnitude of the transport coefficient itself. Taking approximations that

$$\mu_N/\chi_N \simeq 1 \tag{10.34}$$

and

$$\mu_{e.N}/\chi_N \simeq 1 \tag{10.35}$$

hold, a simpler form of the anomalous transport coefficient is given from equation (10.33) as $\chi_N = F(s,\alpha)\alpha^{3/2}(c/a\omega_p)^2$, or in the dimensional form as

$$\chi = F(s,\alpha)q^2(-R\beta')^{3/2}\left(\frac{c}{\omega_p}\right)^2\frac{v_A}{R} \tag{10.36}$$

By use of the eigenfunction of equation (10.10), the ratios of nonlinear transfer rates are also calculated. The result is similar to the one for current diffusive interchange mode turbulence.

In a framework of the mean field approximation, a nonlinear transfer rate is used to evaluate transport coefficients. The function $F(s, \alpha)$ is fitted as $2.5/\sqrt{s}$ in the strong shear limit, and approximated as equation (10.30) in the weak shear limit. Figure 10.5 illustrates the dependence of anomalous transport coefficient on magnetic shear strength. The contour of thermal conductivity is plotted on a plane of pressure gradient and magnetic shear (figure 10.6).

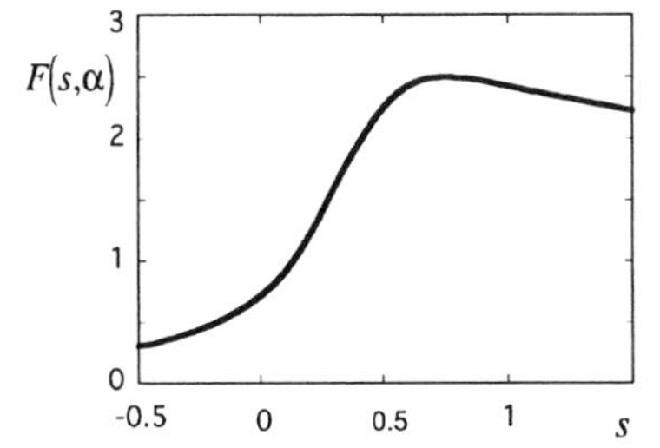

Figure 10.5. Geometrical factor $F(s, \alpha)$ in the transport coefficient.

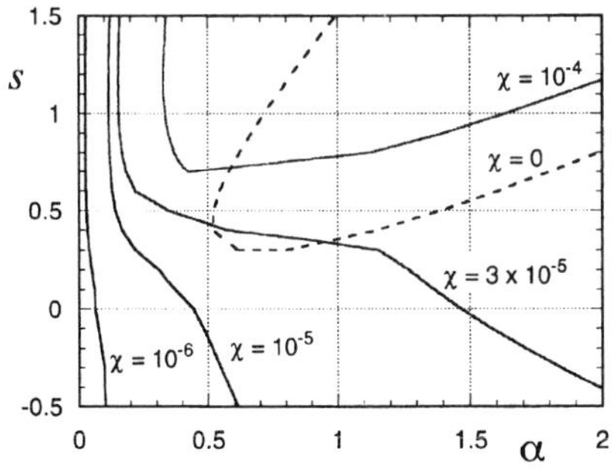

Figure 10.6. Contour of transport coefficient on (s–α)-plane, which is obtained from numerical solution of the nonlinear current diffusive ballooning mode. The normalized value $\hat{\chi} = \tau_{Ap} a^{-2} \chi$ is shown. For a fixed pressure gradient, thermal diffusivity becomes smaller for weaker magnetic shear. The dashed line indicates an instability boundary for the linear ideal MHD ballooning mode.

A current profile has an impact to modify the transport coefficients. If a magnetic shear is very weak (or negative), coefficient $F(s, \alpha)$ becomes small. Influences of pressure profile and current profile are two keys for the structural formation in confined plasmas. Prediction from a theoretical formula and detailed comparison by use of a transport code is explained in chapter 24.

10.4.2 Characteristics of Fluctuations

The least stable mode gives a typical wave number for self-sustained turbulence. Substituting equations (10.31), (10.34)–(10.36) into equation (10.25), a characteristic mode number is given in the dimensional form as

$$k_\perp = \frac{u(s)}{\sqrt{\alpha}} \frac{\omega_p}{c} \tag{10.37}$$

where numerical coefficient $u(s)$ is close to unity in the low shear limit and is about 0.1–1 for the parameter of $s \leq 1$ [10.4]. In a framework of the mean field approximation, a correlation length ℓ_{corr} is given by the relation $\ell_{corr} \simeq k_\perp^{-1}$. This result shows that a characteristic scale-length is given by a collisionless skin depth, and that the scale-length becomes longer as the pressure gradient increases. The fluctuation level is also given as

$$\frac{e}{T}\sqrt{\langle\phi\rangle^2} \simeq \frac{eB}{T}\chi = F(s,\alpha)q^2(-R\beta')^{3/2}\left(\frac{c}{\omega_p}\right)^2 \frac{v_A}{R}\frac{eB}{T} \tag{10.38}$$

showing that the fluctuation level is enhanced as the pressure gradient is increased. The dependence on pressure gradient is the same as the case of CDIM turbulence: the difference appears in geometrical factors.

Equations (10.36), (10.37) and (10.38) also illustrate an important role of pressure gradient for the occurrence of turbulent transport. A normalized pressure gradient α plays a role of order parameter, that determines fluctuations and turbulent-driven transport. As is the case of CDIM turbulence, power laws of forms

$$\chi \propto \alpha^{3/2} \tag{10.39}$$

$$\sqrt{\langle\phi^2\rangle} \propto \alpha^{3/2} \tag{10.40}$$

$$\ell_{corr} \propto \alpha^{1/2} \tag{10.41}$$

$$\tau_{corr} \propto \alpha^{-1/2} \tag{10.42}$$

are obtained. This also represents features of the nonlinear nonequilibrium transport property.

Appendix 10A Modes not Localized at $\eta \simeq 0$

The potential $-H(\eta)$ has many minima in addition to that at $\eta \simeq 0$ as is shown in figure 10.2. All modes are not necessarily localized at $\eta \simeq 0$, and an approximation for a strongly localized mode may not always hold. A minimum of potential $-H(\eta)$ is found at $\eta = \eta_j$, which satisfies

$$s\eta_j = (1-s)\tan\eta_j. \tag{10A.1}$$

A case of finite magnetic shear, $s \simeq O(1)$, is considered. An approximate relation $\eta_j \simeq (2j+1/2)\pi$ holds for the case of $s \simeq 1$. Expanding potential near $\eta = \eta_j$, we have

$$H(\eta) \simeq s\eta_j \left\{-1 + \frac{(\eta - \eta_j)^2}{2}\right\}. \tag{10A.2}$$

Eigenvalue equation (10.21) is approximated as

$$\frac{d^2}{du^2}\hat{\phi}(u) + \left\{\alpha\lambda_N\chi_N^{-1}n^2q^2s\eta_j\left(1 - \frac{u^2}{2}\right) - \mu_N\lambda_N n^6q^6 f(\eta_j)^6\right\}\hat{\phi}(u) = 0 \tag{10A.3}$$

where $u = \eta - \eta_j$. The eigenvalue is obtained as

$$\Im_b = (\tfrac{5}{4})^{5/3}s\eta_j \qquad \text{(for the } j\text{th solution).} \tag{10A.4}$$

A mode which is localized at $\eta = \eta_j$ is more unstable than that localized at $\eta \simeq 0$, if the condition $\Im_b(j\text{th solution}) < \Im_b(0\text{th solution})$ is satisfied. From equations (10.32) and (10A.4), the jth mode at $\eta = \eta_j$ is less stable than the fundamental mode if the condition

$$(\tfrac{5}{4})^{5/3}s\eta_j < F(s,\alpha)^{-2/3} \tag{10A.5}$$

is satisfied. This condition is rewritten as

$$|s| < \left(\frac{4}{5}\right)^{5/3}\frac{1}{(2j+1/2)\pi}F(s,\alpha)^{-2/3} \tag{10A.6}$$

where $\eta_j \simeq (2j+1/2)\pi$ is used.

Equation (10A.6) can be satisfied in an extremely weak shear limit, $|s| < 0.05$. The marginal stability condition is obtained as (10A.4) for a mode localized at $\eta = \eta_j$. That is, the analysis of the localized mode, $\eta \simeq 0$, is acceptable except for an extremely low shear case, $|s| < 0.05$. The turbulent transport coefficient can be large if magnetic shear vanishes in the absence of a magnetic well.

REFERENCES

[10.1] Kittel C 1996 *Introduction to Solid State Physics* 7th edn (New York: Wiley) ch 7

[10.2] Dewar R L and Glasser A H 1983 *Phys. Fluids* **26** 3038
Hazeltine R D and Newcomb A 1990 *Phys. Fluids* B **2** 7

[10.3] Connor J W, Hastie R J and Martin T J 1985 *Plasma Phys. Control. Fusion* **27** 1524

[10.4] Yagi M, Itoh K, Itoh S-I, Fukuyama A and Azumi M 1993 *Phys. Fluids* B **5** 3702

[10.5] Yagi M, Itoh K, Itoh S-I, Fukuyama A and Azumi M 1994 *J. Phys. Soc. Japan* **63** 10

Chapter 11

Resistive Turbulence

Destabilization of the interchange mode and ballooning mode by classical resistivity is possible [2.10, 2.11]. Fluctuations and turbulent transport for a case of resistive turbulence are briefly discussed.

11.1 Current Diffusive Limit and Resistive Limit

The equation for a dressed test mode includes destabilizing mechanisms from both current diffusion and resistivity, as is shown in equation (9.16). The contribution of current diffusivity dominates over resistive effects if the condition

$$\lambda_N k_\perp^2 \gg \eta_\parallel \tag{11.1}$$

is satisfied. (The same normalization is used as in chapter 9.) Evaluations are made for current diffusive mode as $\lambda_N \sim s^{-2}(c/a\omega_p)^4 G_0^{3/2}$ and $k_\perp \sim (a\omega_p/c)sG_0^{-1/2}$ (for CDIM) and $\lambda_N \sim (c/a\omega_p)^4\alpha^{3/2}$ and $k_\perp \sim (a\omega_p/c)\alpha^{-1/2}$ (for CDBM). If these relations are substituted into equation (11.1), the condition (11.1) is rewritten as

$$\eta_\parallel \ll (c/a\omega_p)^2 G_0^{1/2} \qquad \text{(CDIM)} \tag{11.2-1}$$

$$\eta_\parallel \ll (c/a\omega_p)^2 \alpha^{1/2} \qquad \text{(CDBM).} \tag{11.2-2}$$

Equation (11.2) is usually satisfied in present experiments.

In low temperature plasmas which satisfy the conditions below

$$\eta_\parallel \gg (c/a\omega_p)^2 G_0^{1/2} \qquad \text{(resistive IM)} \tag{11.3-1}$$

$$\eta_\parallel \gg (c/a\omega_p)^2 \alpha^{1/2} \qquad \text{(resistive BM)} \tag{11.3-2}$$

parallel resistivity could be more important than current diffusivity. It is also noted that when a long wavelength mode is studied, resistivity can be important. Resistive modes are studied for the interchange mode and ballooning mode.

11.2 Resistive Interchange Mode Turbulence

When the effect of resistivity is larger than that of current diffusivity, equation (9.16) is simplified as

$$\left\{ \frac{\partial}{\partial k_x} \frac{s^2 k_y^2}{\eta_\parallel} \frac{\partial}{\partial k_x} + \frac{k_y^2 G_0}{\bar{\chi} k_\perp^2} - \bar{\mu} k_\perp^4 \right\} \phi(k_x; k_y) = 0. \tag{11.4}$$

This equation describes a stationary state of resistive turbulence. (As in chapter 9, this equation is an eigenvalue equation with respect to a variable k_x, and k_y is a parameter.) Under an assumption that $\eta_\parallel$ is uniform in space and by use of the normalized variable $z = k_x/k_y$, equation (11.4) is rewritten as

$$\left\{ \frac{\mathrm{d}^2}{\mathrm{d}z^2} + \frac{\eta_\parallel G_0}{s^2 \bar{\chi}} \frac{1}{(1+z^2)} - \frac{\eta_\parallel \bar{\mu} k_y^4}{s^2} (1+z^2)^2 \right\} \phi(z) = 0. \tag{11.5}$$

The eigenvalue is obtained by WKB method as

$$\int_0^{z_c} \sqrt{\frac{\eta_\parallel G_0}{s^2 \bar{\chi}} \frac{1}{(1+z^2)} - \frac{\eta_\parallel \bar{\mu} k_y^4}{s^2} (1+z^2)^2}\, \mathrm{d}z = \frac{\pi}{4}. \tag{11.6}$$

Coefficient $\eta_\parallel \bar{\mu} k_y^4 s^{-2}$ is much smaller than $\eta_\parallel G_0 s^{-2} \bar{\chi}^{-1}$, so that a turning point z_c satisfies the condition $|z_c| \gg 1$, and is given as

$$z_c \simeq \left(\frac{G_0}{\bar{\chi} \bar{\mu} k_y^4} \right)^{1/6}. \tag{11.7}$$

The first term in the integrand of equation (11.6) is larger than the second term in the region $|z| < z_c$. Therefore, the second term in the integrand is neglected, and equation (11.6) is approximated as

$$\int_0^{z_c} \sqrt{\frac{\eta_\parallel G_0}{s^2 \bar{\chi}} \frac{1}{(1+z^2)}}\, \mathrm{d}z = \frac{\pi}{4}.$$

The integral is performed, and an eigenvalue is given as [11.1]

$$\sqrt{\frac{G_0 \eta_\parallel}{s^2 \bar{\chi}}} \ln \left\{ 2 \left(\frac{G_0}{\bar{\mu} \bar{\chi} k_y^4} \right)^{1/6} \right\} = \frac{\pi}{4}. \tag{11.8}$$

The critical condition (11.8) is also rewritten as

$$\frac{G_0 \eta_\parallel}{s^2 \bar{\chi}} = C_r \equiv \left[\frac{\pi}{4 \ln\{2(G_0 \bar{\mu}^{-1} \bar{\chi}^{-1} k_y^{-4})^{1/6}\}} \right]^2. \tag{11.9}$$

The right hand side of equation (11.9) is an increasing function of a poloidal mode number k_y. Longer wavelength modes are less stable than shorter wavelength modes. C_r is a weakly varying function of G_0, and coefficient C_r can be approximated as a constant.

From equation (11.9), the turbulent transport coefficient is derived as

$$\chi_N + \chi_c = \frac{\eta_\parallel}{C_r s^2} G_0. \tag{11.10}$$

Equation (11.10) shows that the turbulent transport coefficient is an increasing function of gradient parameter G_0. The gradient parameter is also an order parameter in the limit of resistive plasmas.

11.3 Resistive Ballooning Mode

11.3.1 Resistive Ballooning Mode and Transport Coefficient

This framework of self-sustained turbulence is applied to a low temperature plasma with high resistivity. When equation (11.3-2) holds, the driving source of the mode is a term with $\eta_\parallel$ rather than that with λ in a nonlinear dispersion relation (10.10).

A generalized form of nonlinear marginal equation is given from equation (10.10), instead of equation (10.27), as [11.2]

$$\frac{\alpha \eta_\parallel}{\chi} = \frac{c_1(1 + c_2 \rho_r N^2)}{(1 + \rho_r N^2)^2 (1 - N^4)^2} \left(1 + \frac{2s^2}{c_1} \frac{1 + \rho_r N^2}{1 + c_2 N^2} N^4 \right) \tag{11.11}$$

where ρ_r is the ratio of the effect of the current diffusion to that of the resistivity as

$$\rho_r = \frac{\lambda_N}{\eta_\parallel} \sqrt{\frac{\alpha}{\hat{\chi}_N \hat{\mu}_N}} \tag{11.12}$$

and coefficients are defined as $c_1 = 1/2 + \alpha - s + s^2$, $c_2 = (1/2 + \alpha - s)/c_1$.

Equation (11.11) determines the least stable mode and transport coefficient, as is shown in figure 11.1. In a high temperature limit, $\rho_r \to \infty$ (i.e., current diffusive limit) equation (11.11) is reduced to equation (10.27). If temperature is low and the relation $\rho_r \to 0$, i.e., equation (11.3) holds, resistivity rather than current diffusivity determines the growth of the mode. The analysis reduces to that of the resistive ballooning mode. In such a case, the transport coefficient is given as

$$\chi_N = 2\alpha \eta_\parallel \tag{11.13}$$

or in a dimensional form as

$$\chi = \left(\frac{4\varepsilon r}{L_p} \right) \nu_{ei} \rho_{pe}^2 \tag{11.14}$$

where L_p is a pressure gradient scale-length, $L_p = -\beta/\beta'$, ν_{ei} is the electron–ion collision frequency and ρ_{pe} is the electron poloidal gyro-radius.

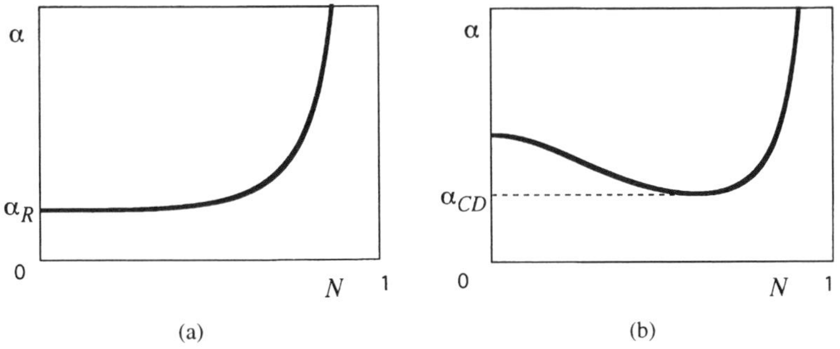

Figure 11.1. Nonlinear stability boundary as a function of mode number of test mode N. The resistive limit is given in (a), and the current diffusive limit is shown in (b). In (a), condition $\alpha = \alpha_R$ gives equation (11.13). In (b), $\alpha = \alpha_{CD}$ provides equation (10.33).

11.3.2 Connection to Pseudo-Classical Transport and Bohm Diffusion

Pseudo-classical Transport. The formula (11.14) is very close to that for pseudo-classical transport which has been proposed by Yoshikawa [11.3]

$$\chi_{pc} = \nu_{ei}\rho_{pe}^2. \tag{11.15}$$

The difference between them is that theoretical formula (11.14) contains a dependence on pressure gradient parameter L_p explicitly. If one allows a condition $L_p \propto r$, a simpler formula (11.15) could be deduced. Change to the resistive turbulence state (or pseudo-classical confinement) from current diffusive turbulence takes place when the condition

$$\eta_\parallel \sim (c/a\omega_p)^2 G_0^{1/2} \qquad \text{(CDIM)} \tag{11.16-1}$$

$$\eta_\parallel \sim (c/a\omega_p)^2 \alpha^{1/2} \qquad \text{(CDBM)} \tag{11.16-2}$$

is satisfied. A comparison study with experimental observations is illustrated in figure 11.2.

Bohm Diffusion. Although various mechanisms could give rise to Bohm diffusion [11.4],

$$\chi \simeq \frac{1}{16}\frac{T_e}{eB} \tag{11.17}$$

a connection to Bohm diffusion can also be drawn. The relation between fluctuation level and diffusivity is given in a strong turbulent limit, e.g., equation (10.38). It is written as

$$\frac{e}{T}\sqrt{\langle\tilde{\phi}^2\rangle} \simeq \frac{eB}{T}\chi_N. \tag{11.18}$$

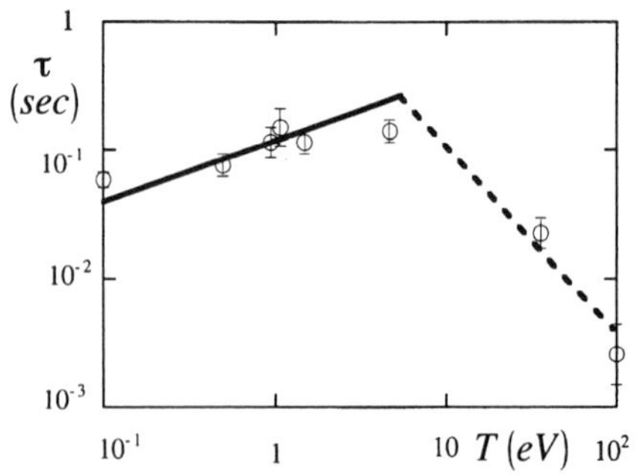

Figure 11.2. The result is compared to experiments in a spherator [11.3], where a transition between pseudo-classical transport and neo-Bohm transport was found to occur around $T = 10$ eV. For parameters of experiments, a transition between pseudo-classical transport and L-mode confinement is predicted to occur at $T = 8$ eV from equation (11.16).

The upper bound of the fluctuation level, $|\tilde{n}/n| \leq 1$, or $|e\tilde{\phi}/T| \leq 1$, gives an upper bound of χ from equation (11.18) as

$$\chi \simeq \mu \leq \frac{T_e}{eB} \tag{11.19}$$

except for a numerical coefficient of order unity. The dependence of

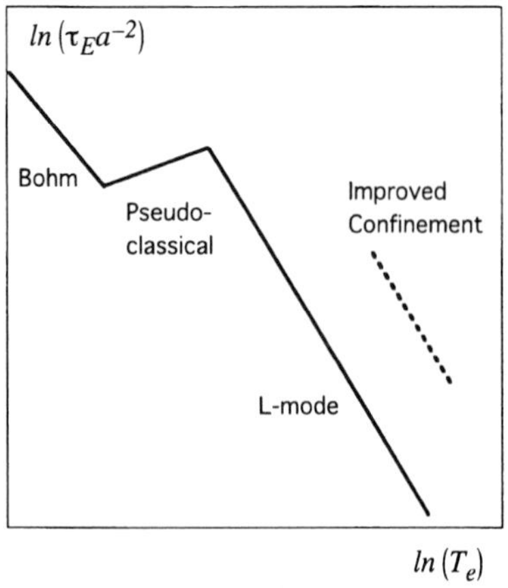

Figure 11.3. Change of energy confinement time as a function of increment in plasma temperature.

equation (11.17) is recovered in this extremely strong turbulent limit.

A turnover from equation (11.14) to equation (11.17) takes place when the condition

$$\left(\frac{4\varepsilon r}{L_p}\right)\nu_{ei}\rho_{pe}^2 > \frac{T}{eB} \tag{11.20}$$

is satisfied. Bohm diffusion is predicted to occur if the condition (11.20) holds. The condition (11.20) is satisfied in low temperature case, if other parameters are fixed. Bohm-like diffusion is expected to appear in low temperature plasmas.

These results, equations (10.36), (11.14), (11.19), show that confinement characteristics change in accordance with a decrease of temperature, from L-mode to Bohm confinement via pseudo-classical confinement. As temperature becomes high and pressure gradient increases, thermal conductivity varies through a parameter α. For a simple variation of $\alpha \propto T_e$ (constant density and plasma size), the characteristic time

$$\tau_E a^{-2} \simeq \chi_{turb}^{-1} \tag{11.21}$$

changes together with a temperature increment. The historical development of confinement characteristics in figure 11.3, from 'Bohm' to 'L-mode' through 'pseudo-classical' confinement, is also drawn from an approach of self-sustained turbulence.

REFERENCES

[11.1] Carreras B A *et al* 1983 *Phys. Rev. Lett.* **50** 503
Diamond P H and Carreras A B 1987 *Comments Plasma Phys. Control. Fusion* **10** 271

[11.2] Itoh K, Itoh S-I, Yagi M, Fukuyama A and Azumi M 1993 *Phys. Fluids* B **5** 3299

[11.3] Yoshikawa S 1973 *Nucl. Fusion* **13** 433

[11.4] Bohm D 1949 *Characteristics of Electrical Discharges in Magnetic Fields* ed A Guthrie and R K Wakerling (New York: Mcgraw-Hill) ch 2

Chapter 12

Subcritical Excitation

Analyses on current diffusive modes (the interchange mode as well as the ballooning mode) have shown that strong turbulence could be realized by a balance of nonlinear excitation and a nonlinear damping. The fluctuation level and associated transport coefficient depend weakly on the distance from the linear stability boundary in parameter space. This suggests that fluctuations can grow as subcritical turbulence [12.1]. Subcritical turbulence has been studied in fluid dynamics, and the conventional method by use of amplitude expansion often encounters the difficulty of poor convergence with respect to truncation of modes [12.2]. To analyse subcritical turbulence, analysis near thermal equilibrium is insufficient. A theory of subcritical turbulence in plasmas must be advanced in order to understand fluctuations and anomalous transport in confined plasmas.

In this chapter, some characteristic features of subcritical excitation of current diffusive interchange mode turbulence are illustrated: (1) large amplitude fluctuation exists in a parameter regime where linear stability is predicted, (2) there is a backward bifurcation near the linear stability boundary and (3) transition from a collisional transport to turbulent transport takes place at a critical gradient $G_0 = G_*$.

12.1 Coexistence of the Collisional Dissipation

12.1.1 Renormalization Relation

The transport coefficients μ_c, λ_c, χ_c, which are contributions from collisional diffusion, are small but could be important in a quiescent plasma. Coexistence of collisional transport and turbulent transfer modifies renormalization relations and eigenvalues.

A stationary solution is studied and a limit of $\gamma \to 0$ is taken. Relations

$$\gamma_{u1} \approx (\mu_N + \mu_c)k_{1\perp}^2 \qquad \gamma_{j1} \approx (\mu_{eN} + \mu_{ec})k_{1\perp}^2 \qquad \gamma_{p1} \approx (\chi_N + \chi_c)k_{1\perp}^2 \tag{12.1}$$

and

$$K_1 = (\mu_N + \mu_c)k_{1\perp}^2(1 + C) \tag{12.2}$$

are obtained, where C is defined by

$$C = k_{\perp 1}^2 \gamma_{u1}^{-1} \gamma_{j1}^{-1} \xi k_{\parallel 1}^2 - k_{\perp 1}^{-2} \gamma_{u1}^{-1} \gamma_{p1}^{-1} G_0 k_{\theta 1}^2. \tag{12.3}$$

In this limit, the relations between turbulent transfer rates and fluctuation amplitude are given as

$$\mu_N(\mu_N + \mu_c) = \overline{\tilde{\phi}^2} \tag{12.4}$$

$$\mu_{eN}(\mu_{eN} + \mu_{ec}) = P^2\overline{\tilde{\phi}^2} \tag{12.5}$$

$$\chi_N(\chi_N + \chi_c) = Q^2\overline{\tilde{\phi}^2} \tag{12.6}$$

where the normalized fluctuation amplitude is defined as

$$\overline{\tilde{\phi}^2} \equiv \sum_{k_1} \frac{|\phi_1|^2}{2(1 + C)}. \tag{12.7}$$

Coefficients P and Q (which are related to Prandtl numbers) are defined as

$$P^2 \equiv \left\{ \sum_{k_1} \frac{|\phi_1|^2}{(1 + C)} (1 - k_{\perp 1}^{-2} \gamma_{u1}^{-1} \gamma_{p1}^{-1} k_{\theta 1}^2 G_0) \right\} \left\{ \sum_{k_1} \frac{|\phi_1|^2}{(1 + C)} \right\}^{-1} \tag{12.8}$$

$$Q^2 \equiv \left\{ \sum_{k_1} \frac{|\phi_1|^2}{(1 + C)} (1 + k_{\perp 1}^{-2} \gamma_{u1}^{-1} \gamma_{j1}^{-1} \xi k_{\parallel 1}^2) \right\} \left\{ \sum_{k_1} \frac{|\phi_1|^2}{(1 + C)} \right\}^{-1}. \tag{12.9}$$

P and Q are the ratios of different (normalized) moments of the turbulent fluctuation spectrum. Their variations are slow compared to that of the turbulence level, and are approximated as constants. In a strong turbulence limit, $\mu_{eN}/\mu_N = P$ and $\chi_N/\mu_N = Q$ are found to be close to unity as in preceding chapters. Figure 12.1 illustrates the turbulent transfer rate as a function of fluctuation amplitude $\sqrt{\overline{\tilde{\phi}^2}}$. In the large amplitude limit, a relation $\mu_N \simeq \sqrt{\overline{\tilde{\phi}^2}}$ holds. When the level is low, a quadratic relation of quasilinear form $\mu_N \simeq \overline{\tilde{\phi}^2}/\mu_c$ is deduced. These limits are discussed in the heuristic argument of equations (4.11) and (4.8).

12.1.2 Nonlinear Marginal Stability Condition

The marginal stability condition for the least stable mode (CDIM) was derived as

$$\mathfrak{I} = \mathfrak{I}_c(k_y) \tag{12.10}$$

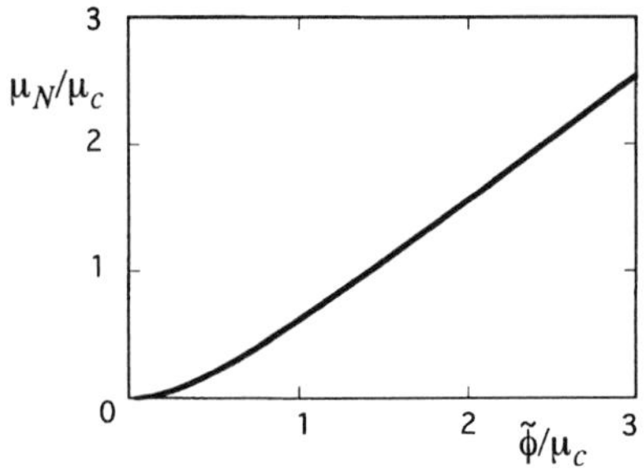

Figure 12.1. Nonlinear transfer rate μ_N as a function of fluctuation amplitude $\sqrt{\overline{\tilde{\phi}^2}}$. In the strong turbulence limit, an approximate relation $\mu_N \sim \tilde{\phi}$ holds.

where a normalized pressure gradient (Itoh number) is redefined as

$$\Im \equiv \frac{G_0}{s^{4/3}} \frac{(\lambda_N + \lambda_c)^{2/3}}{(\chi_N + \chi_c)(\mu_N + \mu_c)^{1/3}}. \tag{12.11}$$

An explicit form of $\Im_c(k_y)$ is given in equation (9.33). The minimum value of $\Im_c(k_y)$, a critical Itoh number, is of the order of unity. Equations (12.4)–(12.6) and (12.10) determine fluctuation level and turbulent transport coefficient as a function of equilibrium pressure gradient, i.e., $\tilde{\phi}(G_0)$ and $\chi_N(G_0)$, respectively.

12.2 Subcritical Excitation

12.2.1 Linear Threshold and Backward Bifurcation

In order to examine the bifurcation nature of a mode, we consider a neutral condition in the vicinity of a linear stability boundary. A linear stability is obtained under a limit of $\tilde{\phi} \to 0$ as

$$G_0 \le G_c \tag{12.12}$$

where the critical pressure gradient is given from equation (12.10) with $\mu_N = \lambda_N = \chi_N = 0$ as

$$G_c = \Im_c (s a \omega_p / c)^{4/3} \chi_c \mu_c^{1/3} \chi_{ec}^{-2/3}. \tag{12.13}$$

Expansion of equation (12.11) near $G_0 \simeq G_c$ with respect to fluctuation amplitude

$$G_0 \simeq G_c + \left(\partial G_0 / \partial \overline{\tilde{\phi}^2}\right) \overline{\tilde{\phi}^2} + \dots \tag{12.14}$$

gives the amplitude $\tilde{\phi}^2$ near the marginal condition as

$$\overline{\tilde{\phi}^2} = \frac{1}{\partial G_0 / \partial \overline{\tilde{\phi}^2}} (G_0 - G_c). \tag{i2.15}$$

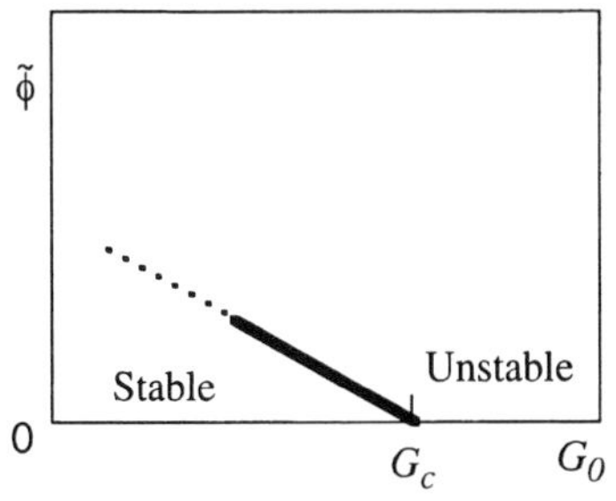

Figure 12.2. Linear stability boundary G_c and backward bifurcation.

From equation (12.11), the derivative is calculated as

$$\frac{\partial G_0}{\partial \overline{\tilde{\phi}^2}} = \left(\frac{\partial \chi_N / \partial \overline{\tilde{\phi}^2}}{\chi_c + \chi_N} + \frac{\partial \mu_N / \partial \overline{\tilde{\phi}^2}}{3(\mu_c + \mu_N)} - \frac{2\partial \mu_{eN} / \partial \overline{\tilde{\phi}^2}}{3(\mu_{ec} + \mu_{eN})} \right) G_0. \tag{12.16}$$

Noting equations (12.4)–(12.6) and taking the limit of $\tilde{\phi} \to 0$, we have

$$\frac{\partial G_0}{\partial \overline{\tilde{\phi}^2}} = \left(\frac{Q^2}{\chi_c^2} + \frac{1}{3\mu_c^2} - \frac{2P^2}{2\mu_{ec}^2} \right) G_c \tag{12.17}$$

near $G_0 \simeq G_c$. For collisional diffusion, conditions $\mu_c \approx \chi_c$ and $\mu_{ec} \ll \chi_c$ usually hold, and the relation

$$\mu_{e.c}/\chi_c \sim \sqrt{m_e/m_i} \tag{12.18}$$

also holds, where m_e/m_i is the mass ratio. Under this circumstance, equation (12.17) provides an approximate relation $\partial G_0/\partial \overline{\tilde{\phi}^2} = -(2P^2/3\mu_{ec}^2)G_c$. Using this relation, equation (12.15) is rewritten as

$$\overline{\tilde{\phi}^2} = -\frac{3\mu_{ec}^2}{2P^2} \left(\frac{G_0}{G_c} - 1 \right). \tag{12.19}$$

This result demonstrates the existence of backward bifurcation ($\partial \overline{\tilde{\phi}^2}/\partial G_0 < 0$). A small but finite amplitude of ϕ is expected to exist in a region below the critical pressure gradient, $G_0 < G_c$. Figure 12.2 schematically illustrates the backward bifurcation.

12.2.2 Subcritical Turbulence

A branch of large fluctuation amplitude is also obtained from equations (12.4)–(12.6) and (12.10). In the large amplitude limit, $\chi_N \gg \chi_c$, equations (12.4)–

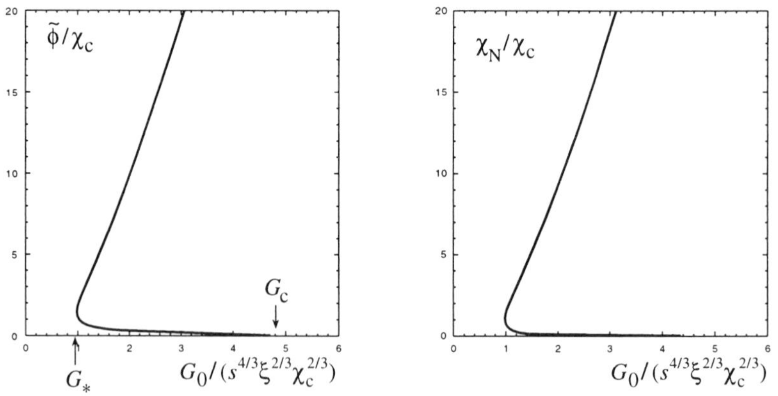

Figure 12.3. Fluctuation level (left) and turbulent transport coefficient (right) as a function of the pressure gradient. Strong turbulence exists below the critical pressure gradient against the linear instability, $G_0 < G_c$. Transition to the turbulent state takes place at $G_0 = G_*$. The dotted line shows $\chi = \chi_c$.

(12.6) and (12.10) provide a relation

$$\chi_N \simeq \sqrt{\bar{\tilde{\phi}^2}} \simeq \frac{G_0^{3/2}}{s^2}\left(\frac{c}{a\omega_p}\right)^2 \tag{12.20}$$

where the typical mode number scales as $k_\perp \sim G_0^{-1/2} sa\omega_p/c$. This branch connects to the strong turbulence limit of self-sustained turbulence. The lower branch (12.19) and upper branch (12.20) are found to merge at a particular pressure gradient, $G_0 = G_*$ [12.3]. The merging point is given by a singularity condition, $\partial G_0/\partial\tilde{\phi} = 0$. This condition is satisfied at $\chi_N \simeq \chi_c - \mu_{ec}$ in a limit of $\chi_c \sim \mu_c \gg \mu_{ec}$ where a critical pressure gradient is given as

$$G_0 = G_* \cong \Im_c (2sa\omega_p/c)^{4/3} \chi_c^{2/3}. \tag{12.21}$$

At the critical pressure gradient G_*, the turbulent transport coefficient is expected to be of the same order as collisional transport coefficient. Figure 12.3 illustrates a theoretical prediction of fluctuation level as a function of pressure gradient, G_0. Explicit multifold forms of $\tilde{\phi}(G_0)$ and $\chi(G_0)$ are seen. The lower amplitude branch is thermodynamically unstable. A subcritical excitation of turbulence is predicted to occur if G_0 exceeds the critical value G_*. Anomalous transport is also induced below the linear stability boundary G_c.

A turbulent state appears as a bifurcation from collisional to turbulent transport at a critical pressure gradient, $G_0 \geq G_*$. The critical gradient, G_*, is expressed in experimental variables as

$$\frac{a}{L_p^*} = \frac{\varepsilon^2}{2\kappa}\left(\frac{s^2 g m_i}{m_e}\right)^{2/3}\left(\frac{a\nu_i}{v_{thi}}\right)^{2/3} \tag{12.22}$$

where $1/L_p \equiv -\beta'/\beta$, g is a numerical factor introduced as

$$\chi_c = g\nu_i\rho_i^2 \tag{12.23}$$

and ν_i, ρ_i and v_{thi} are collision frequency, gyro-radius and thermal velocity of ions.

The subcritical excitation of turbulence illustrates a new feature of its appearance. The transition at critical gradient $G_0 = G_*$ is an important concept of self-sustained turbulence. This implies that transport obeys a rule of collisional transport only in the region of $G_0 < G_*$. An identification of the region in which the linear response theory is valid is important from the viewpoint of transport in systems far from equilibrium.

REFERENCES

[12.1] Landau L D and Lifshitz E M 1987 *Fluid Mechanics*, 2nd edn (Oxford: Pergamon) section 26

Jackson E A 1989 *Perspectives of Nonlinear Dynamics* vol 1 (Cambridge: Cambridge University Press)

[12.2] Herbert T 1980 *AIAA J.* **18** 243

[12.3] Itoh K, Itoh S-I, Yagi M and Fukuyama A 1996 *J. Phys. Soc. Japan* **65** 2749

Chapter 13

Spectrum of Subcritical Turbulence in an Inhomogeneous Plasma

Discussions in the preceding chapters are made based upon a mean field approximation, in which nonlinear transfer rates are unique for every test mode. This approximation is useful for an analytic insight, but is not inevitable. If the mean field approximation is not employed, a spectrum of background turbulence is obtained. In this chapter, the study of the spectrum of fully developed turbulence with subcritical nature is explained, taking the interchange mode turbulence as an example.

An energy spectrum of the form $E(k_\perp) \propto G_0 k_\perp^{-3}$ is obtained in the energy containing region. Turbulent kinetic energy, $W_{\nabla\phi} \propto G_0^2$, and transport coefficient, $\chi_T \propto G_0^{3/2}$, are found to be nonlinear functions of G_0, presenting a typical example of nonlinear transport far from equilibrium.

13.1 Equations for Dressed Test Modes

13.1.1 Recurrent Formula

A set of basic equations is expressed for a test mode in terms of a renormalized dielectric tensor. In a stationary state, $\gamma \to 0$, it is given for CDIM turbulence

$$\boldsymbol{L}_k \boldsymbol{f}_k = 0 \tag{13.1-1}$$

$$\boldsymbol{L}_k = \begin{pmatrix} -(\mu_{N.k} + \mu_c)\nabla_\perp^4 & -\nabla_\parallel & -\kappa_x \partial/\partial y \\ \nabla_\parallel & -(\lambda_{N.k} + \lambda_c)\nabla_\perp^2 & 0 \\ (-\mathrm{d}p_0/\mathrm{d}x)\partial/\partial y & 0 & -(\chi_{N.k} + \chi_c)\nabla_\perp^2 \end{pmatrix} \tag{13.1-2}$$

with $f^T = (\phi_k, J_{\parallel k}, p_k)$. As in preceding chapters, only diagonal terms are kept in nonlinear transfer rates. The renormalized dielectric tensor contains nonlinear transfer rates, and is expressed as effective diffusion coefficients for a test mode, $\{\mu_{N.k}, \lambda_{N.k}, \chi_{N.k}\}$, as is explicitly given as equations (8.38)–(8.41).

Equations (8.38)–(8.41) form a recurrent formula for renormalized coefficients $\{\mu_{N.k}, \lambda_{N.k}, \chi_{N.k}\}$. When we do not use a mean field approximation, coefficients $\{\mu_{N.k}, \lambda_{N.k}, \chi_{N.k}\}$ are not constant, but depend on the wave number of the test mode. In a stationary state, equation (13.1) is satisfied for an arbitrary choice of test mode. The nonlinear marginal condition for an arbitrary test mode provides conditions to determine the spectrum of background turbulence.

If we look for solutions with a power dependence of $\phi(k)$ and $\mu_{N.k}$ as

$$|\phi(k)|^2 \propto k^{-\alpha} \tag{13.2-1}$$

and

$$\mu_{N.k} \propto k^{-\beta} \tag{13.2-2}$$

the renormalized relation

$$\mu_{N.k} = \sum_{k_1 > k} \frac{1}{2K_2} |k_{\perp 1}\phi_1|^2 \tag{13.3}$$

gives a relation

$$\alpha = 2\beta + d \tag{13.4}$$

where d is a dimensional number, and $d = 2$ for two-dimensional turbulence.

13.1.2 Energy Containing Region

The wave number space (k-space) of fluctuations is divided into three; i.e., the energy containing region, $k_0 < k_\perp < k_b$, the inertial range if it exists, $k_b < k_\perp < k_\eta$, and the dissipation range $k_\eta < k_\perp$. In an energy containing region, the nonlinear transfer rate to higher k modes is expressed in terms of an effective diffusion operator $\{\mu_{N.k}, \lambda_{N.k}, \chi_{N.k}\}$. They are much larger than those by binary collisions in turbulent plasmas.

13.1.3 Energy Spectrum

Energy spectrum associated with turbulent plasma motion, $E(k_\perp)$, is often introduced as

$$\sum_k w(k_\perp)|k_\perp \phi_k|^2 = \int \mathrm{d}k_\perp w(k_\perp) E(k_\perp) \tag{13.5-1}$$

where $w(k_\perp)$ is an arbitrary weighting function. E_θ and E_J are also introduced for pressure and current perturbations as

$$\sum_k w(k_\perp)|p_k|^2 = \int \mathrm{d}k_\perp w(k_\perp) E_p(k_\perp)$$

$$\text{and} \sum_k w(k_\perp)|J_{\parallel k}|^2 = \int \mathrm{d}k_\perp w(k_\perp) E_J(k_\perp). \tag{13.5-2}$$

13.1.4 Dissipation and Flux

Dissipations in turbulent plasma, ε^c, ε_J^c and ε_p^c are given as

$$\varepsilon^c = V^{-1}\mu_c \int \mathrm{d}V|\nabla_\perp^2\phi|^2 \qquad \varepsilon_J^c = V^{-1}\lambda_c \int \mathrm{d}V|\nabla_\perp J_\parallel|^2$$

$$\varepsilon_p^c = V^{-1}\chi_c \int \mathrm{d}V|\nabla_\perp p|^2 \tag{13.6}$$

where V is the local volume of the plasma. An average heat flux in the direction of the temperature gradient, q, and Joule heating per unit volume, Q_J, are also calculated as

$$q = V^{-1}\int \mathrm{d}V\, p^*\partial\phi/\partial x \text{ and } Q_J = -V^{-1}\int \mathrm{d}V\, J^*\nabla_\parallel\phi \tag{13.7}$$

where * denotes complex conjugate. In a stationary state, conservation relations are obtained as

$$\varepsilon_p^c = -(\mathrm{d}p_0/\mathrm{d}x)q \tag{13.8-1}$$

$$\varepsilon_J^c = Q_J \text{ and } \varepsilon^c + \varepsilon_J^c = \kappa_x(\mathrm{d}p_0/\mathrm{d}x)^{-1}\varepsilon_p^c. \tag{13.8-2}$$

13.2 Marginal Conditions and Integral Equations

13.2.1 Nonlinear Marginal Conditions

Equation (13.1) describes dynamics of a test mode, which is shielded by background turbulence. The nonlinear marginal stability criterion dictates a condition that $(\mu_{N.k}, \lambda_{N.k}, \chi_{N.k})$ must satisfy. When the k-dependences of these terms are derived, then the spectral function $E(k)$ is obtained from equations (8.38)–(8.41).

From equation (13.1), the nonlinear marginal stability condition is derived. Namely, an equation $\det \boldsymbol{L} = 0$ gives a nonlinear dispersion relation for the existence of a nontrivial solution of the dressed (shielded) test mode as

$$\frac{k_\parallel^2}{\lambda_{N.k}k_\perp^2}\phi_k + \mu_{N.k}k_\perp^4\phi_k - \frac{G_0 k_y^2}{\chi_{N.k}k_\perp^2}\phi_k = 0. \tag{13.9}$$

Equation (13.9) provides a nontrivial solution if the relation between eigenvalues $(\mu_{N.k}, \lambda_{N.k}, \chi_{N.k})$ is satisfied. We take

$$\phi_k = 0 \text{ in the region } k < k_0. \tag{13.10}$$

The dispersion relation (13.9) is solved, and the k-dependences of $(\mu_{N.k}, \lambda_{N.k}, \chi_{N.k})$ are derived. For a fixed G_0, it is shown that a scaling relation like equation (9.34)

$$\mu_N(k), \lambda_N(k), \chi_N(k) \propto k_\perp^{-2} \tag{13.11}$$

is satisfied as an asymptotic form. For given ratios $k_y^2/k_\perp^2$ and $k_\parallel^2/k_\perp^2$, equation (13.9) becomes independent of k if equation (13.11) is satisfied. In equation (13.11) and in the following, notation like $\mu_N(k)$ is also used in order to illuminate the k-dependence of nonlinear transfer rates $(\mu_{N.k}, \lambda_{N.k}, \chi_{N.k})$. Noting this relation, we write

$$\mu_N(k_\perp) = \mu_N(k_0)\frac{k_0^2}{k_\perp^2} \qquad \lambda_N(k_\perp) = \lambda_N(k_0)\frac{k_0^2}{k_\perp^2} \qquad \chi_N(k_\perp) = \chi_N(k_0)\frac{k_0^2}{k_\perp^2}. \tag{13.12}$$

Substituting equation (13.12) into equation (13.9), we have a relation

$$\frac{1}{\lambda_N(k_0)k_0^2}\frac{k_\parallel^2}{k_\perp^2} + \mu_N(k_0)k_0^2 - \frac{G_0}{\chi_N(k_0)k_0^2}\frac{k_y^2}{k_\perp^2} = 0. \tag{13.13}$$

The renormalized equation (13.1) is derived with an assumption of the isotropy of turbulence, i.e., $k_y^2/k_\perp^2 \simeq 1/2$. With this fact in mind, approximate evaluations of the nonlinear transfer rates $\mu_N(k_0)$ etc are obtained. In a stationary state, the driving term, i.e., the third term of equation (13.13), becomes of the same order as the nonlinear damping, which is represented by the effective ion viscosity. One has an estimate $\mu_N(k_0)k_0^2 \simeq G_0(2\chi_N(k_0)k_0^2)^{-1}$, i.e.,

$$\mu_N(k_0)\chi_N(k_0)k_0^4 \simeq \frac{G_0}{2}. \tag{13.14}$$

13.2.2 Integral Equations

From relations (13.12) and (13.14), an integral equation for the spectral function is derived. In order to have an analytic insight, we study the limit where the nonlinear decorrelation term $\mu_N k_\perp^2$ is considered to be dominant in K_2. In this case, the renormalization relation (8.38) is simplified as [13.1]

$$\mu_N(k) = \int_k^{k_b} \mathrm{d}k_1 \frac{E(k_1)}{\mu_N(k+k_1)(k+k_1)^2}. \tag{13.15}$$

Other nonlinear transfer rates, μ_{eN} and χ_N, satisfy similar relations, and we have

$$\lambda_N \simeq \xi^{-1}\mu_N \text{ and } \chi_N \simeq \mu_N. \tag{13.16}$$

By use of the relation $\chi_N \simeq \mu_N$, equation (13.14) provides an estimate of the nonlinear transfer rate as

$$\mu_N(k_0)k_0^2 \simeq \frac{1}{\sqrt{2}}G_0^{1/2}. \tag{13.17}$$

Employing equations (13.12), (13.15) and (13.17), the integral equation which $E(k_\perp)$ must satisfy is given as

$$\int_{k_\perp}^{k_b} \mathrm{d}k_1\, E(k_1) = \frac{1}{2}G_0 k_\perp^{-2}. \tag{13.18}$$

13.3 Spectrum

13.3.1 Spectral Functions

Integral equation (13.18) is solved as

$$E(k_\perp) = G_0 k_\perp^{-3} \tag{13.19}$$

in the energy containing region ($k_\perp \ll k_b$). The power spectral function has $k_\perp^{-3}$-dependence as well as G_0-dependence. Current and pressure spectral functions are obtained as

$$E_J(k_\perp) = \xi G_0 k_\perp^{-3} \text{ and } E_p(k_\perp) = (\mathrm{d}p_0/\mathrm{d}x)^2 k_\perp^{-3}. \tag{13.20}$$

13.3.2 Turbulence Energy and Dissipation

Integrating $E(k_\perp)$ over the energy containing region, the kinetic energy of fluctuation, $W_{\nabla\phi}$, is given as

$$W_{\nabla\phi} = k_0^{-2} G_0 \tag{13.21}$$

in the limit of $k_0 \ll k_b$. Fluctuation energy is found to be of the order of $k_0^{-2}G_0$. Corresponding terms in current and pressure fluctuations, W_J and W_p, are also given as

$$W_J = \xi k_0^{-2} G_0 \text{ and } W_p = (\mathrm{d}p_0/\mathrm{d}x)^2 k_0^{-2} \tag{13.22}$$

satisfying an approximate relation

$$W_{\nabla\phi} : W_J : W_p \sim 1 : \xi : (\mathrm{d}p_0/\mathrm{d}x)\kappa_x^{-1} \tag{13.23}$$

where $\kappa_x = G_0(\mathrm{d}p_0/\mathrm{d}x)^{-1}$ holds.

Dissipation terms are calculated by use of spectral functions. We have

$$\varepsilon = \frac{1}{\sqrt{2}k_0^2} G_0^{3/2} \qquad \varepsilon_J = \frac{1}{\sqrt{2}k_0^2} G_0^{3/2} \qquad \varepsilon_p = \frac{1}{\sqrt{2}k_0^2}\left(\frac{\mathrm{d}p_0}{\mathrm{d}x}\right)^2 G_0^{1/2}. \tag{13.24}$$

The average heat flux is calculated from fluctuation spectra as $q_x = |\mathrm{d}p_0/\mathrm{d}x|^{-1}\varepsilon_p$. If one introduces the turbulent transport coefficient as $q_x = -\chi_T(\mathrm{d}p_0/\mathrm{d}x)$, χ_T is expressed as $\chi_T = (\mathrm{d}p_0/\mathrm{d}x)^{-2}\varepsilon_p$. From equation (13.24), χ_T is given as

$$\chi_T = \frac{1}{\sqrt{2}k_0^2} G_0^{1/2}. \tag{13.25}$$

13.3.3 Mode Number

Evaluation of the lower bound mode number k_0 closes the analysis. Relation (13.11) holds as an asymptotic behaviour in the high k region, and is not valid below k_0. The value of k_0 is estimated from a typical mode number which gives the least stable mode. As is discussed in chapter 9, the least stable mode satisfies the relation

$$k_0 \sim s^{1/3}(2\mu_N(k_0)\lambda_N(k_0))^{-1/6} \tag{13.26-1}$$

or

$$k_0 \sim \xi^{1/2} s\, G_0^{-1/2}. \tag{13.26-2}$$

Equation (13.26-2) is substituted into equation (13.25), and the turbulent thermal conductivity is expressed in terms of global parameters as

$$\chi_T \simeq \frac{1}{\sqrt{2}\xi s^2} G_0^{3/2}. \tag{13.27}$$

The turbulence level is also obtained by substitution of k_0 into equations (13.21) and (13.22) as

$$W_{\nabla\phi} \simeq \xi^{-1} s^{-2} G_0^2 \qquad W_j \simeq s^{-2} G_0^2 \qquad W_p \simeq (\mathrm{d}p_0/\mathrm{d}x)^2 \xi^{-1} s^{-2} G_0. \tag{13.28}$$

Turbulent magnetic energy W_B could be more easily observed compared to turbulent current level W_J. Noting a relation $B_k^2 = k_\perp^2 J_k^2$, the spectral function of the magnetic perturbation is given as

$$E_B(k_\perp) = k_\perp^2 E_J(k_\perp). \tag{13.29}$$

The energy level of magnetic fluctuation, W_B, is given as

$$W_B \simeq \xi^{-1} s^{-4} G_0^3. \tag{13.30}$$

When the driving parameter is large,

$$G_0 > s^2 \tag{13.31}$$

magnetic energy W_B dominates over kinetic energy $W_{\nabla\phi}$.

By use of equations (13.17) and (13.26-2), the effective nonlinear transfer rate $\chi_N(k_0)$ for the least stable mode is explicitly calculated. It is found to be nearly equal to the turbulent transport coefficient χ_T

$$\chi_T \sim \chi_N(k_0). \tag{13.32}$$

This result gives a basis for previous analyses in which mean field approximation has been employed.

13.4 Summarizing

Study of the spectrum by use of a dressed text mode is summarized. Parameter G_0 plays a role in characterizing the nonequilibrium nature of magnetized plasmas. The nonlinear marginal stability condition is solved for a given value of G_0, yielding a power law for nonlinear transfer rates,

$$\mu_N(k_\perp) \sim \sqrt{G_0}k_\perp^{-2} \qquad \lambda_N(k_\perp) \sim \xi^{-1}\sqrt{G_0}k_\perp^{-2} \qquad \chi_N(k_\perp) \sim \sqrt{G_0}k_\perp^{-2}.$$

The integral equation for spectral functions is solved as

$$E(k_\perp) \sim G_0 k_\perp^{-3}$$

and so on. Turbulent kinetic energy is also obtained as

$$W_{\nabla\phi} \simeq \xi^{-1}s^{-2}G_0^2 \qquad W_J \simeq s^{-2}G_0^2$$

$$W_B \simeq \xi^{-1}s^{-4}G_0^3 \qquad W_p \simeq (\mathrm{d}p_0/\mathrm{d}x)^2\xi^{-1}s^{-2}G_0.$$

Decorrelation length and time are estimated as

$$\ell_{corr} \simeq \xi^{-1/2}s^{-1}G_0^{1/2} \text{ and } \tau_{corr}^{-1} \simeq G_0^{1/2}$$

respectively. The turbulent transport coefficient is obtained

$$\chi_T \simeq \xi^{-1}s^{-2}G_0^{3/2}.$$

Turbulence is nonlinearly self-sustained, and the result does not depend on collisional dissipation in the strong turbulence limit.

A result like $\chi_T \simeq \xi^{-1}s^{-2}G_0^{3/2}$ shows that heat flux in the direction of the gradient is a highly nonlinear function of the gradient. If it is expressed as $\chi_T = \ell_{corr}^2\tau_{corr}^{-1}$, the $G_0^{3/2}$-dependence of χ_T is contributed to by an amount of $G_0^{1/2}$ due to temporal decorrelation rate, and of G_0 due to spatial decorrelation length. This nonlinear dependence of χ_T shows the characteristics of a system far from thermal equilibrium. The present method is demonstrated to be useful in studying nonlinear–nonequilibrium transport under the presence of subcritical excitation of turbulence.

Note that in the strong gradient limit, $s < \xi^{-1/2}G_0^{1/2}$, the decorrelation length is limited by system size. In such a limit, the conductivity is bounded by $\chi_T \simeq G_0^{1/2}$.

REFERENCES

[13.1] Itoh S-I and Itoh K 1998 *Plasma Phys. Control. Fusion* **40** 1729
Itoh S-I and Itoh K 1997 *J. Phys. Soc. Japan* **66** 1571

Chapter 14

Nonlinear Simulation

Analytic theory suggests that subcritical turbulence exists in the system of an inhomogeneous magnetized plasma. A direct simulation of nonlinear equations confirms this theoretical picture. Results of the nonlinear simulation are shown after [7.8] and [14.1].

14.1 Model Geometry and Set of Equations

A model geometry is a plasma slab, which is shown in figure 14.1. Cartesian coordinates (x, y, z) are used, and a strong magnetic field is in the z-direction. The magnetic field has a shear as is given by equation (9.5). Nonlinear simulation is performed for a two-dimensional turbulence. The fluctuating quantities have a form

$$\tilde{\phi}(x, y, z; t) = \sum_{k_y} \phi_{k_y}(x; t) \exp(\mathrm{i}k_y y). \tag{14.1}$$

The wave number along the magnetic field line, $k_\parallel$, is given as $k_\parallel = sxk_y$.

A set of model equations is written in a normalized form as

$$\frac{\mathrm{d}}{\mathrm{d}t}\begin{pmatrix} \nabla_\perp^2\phi \\ j \\ p \end{pmatrix} + \begin{pmatrix} -\mu_c\nabla_\perp^4 & -\mathrm{i}k_y sx & \mathrm{i}k_y G_0 \\ \mathrm{i}k_y sx & -\lambda_c\nabla_\perp^2 & 0 \\ \mathrm{i}k_y & 0 & -\chi_c\nabla_\perp^2 \end{pmatrix}\begin{pmatrix} \phi \\ j \\ p \end{pmatrix} = 0 \tag{14.2}$$

$$\frac{\mathrm{d}}{\mathrm{d}t} = \frac{\partial}{\partial t} + [\phi,\ \ ] \tag{14.3}$$

where the Poisson bracket $[\phi,\ \ ]$ represents a convective nonlinear interaction. For simplicity of the expression, length and time are normalized to collisionless skin depth and Alfvén transit time τ_{Ap}, respectively, in this chapter. Perturbation of pressure is normalized to $c\omega_p^{-1}\,\mathrm{d}p_0/\mathrm{d}x$ in equation (14.2). In order to illuminate the nonlinear mechanism of an instability, an electrostatic approximation is used, and an inductive electric field (time derivative of the vector potential) is neglected in the Ohm law.

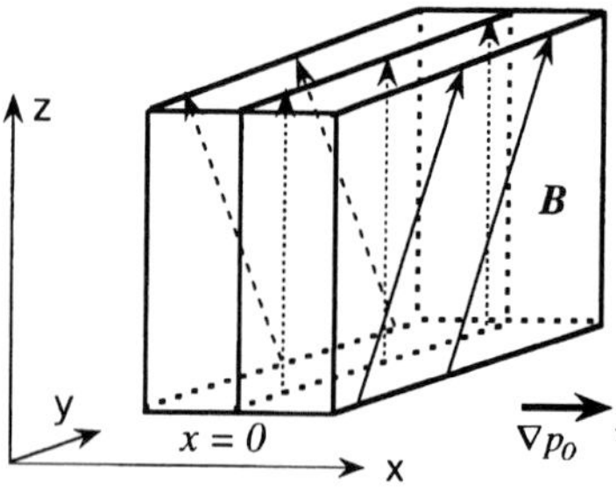

Figure 14.1. Geometry of the simulation.

In evaluating fluctuation quantities, we define fluctuation spectra of electric field, $W_{\nabla\phi}$, current, W_j, and pressure, W_p, as

$$W_{\nabla\phi}(k_y) = \frac{1}{2L_x}\int_0^{L_x} \mathrm{d}x |\nabla_\perp \phi(x, k_y)|^2 \tag{14.4}$$

$$W_j(k_y) = \frac{1}{2L_x}\int_0^{L_x} \mathrm{d}x |j(x, k_y)|^2 \tag{14.5}$$

and

$$W_p(k_y) = \frac{1}{2L_x}\int_0^{L_x} \mathrm{d}x |p(x, k_y)|^2, \tag{14.6}$$

respectively. Fluctuation levels are expressed by $\langle W_{\nabla\phi}\rangle$, $\langle W_j\rangle$ and $\langle W_p\rangle$, where an average $\langle\,\rangle$ denotes a summation over poloidal mode numbers, e.g.,

$$\langle W_p\rangle = \sum_{k_y} W_p(k_y). \tag{14.7}$$

By use of these quadratic forms, the energy conservation law is derived from equation (14.2) as

$$\frac{\partial}{\partial t}(\langle W_{\nabla\phi}\rangle + \langle W_j\rangle + \langle W_p\rangle) = (G_0+1)q_x - \langle \mu_c|\nabla_\perp^2\phi|^2\rangle - \langle\lambda_c|\nabla_\perp j|^2\rangle - \langle\chi_c|\nabla_\perp p|^2\rangle \tag{14.8}$$

where

$$q_x = -\langle \mathrm{i}k_y p^* \phi\rangle \tag{14.9}$$

denotes the average energy flux in the direction of the pressure gradient. Equation (14.8) is a simplified form of equation (7.33). This relation suggests that a stationary state is realized by the balance between the power released by plasma expansion and dissipation through collisional damping. The nonlinearities, $[\phi,\quad]$, transfer wave energy between different modes and act to cause effective diffusion in a spectrum, but real dissipation occurs through collisional transport.

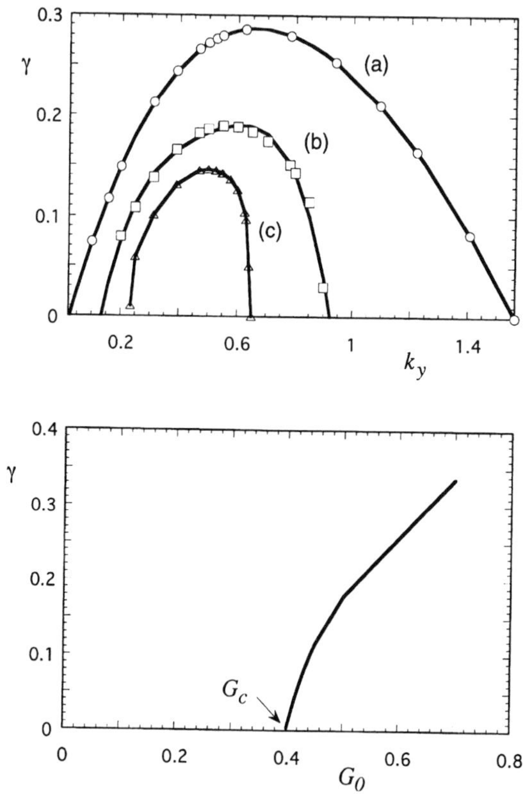

Figure 14.2. Linear growth rate (top) as a function of mode number k_y for fixed pressure gradient G_0. ((a) $\lambda_c = 0.2$, (b) $\lambda_c = 0.02$, (c) $\lambda_c = 0$ for $G_0 = 0.5$.) The maximum linear growth rate (bottom) as a function of pressure gradient G_0. (Other parameters are: $\mu_c = \chi_c = 0.2$ and $s = 0.5$.)

14.2 Linear Instability

This set of equations predicts linear instability. If the pressure gradient is large enough, reactive instability appears and the unstable mode is called an interchange mode (see section 6.2.1). In the limit of small pressure gradient, a dissipative instability exists owing to the current diffusivity, which impedes free motion of electrons along the magnetic field line. Besides current diffusivity, other elements of transport processes influence a linear growth rate.

Figure 14.2 (top) illustrates the linear growth rate as a function of mode number k_y. It is shown that a mode becomes linearly unstable in the range of $k_y \simeq 1/2$. The largest growth rate has a dependence on a magnitude of driving parameter G_0, which is pressure gradient multiplied by gradient of magnetic field strength, equation (8.21). Figure 14.2 (bottom) shows the largest growth rate as a function of the driving parameter G_0. For parameters of $s = 0.5$, $\mu_c = \chi_c = 0.2$ and $\lambda_c = 0.01$, a critical value is given as

$$G_c \simeq 0.4 \tag{14.10}$$

below which linear modes are stable.

14.3 Nonlinear Simulation

Nonlinear simulation is performed for a system in a box (L_x, L_y). Boundary conditions are: (1) fluctuations vanish at $x = \pm L_x/2$ and (2) fluctuations are periodic at $y = 0$ and $y = L_y$. In the following, parameters $L_x = 80$ and $L_y = 2\pi$ are employed. $M = 64$ modes are taken in k_y-space ($k_{y,min} = 10/64$ and $k_{y,max} = 10$).

14.3.1 Nonlinear Instability

Figure 14.3 shows the temporal evolution of fluctuations and its nonlinear growth (solid line). In the small amplitude limit, perturbations grow in accordance with a linear growth rate. However, at a time of $t \simeq 35$, when the amplitude exceeds a certain threshold value,

$$\langle W_{\nabla\phi} \rangle = W_{thr} \sim 10^{-4} \tag{14.11}$$

the growth rate starts to increase. In the time range of $35 < t < 50$, the growth rate becomes larger as amplitude increases. This is nonlinear destabilization through electron dynamics. The dashed line shows a result of a reference simulation, in which the $[\phi, j]$ term in equation (14.2) is omitted. (This reference case is called the 'linear Ohm's law' for short.) In the case of the linear Ohm's law, only linear growth and nonlinear stabilization are observed, showing a conventional picture of nonlinear development.

Figure 14.4 illustrates the growth rate as a function of fluctuation amplitude for the case of figure 14.3. The growth rate first increases as amplitude grows. Nonlinear destabilization is clearly observed.

The level of fluctuations for nonlinear growth is compared to theory. The current diffusivity, λ_N, is estimated as

$$\lambda_N \simeq \frac{\sqrt{\langle W_{\nabla\phi} \rangle}}{k_y}. \tag{14.12}$$

(Note the normalization.) For values of $\bar{\lambda} = \lambda_N + \lambda_c$ and $k_y = 0.6$, λ_N is estimated as 1.5×10^{-2} and becomes larger than λ_c for this case. The growth rate, estimated by $\bar{\lambda} = \lambda_N + \lambda_c$, noticeably deviates from a linear estimation. A theoretical prediction by use of one point renormalization gives a good approximation on a transition from the linear growth to nonlinear instability.

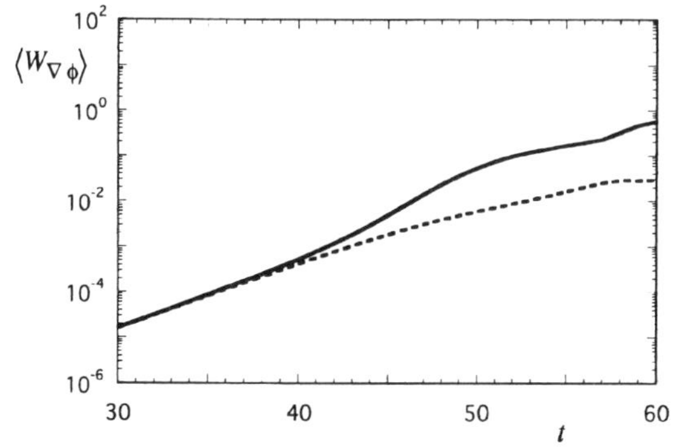

Figure 14.3. Fluctuating electric field energy versus time, showing nonlinear growth due to an electron nonlinearity at $t > 35$ (solid line). Parameters are $\mu_c = \chi_c = 0.2$, and $\lambda_c = 0.01$ and $s = G_0 = 0.5$. The dashed line indicates a simulation, for which convective nonlinearity $[\phi, J]$ in Ohm's law is omitted. (Quoted from [7.8].)

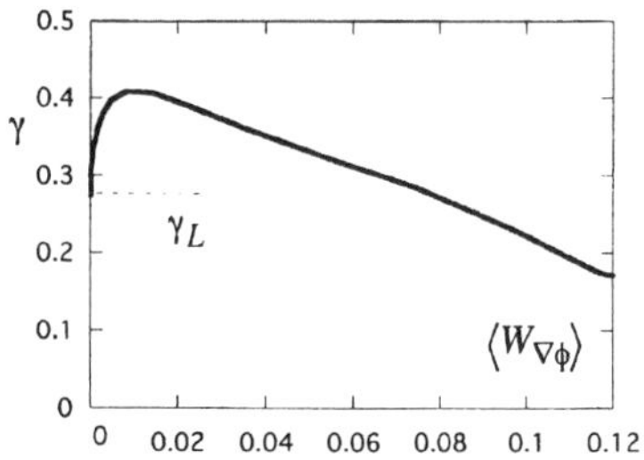

Figure 14.4. Growth rate is shown as a function of fluctuation amplitude. (Simulation result of solid line in figure 14.3.)

14.3.2 Higher Saturation Level

Simulation in a longer time scale is shown in figure 14.5. At the time of $t \sim 80$, an inverse cascade takes place, and the level shows a transient dip. When t exceeds 100, a saturated state is realized. The dashed line (b) shows the reference case of linear Ohm's law (other parameters are common). The saturation level is smaller in the case of linear Ohm's law. The case for an increased value of λ_c ($\lambda_c = 0.2$) is shown by a line (c). For this value of λ_c, the maximum linear growth rate becomes about twice as large, while a corresponding value of $k_\perp$ shifts by less than 20% from that in the case (a). Comparing results of (a) and (c), we see that the saturation level is not sensitive to the linear growth rate in current diffusive interchange mode turbulence.

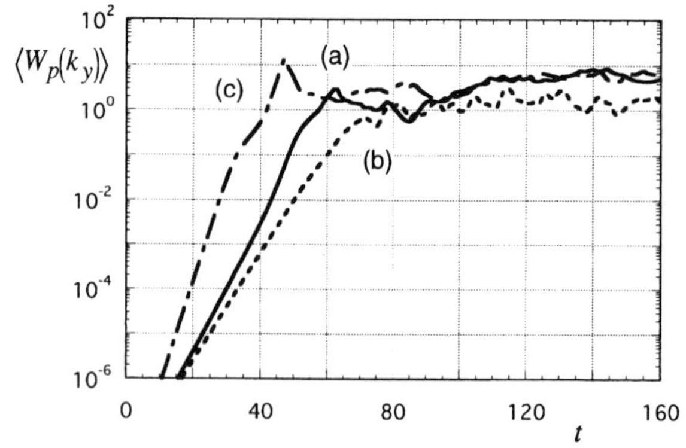

Figure 14.5. Evolution of fluctuating pressure energy. (a) $\lambda_c = 0.01$ and (b) $\lambda_c = 0.01$ with linear Ohm's law and (c) $\lambda_c = 0.2$. The case (c) has a larger linear growth rate than the case (a). Nevertheless, the saturation level is less affected by the linear growth rate in (a) and (c). Other parameters are the same as in figure 14.3.

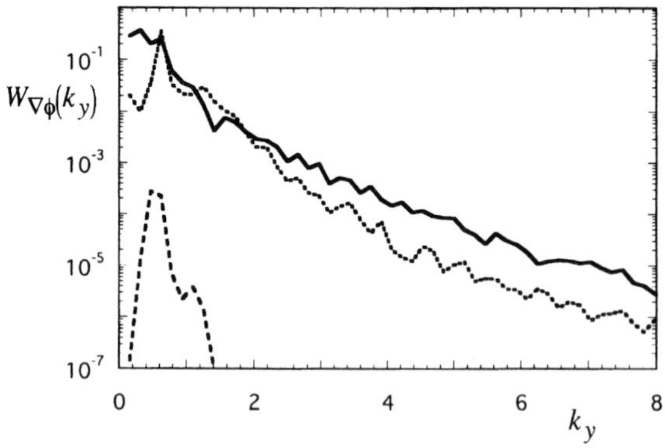

Figure 14.6. Power spectrum of electric field energy for nonlinear Ohm's law. The dashed line corresponds to a linear phase ($t = 40$), and the dotted line is in a phase of nonlinear growth ($t = 60$). The solid line shows a stationary state $t = 200$.

The power spectrum is illustrated in figure 14.6. In a linearly growing phase, a narrow spectrum is observed (dashed line), representing linearly unstable modes. When a nonlinear instability takes place, a strong cascade to a shorter wavelength mode is observed.

The spectrum of fluctuations is studied. A nonlinear growth takes place in the range of $k_y \sim 0.6$, and a spectrum extends to higher mode numbers. This extension (normal cascade) is characteristic of the nonlinear Ohm's law. In a stationary stage, the largest amplitude mode is observed in longer wavelength modes. This is due to an inverse cascade from nonlinearly excited modes of $k_y \sim 0.6$. In the region of $0.3 < k_y < 1.5$, the spectrum is roughly fitted to

$$W_{\nabla\phi}(k_y) \propto k_y^{-3}. \tag{14.13}$$

Above $k_y > 2$, collisional dissipation dominates and $W_{\nabla\phi}$ is cut off as $W_{\nabla\phi}(k_y) \propto \exp(-k_y)$.

The largest amplitude mode is found in longer wavelength modes. In this particular study, the largest amplitude mode is obtained at $k_y = 10/64$. This suggests that the wavelength at the spectrum peak is longer than the collisionless skin depth, though nonlinear interactions in the range of $k_y \sim 1$ cause nonlinear growth and higher saturation. The inverse cascade has been discussed in the literature [14.2].

14.3.3 Subcritical Excitation

Insensitivity of the turbulence level to a linear stability condition becomes more prominent if simulation is performed for a linearly stable case. If the value of G_0 is reduced to 0.3 and other parameters are fixed, all modes become linearly stable (as is shown in figure 14.2 (bottom)). Even in this case, growth of fluctuation occurs in the system of nonlinear Ohm's law. If an initial amplitude of fluctuation level exceeds a threshold value, fluctuations start to grow to a very high level. Even if a control parameter G_0 remains in a linearly stable regime, the fluctuation level approaches those for linearly unstable cases (figure 14.7). When the fluctuation level is above the threshold value, $\langle W_p \rangle > W_{p.thr} \simeq 10^{-3}$, fluctuation develops to a turbulent state with amplitude of $\langle W_p \rangle \simeq 6$. The power spectrum extends to the higher k region, if nonlinear excitation takes place. When the initial amplitude is below the threshold value, the system remains stable and the fluctuation level does not increase. In such a case (line (4) in figure 14.7), a cascade to shorter wave modes does not occur, and the fluctuation spectrum is shown in figure 14.7 (bottom).

Through simulations, it has become clear that the observed growth rate, $\langle W_p \rangle^{-1}(\partial \langle W_p \rangle / \partial t)$, can be an increasing function of the fluctuation level.

The steady state fluctuation amplitude is illustrated as a function of pressure gradient in figure 14.8. The linear growth is realized only in a very low level of initial amplitude (about 10^{-3} times the saturation level or less). The dynamical evolution of fluctuations is governed by nonlinear growth; the growth rate depends on the fluctuation level as is illustrated in previous sections.

The nonlinear simulation confirms a theoretically predicted feature of subcritical excitation of turbulence in inhomogeneous plasmas.

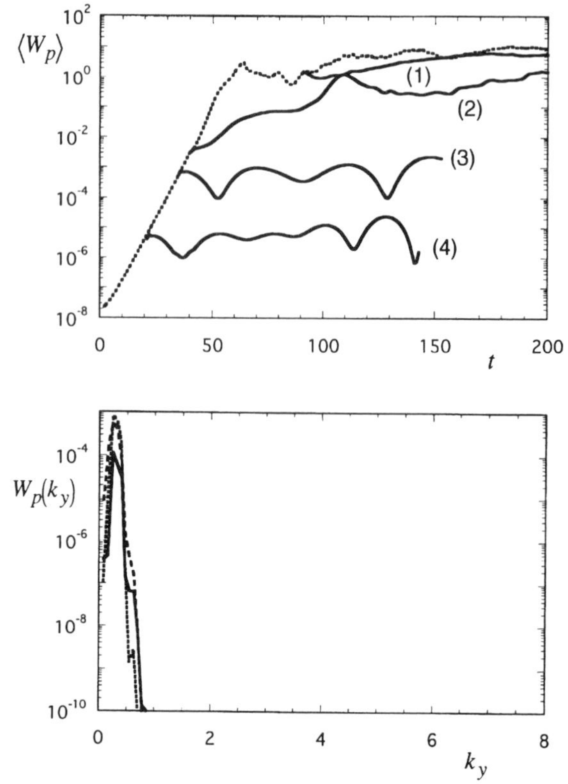

Figure 14.7. Fluctuating pressure energy against time for the linearly stable case $G_0 = 0.3$. (Other parameters are the same as in figure 14.2) Four cases (1)–(4), which have different initial values of fluctuation level, are plotted. The dotted line indicates the case of $G_0 = 0.5$ for the reference. The power spectrum of the case (4) is illustrated below. Dashed line, dotted line and solid line correspond to $t = 20$, $t = 40$ and $t = 140$, respectively. Energy transfer to the shorter wavelength region is not observed in this case.

14.3.4 Gradient–Flux Relation

Nonlinear simulation also confirms the dependence of saturation level on pressure gradient. In the previous section it is shown that the static potential fluctuation scales as $\phi \propto G_0^{3/2}$, and the typical scale-length of fluctuations behaves as $\langle k_\theta \rangle \propto G_0^{-1/2}$. This indicates that electric field energy $\langle W_{\nabla\phi} \rangle$, which is in proportion to $\phi^2 k_\theta^2$, has a dependence

$$\langle W_{\nabla\phi} \rangle \propto G_0^2. \tag{14.14}$$

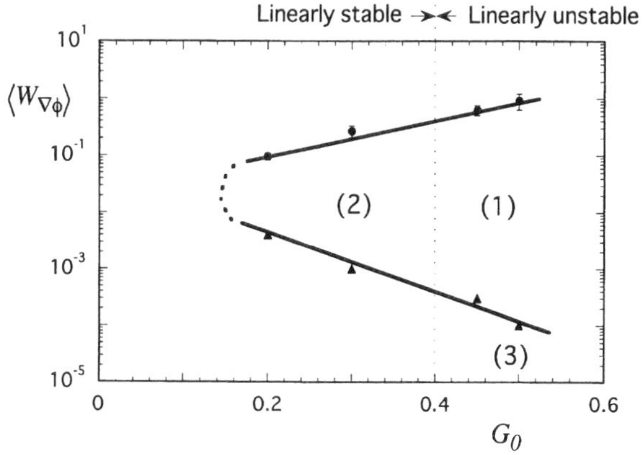

Figure 14.8. Nonlinear stability boundary in the $\langle W_{\nabla\phi}\rangle$–$G_0$-plane. Parameters are the same as in figure 14.2. The regions (1) and (2) represent nonlinearly unstable regions with $\gamma_L > 0$, $\gamma_N \neq 0$ and $\gamma_L < 0$, $\gamma_N \neq 0$, respectively. The region (3) represents a linearly unstable region. The upper line represents the stationary state.

The upper solid line of figure 14.8 satisfies this relation.

A turbulent-driven energy flux is computed and is compared to the theoretical prediction. Since simulation is performed in a 2D model, an associated flux is localized in the vicinity of the rational surface. By introduction of a localization width ℓ_x, the space average of the heat flux is defined within this localization length ℓ_x as

$$\langle\langle q_x \rangle\rangle = \ell_x^{-1} \int_{-\ell_x/2}^{\ell_x/2} \mathrm{d}x\, \bar{q}_x. \tag{14.15}$$

The localization width ℓ_x is defined such that $\bar{q}_x > 0.1\, \bar{q}_{x\,max}$ holds for $|x| \leq \ell_x$, where

$$\bar{q}_x = L_y^{-1} \int_0^{L_y} q_x\, \mathrm{d}y$$

and $q_x = -\mathrm{i}k_y p^* \phi$. In the saturation stage, a time average in an interval of $100 < t < 200$ is calculated. The turbulent transport coefficient $\chi_{turb} \equiv -\langle\langle q_x \rangle\rangle / \nabla p$ is obtained as $\chi_{turb} = 3$ for parameters of figure 14.2 ($s = 0.5$, $G_0 = 0.5$). In the large gradient limit, the simulation result is fitted to

$$\chi_{turb} \simeq 10\, G_0^{3/2} \qquad \text{(case of } s = 1/2). \tag{14.16}$$

Below a critical gradient G_*, $G_0 < G_*$, the turbulent transport coefficient remains small compared with transport resulting from thermal fluctuations. Figure 14.9 compares an analytic formula with the result of nonlinear simulation.

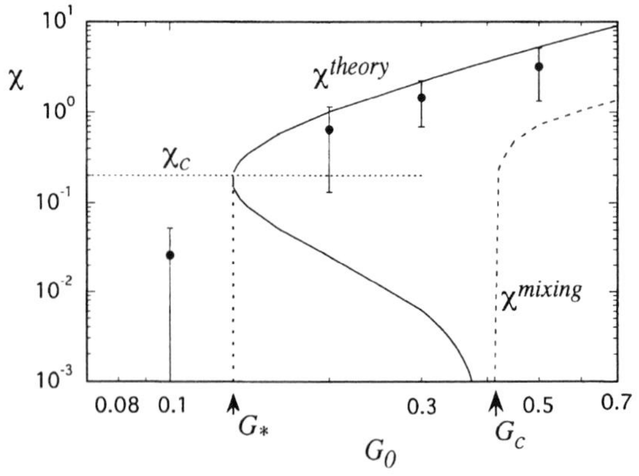

Figure 14.9. Comparison of analytic formula (solid line) with a numerical simulation (dot with error bar). (χ^{mixing} denotes a quasilinear estimate $\gamma_L k_\perp^{-2}$.)

This result confirms theoretical predictions: the transition from collisional transport to turbulent transport takes place at $G_0 = G_*$. An asymptotic relation $\chi_{turb} \propto G_0^{3/2}$ holds in the strong gradient limit.

Through this simulation study, it is shown that the nonlinear dynamics of electron motion in the presence of turbulence is important in plasma turbulence and turbulent transport. Small but finite electron mass introduces an essential aspect of structural formation in plasmas. Acceleration of growth rate in the presence of the electron inertia effect has also been observed in simulations of global MHD instabilities [14.3].

Appendix 14A Nonlinear Decorrelation

Nonlinear simulation demonstrates that nonlinear interactions cause strong decorrelation if the fluctuation amplitude exceeds a threshold.

Figure 14A.1 illustrates results of numerical simulation with two different codes [14.4]. Parameters are chosen to be identical in two computations. (Precise descriptions of the two codes are given in [14.4].) In an early phase of simulation, the two codes provide agreement. However, in a latter phase, a noticeable difference appears. This difference in result is owing to a nonlinear decorrelation. In numerical simulations, a round-off error cannot be eliminated. An observation of the round-off error gives an insight into turbulence in plasmas.

A difference of the result of two computations

$$\delta W \equiv W_{\nabla\phi}(\text{code A}) - W_{\nabla\phi}(\text{code B}) \tag{14A.1}$$

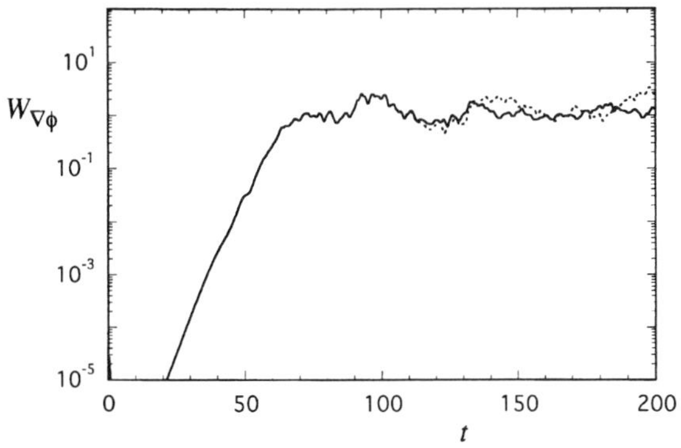

Figure 14A.1. Temporal evolutions of $W_{\nabla\phi}$ are drawn for two codes A and B (solid and dashed lines). Two physically identical codes with the same initial condition show a considerable deviation in solutions after the nonlinear growth of the mode. (Quoted from [14.4].)

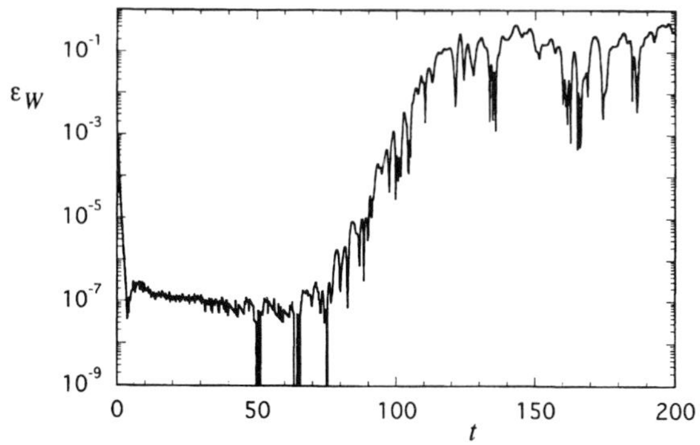

Figure 14A.2. The time trace of the relative deviation for the case of figure 14.10 is plotted. In the linear phase, the difference which is caused by the initial noise remains constant. Deviation grows associated with the nonlinear growth and saturation of fluctuations. (Quoted from [14.4].)

is evaluated, and a relative difference of the results of two codes

$$\varepsilon_W \equiv \frac{|W_{\nabla\phi}(\text{code A}) - W_{\nabla\phi}(\text{code B})|}{W_{\nabla\phi}(\text{code A}) + W_{\nabla\phi}(\text{code B})} \tag{14A.2}$$

is illustrated in figure 14A.2. In an early phase of calculation, $0 < t < 50$, the relative difference does not grow. It grows exponentially in the time period of

$60 < t < 110$. Saturation at around $120 < t$ is observed. In the early phase, fluctuation energy grows following a linear growth rate. Two computations give the linear growth rate with sufficient accuracy, and an initial difference due to round-off error is preserved. On the other hand, for $t > 70$ the noise grows nonlinearly according to the nonlinear decorrelation rate of the system.

The nonlinear growth rate of the deviation $|\delta W|$ is estimated in the form of

$$|\delta W| \propto \exp(\gamma_{ac} t) \tag{14A.3}$$

with

$$\gamma_{ac} \simeq 0.4. \tag{14A.4}$$

In this system, the typical mode number of the turbulence is of the order of $k_\perp \sim 0.3$. The nonlinear decorrelation rate of such a mode due to nonlinear coupling is estimated to be $\gamma_{NL} \sim \chi_{NL} k_\perp^2 \sim 0.3$. A positive Lyapunov exponent is also observed. This suggests that the noise due to the round-off errors also grows under the nonlinear interactions of the turbulence mode.

It is well known in a low order model system (like a Lorenz model) that nonlinear interactions give a positive Lyapunov exponent and introduce a finite decorrelation time [14.5]. Nonlinear simulation of plasma turbulence demonstrates that the nonlinear decorrelation rate plays an essential role in the dynamics of fluctuations as well as in turbulent transport.

REFERENCES

[14.1] Yagi M 1995 *J. Plasma Fusion Res.* **71** 1123

[14.2] Hasegawa A 1985 *Adv. Phys.* **34** 1

[14.3] Acceleration of the growth rate in the presence of electron inertia has also been observed in

Biskamp D and Drake J F 1994 *Phys. Rev. Lett.* **73** 971

Ottaviani M and Polcelli F 1993 *Phys. Rev. Lett.* **71** 3802

[14.4] Yagi M *et al* 1997 *Chaos* **7** 198

[14.5] See, e.g., Lorenz E N 1993 *The Essence of Chaos* (Seattle, WA: University of Washington Press) ch 4

Chapter 15

Scale Invariance

The method of scale invariance is a clear approach to understanding the nature of turbulence. Pioneering work has been performed by Kolmogorov, in which characteristics of homogeneous turbulence are investigated [15.1]. Efforts have been made for plasma turbulence [15.2–5]. Analysis based on a scale invariance method is briefly illustrated.

15.1 Kolmogorov Inertial Range

Kolmogorov's insight captures an essence of turbulence. A complex pattern of flow is realized only if small scale fluctuations are strongly excited. Perturbations of flow, which are excited owing to an inhomogeneity or boundary conditions in a global (observable) structure, in the end, are dissipated by processes of microscopic (molecular) dissipations. A nonlinear cascade, which transfers wave energy to the shorter wavelength regime, allows pumping of short wavelength modes. The latter modes are dissipated by molecular viscosity. The level and spectrum of fluctuations should be consistently determined with the rate of molecular dissipation. A mode number range of fluctuations which only conducts energy to shorter wavelength modes is called an inertial range. This is because they are not pumped or dissipated. A schematic drawing of the energy spectrum of fully developed turbulence is illustrated in figure 15.1

The literature explains the physics of Kolmogorov scaling [15.6]. Let us consider a homogeneous system and assume that dissipated power per unit volume ε is given. Change of scales are applied as

$$x \to \lambda_1 x \qquad \text{and} \qquad t \to \lambda_2 t. \tag{15.1}$$

Under this transformation of variables, viscosity μ (dimension [length^{2}time^{-1}]) and dissipated power per unit volume ε (dimension [length^{2}time^{-2}]) are transformed as

$$\mu \to \lambda_1^2 \lambda_2^{-1} \mu \qquad \text{and} \qquad \varepsilon \to \lambda_1^2 \lambda_2^{-2} \varepsilon. \tag{15.2}$$

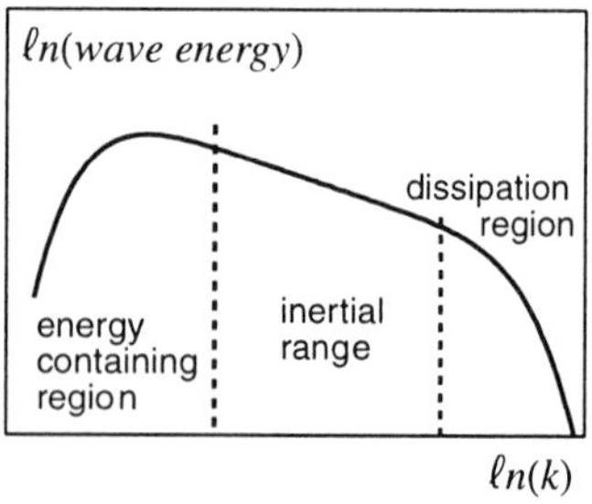

Figure 15.1. Concept of an inertial range, which transfers energy from an energy containing region to a dissipation region.

The energy spectrum of fluctuations $E(k)$ (dimension [length^{3}time^{-2}]) has the transformation

$$E(k) \rightarrow \lambda_1^3 \lambda_2^{-2} E(k). \tag{15.3}$$

Noting that a transformation of mode number is $k \rightarrow \lambda_1^{-1} k$, one finds the following two invariants against the transformation (15.1):

$$\frac{E(k)}{\mu^{5/4}\varepsilon^{1/4}} \quad \text{and} \quad \frac{k}{\mu^{-3/4}\varepsilon^{1/4}}. \tag{15.4}$$

From this result, the spectrum is independent of choice of coordinates if it is expressed as

$$E(k) = \mu^{5/4}\varepsilon^{1/4} F(k\mu^{3/4}\varepsilon^{-1/4}) \tag{15.5}$$

where $F(z)$ is an arbitrary function. The essence is that wave energy is not dissipated by viscosity in the inertial region. That is, $E(k)$ is approximated as being independent of μ. Such a nature requires a relation $F \propto \mu^{-5/4}$. In other words, a functional form of $F(z) \propto z^{-5/3}$ is satisfied. By use of this form factor, one finally has

$$E(k) \propto k^{-5/3}\varepsilon^{2/3}. \tag{15.6}$$

15.2 Scale Invariance Property of Plasma Turbulence

Our interest lies in determining the turbulence level and dissipation rate in the presence of an inhomogeneity. The energy containing region is of direct interest, where global parameters play decisive roles. For the determination of plasma turbulence level, the philosophy of the Kolmogorov method could be applied. Its extension is necessary, because inhomogeneity of a system requires a rescaling of spatial inhomogeneities.

15.2.1 Current Diffusive Turbulence

Let us study current diffusive interchange mode turbulence, which is analysed either by analytic theory or by direct numerical simulation in the preceding chapters. The scale invariance method provides a new insight into a problem. We choose a model equation of chapter 14

$$\frac{\mathrm{d}}{\mathrm{d}t}\begin{pmatrix}\nabla_\perp^2\phi \\ j \\ p\end{pmatrix}+\begin{pmatrix}0 & -\mathrm{i}k_y sx & \mathrm{i}k_y G_0 \\ \mathrm{i}k_y sx & 0 & 0 \\ \mathrm{i}k_y & 0 & 0\end{pmatrix}\begin{pmatrix}\phi \\ j \\ p\end{pmatrix}=0 \tag{15.7}$$

$$\frac{\mathrm{d}}{\mathrm{d}t}=\frac{\partial}{\partial t}+[\phi, \quad]. \tag{15.8}$$

In order to clarify our arguments, collisional dissipation terms are neglected. This implies that the inertial regime and energy containing region are of interest. In equation (15.7), the matrix includes a driving mechanism caused by an inhomogeneity. The Poisson bracket of equation (15.8), which represents a Lagrangean nonlinearity, is a nonlinear operator which causes a cascade of the spectrum.

A variation of scale is applied to equations (15.7) and (15.8). It should be noted that a driving source of fluctuations, which is represented by parameter G_0 in this problem, is also transformed. Let us consider a transformation $x \to \lambda_1 x$ and $t \to \lambda_2 t$ [15.5]. (Parameters λ_1 and λ_2 here are variables to denote scale transformations, not a current diffusivity.) The system of equations (15.7) and (15.8) is found to be invariant under this transformation with $\lambda_1\lambda_2 = 1$, namely,

$$x \to \lambda_1 x \qquad t \to \lambda_1^{-1} t \tag{15.9}$$

with

$$\begin{pmatrix}\phi \\ j \\ p\end{pmatrix} \to \begin{pmatrix}\lambda_1^3\phi \\ \lambda_1^2 j \\ \lambda_1 p\end{pmatrix} \tag{15.10}$$

and

$$G_0 \to \lambda_1^2 G_0. \tag{15.11}$$

Magnitude. The turbulent transport coefficient, D, which has a dimension of $D \sim [\text{length}^2\text{time}^{-1}]$, is transformed as

$$D \to \lambda_1^3 D. \tag{15.12}$$

From equations (15.11) and (15.12), λ_1 is eliminated, and a ratio $D/G_0^{3/2}$ is found to be scale invariant. We see that the dependence

$$D \propto G_0^{3/2} \tag{15.13}$$

is satisfied. In other words, the relation (15.13) must be fulfilled in order to be free from the choice of coordinates. The role of gradient is deduced from the scale invariance of basic equations. A similar argument applies to characteristics of fluctuations, and one has

$$\text{correlation length } \propto G_0^{1/2} \tag{15.14}$$

$$\text{correlation time } \propto G_0^{-1/2} \tag{15.15}$$

$$\text{potential fluctuation level } \propto G_0^{3/2}. \tag{15.16}$$

These results also demonstrate the essential role of inhomogeneity in driving fluctuations.

The role of magnetic shear is also shown. Writing the s-dependence explicitly, equations (15.7) and (15.8) are invariant under a transformation

$$\begin{pmatrix} x \\ t \\ \phi \\ j \\ p \end{pmatrix} \rightarrow \begin{pmatrix} \lambda_1 s^{-1} x \\ \lambda_1^{-1} t \\ \lambda_1^3 s^{-2} \phi \\ \lambda_1^2 s\, j \\ \lambda_1 p \end{pmatrix} \tag{15.17}$$

with equation (15.12). The scale invariance nature leads to the result of

$$\begin{pmatrix} \text{transport coefficient} \\ \text{correlation length} \\ \text{correlation time} \\ \text{potential fluctuation level} \end{pmatrix} \propto \begin{pmatrix} G_0^{3/2} s^{-2} \\ G_0^{1/2} s^{-1} \\ G_0^{-1/2} \\ G_0^{3/2} s^{-2} \end{pmatrix}. \tag{15.18}$$

It should be noted that the gradient parameter G_0 depends on λ_1^2. When the length is magnified (i.e., $\lambda_1 < 1$ holds, figure 15.2), the global gradient is also weakened in a new coordinate system. The gradient parameter G_0 is a combination of pressure gradient and magnetic field gradient. Therefore a dependence of $G_0 \rightarrow \lambda_1^2 G_0$ is deduced.

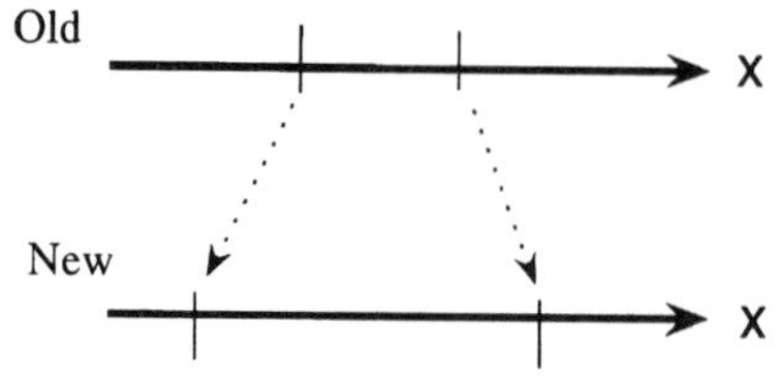

Figure 15.2. Example of scale transformation. In a new coordinate, length is magnified ($\lambda_1 < 1$). In other words, small scale structure is to be discussed.

Spectrum. Constraints for a spectrum are also deduced from a scale-invariant property. From scaling properties of $\tilde{\phi}$, equation (15.16), a quantity $\phi^2 G_0^{-3}$ is invariant under the scale transformation (15.9). The scaling property of k is the inverse of that of correlation length. Equation (15.14) shows that $k\sqrt{G_0}$ is scale invariant. If a scale-invariant quantity $\phi^2 G_0^{-3}$ is expressed as a function of mode number k, it is expressed in terms of a scale-invariant quantity $k\sqrt{G_0}$. That is

$$\tilde{\phi}^2 G_0^{-3} \sim F_{CD}(\sqrt{G_0}k) \tag{15.19-1}$$

or $\tilde{\phi}^2$ satisfies a relation

$$\tilde{\phi}^2 \sim G_0^3 F_{CD}(\sqrt{G_0}k) \tag{15.19-2}$$

where F_{CD} is an arbitrary function.

The dependence of the energy spectrum, $E(k) = \{k_\perp^2 |\phi(k)|^2\} k^{d-1}$, is also deduced. The Fourier component of potential fluctuation $|\phi(k)|^2$ has dimension$[|\phi(k)|^2] \sim$ dimension$[\tilde{\phi}^2]$ dimension$[k^{-d}]$, where d is the space dimension of a system. For a quasi-two-dimensional turbulence in magnetized plasmas, $d = 2$, we have dimension$[E(k)] =$ dimension$[\tilde{\phi}^2 k_\perp]$. Combining this relation with equation (15.19), one obtains

$$E(k_\perp) \sim G_0^3 F_{CD}(\sqrt{G_0}k_\perp)k_\perp. \tag{15.20}$$

Equation (15.20) corresponds to equation (15.5) in the case of Kolmogorov's argument, and is an extension to inhomogeneous and unstable systems. If the energy spectrum is expressed by a power law of mode number as

$$E(k) \propto k^{-\nu} \tag{15.21}$$

the function $F_{CD}(z)$ has a dependence $F_{CD}(z) \propto z^{1-\nu}$. By use of this form, equation (15.20) is rewritten as

$$E(k_\perp) \sim G_0^{5/2-\nu/2} k_\perp^{-\nu}. \tag{15.22}$$

The index ν is calculated if a nonlinear solution is obtained in the energy containing region. The solution of spectrum equation (13.19), which is obtained by the method of the dressed test mode, $E(k_\perp) \propto G_0 k_\perp^{-3}$, satisfies the constraint equation (15.22). The method of the dressed test mode preserves a scale-invariant property.

15.2.2 Collisional Turbulence

In order to illustrate the distinction between subcritical turbulence and supercritical turbulence, the scale invariance method is applied to collisional instability. Supercritical excitation is predicted for the collisional one. Equation (15.7) is replaced by

$$\frac{\mathrm{d}}{\mathrm{d}t}\begin{pmatrix} \nabla_\perp^2 \phi \\ 0 \\ p \end{pmatrix} + \begin{pmatrix} 0 & -\mathrm{i}k_y s x & \mathrm{i}k_y G_0 \\ \mathrm{i}k_y s x & \eta_\parallel & 0 \\ \mathrm{i}k_y & 0 & 0 \end{pmatrix}\begin{pmatrix} \phi \\ j \\ p \end{pmatrix} = 0 \tag{15.23}$$

where the difference appears in Ohm's law. In equation (15.7), parallel current is impeded by electron inertia which is enhanced by turbulent interactions. In contrast, collisional resistivity $\eta_\parallel$ is kept constant in the collisional model, equation (15.23). This set of equations is used to study low temperature and quiescent plasmas.

Let us consider a transformation

$$x \to \lambda_1 x \text{ and } t \to \lambda_2 t. \tag{15.24}$$

Field quantities are transformed as

$$\begin{pmatrix} \phi \\ j \\ p \end{pmatrix} \to \begin{pmatrix} \lambda_1^2 \lambda_2^{-1} \phi \\ [\eta_\parallel^{-1}] \lambda_1^2 \lambda_2^{-1} j \\ \lambda_1 p \end{pmatrix} \tag{15.25}$$

where $[\eta_\parallel^{-1}]$ is the dependence of the conductivity. Note that collisional resistivity $\eta_\parallel$ is determined by local temperature, and is not influenced by the gradient. From the equation of motion, a transformation

$$G_0 \to \lambda_2^{-2} G_0 \tag{15.26}$$

and a relation

$$[\eta_\parallel^{-1}] \lambda_1^2 \lambda_2 = 1 \tag{15.27}$$

are derived. Equation (15.26) is essentially equivalent to equation (15.11).

By use of this scale transformation, the dependence of transport coefficient is derived. Eliminating λ_1 and λ_2 from equations (15.24), (15.26) and (15.27), one has relations

$$\text{correlation length } \propto [\eta_\parallel]^{1/2} G_0^{1/4} \tag{15.28}$$

$$\text{correlation time } \propto G_0^{-1/2}. \tag{15.29}$$

The turbulent transport coefficient, D, has a dimension of $D \sim \text{length}^2 \text{time}^{-1}$, and has the dependence

$$D \propto [\eta_\parallel] G_0. \tag{15.30}$$

This dependence is identical to an analytical estimate equation (11.10) which is deduced by the method of the dressed test mode for resistive plasmas.

Combining this result with one for current diffusive turbulence, a condition where a change from resistive turbulence to current diffusive turbulence takes place is derived. Comparing equations (15.14) and (15.28), one has the condition

$$G_0 > \eta_\parallel^2 \tag{15.31}$$

for current diffusive turbulence to exist. In the weak gradient limit, $G_0 < \eta_\parallel^2$, resistive turbulence dominates. This result agrees with equation (11.16). (Notice the difference of normalization in chapters 11 and 15. Length in the basic equations is normalized to c/ω_p in chapters 14 and 15, while it is normalized to a in chapters 7–13.)

15.2.3 General Feature of Decorrelation Rate

Before closing this chapter, a general feature is discussed. First, the turbulence level is deduced from a Lagrange derivative, equation (15.8), $\mathrm{d}/\mathrm{d}t = \partial/\partial t + [\phi, \ \ ]$. The second term has the dimension of [potential fluctuation][length]$^{-2}$. (Note the normalization of equation (15.7).) If this term determines the rate of time derivative, [time]$^{-1}$, the potential fluctuation satisfies a scaling property [potential fluctuation] $\sim$ [length]^2[time]$^{-1}$. This has the same scaling property as diffusivity has, i.e.,

$$\phi \sim D. \tag{15.32}$$

This is characteristic of a strong turbulence. This relation corresponds to a heuristic argument equation (4.11) as well as to an example of nonlinear analysis equation (10.38). Second, the dependence of decorrelation rate τ_{corr}^{-1} on gradient as

$$\tau_{corr}^{-1} \propto \sqrt{G_0} \tag{15.33}$$

holds in general. These relations (15.32) and (15.33) are satisfied regardless of dissipation mechanisms that impede electron free motion along the magnetic field line. Dissipation mechanisms are influential in determining the correlation length of fluctuations as is shown in equations (15.14) and (15.28).

REFERENCES

[15.1] Kolmogorov A N 1941 *Dokl. Akad. Nauk SSSR* **30** 299

[15.2] Kadomtsev B B 1975 *Fiz. Plazmy* **1** 531 (Engl. transl. *Sov. J. Plasma Phys.* **1** 295)

[15.3] Connor J W and Taylor J B 1977 *Nucl. Fusion* **17** 1047
Connor J W 1988 *Plasma Phys. Control. Fusion* **30** 619

[15.4] Yagi M, Wakatani M and Shaing K C 1988 *J. Phys. Soc. Japan* **57** 117

[15.5] Connor J W 1993 *Plasma Phys. Control. Fusion* **35** 757

[15.6] See, e.g., Tatsumi T 1982 *Fluid Dynamics* (Tokyo: Baihukan) (in Japanese)
Zakharov V E, L'vov V S and Falkovich G 1992 *Kolmogorov Spectra of Turbulence I* (Berlin: Springer)

Chapter 16

Transition Phenomena and Concept of Electric Field Bifurcation

窮則変　変則通　(易経)

In preceding chapters, it is shown that turbulent transport is driven by inhomogeneities in plasmas. Time scales are separated between pressure gradient evolution and turbulence level development, and stationary turbulence is determined. By this time scale separation, the pressure gradient is considered as an order parameter for turbulence, and the nonequilibrium nature of fluctuations could be expressed in a power law, e.g., $\chi \propto |\nabla p_0|^{3/2}$.

In confined plasmas, there have been observed abrupt changes in transport properties as is briefly explained in chapter 2. Introduction of a new time scale is necessary to thoroughly understand these phenomena. In this chapter, a phenomenological description of transition phenomena is explained. The physics picture to treat a fast transition is then discussed.

16.1 Transition Phenomena in Global Structure

A sudden change of plasma flux across a magnetic surface was first found in an experiment on the ASDEX tokamak [16.1]. Figure 16.1 (left) illustrates the time evolution of plasma parameters. When a plasma is subject to a strong external heating, plasma energy increases and saturates. In this figure, it is shown that at a certain time, t_t, the plasma energy and density suddenly start to increase again, even though supplies of energy and particles are unchanged. This sudden increase of energy is associated with an abrupt reduction of loss across the surface, whose evolution is schematically shown in figure 16.1 (right). This type of transition in confined plasmas has been reproduced and confirmed in other experimental devices, and is considered to be a generic feature belonging to plasmas. This is not a peculiar event in a particular circumstance, but is a commonly observable one if a certain condition is satisfied.

Owing to this finding, it has become clear that confined plasmas have multiple states for a given external supply. This new state of confined plasmas

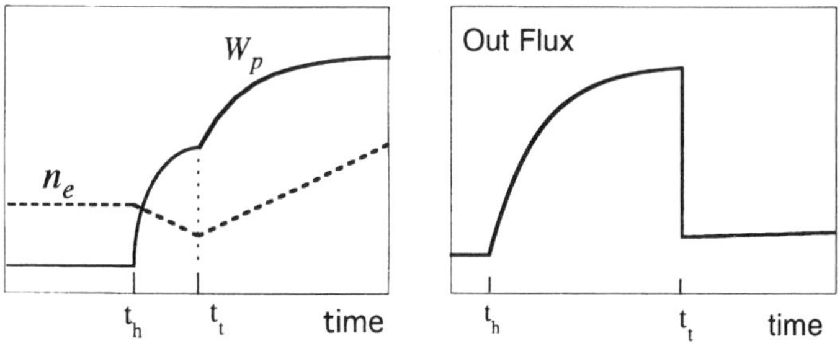

Figure 16.1. Experimental observation of transition of a confined plasma is illustrated. Total plasma energy and density suddenly start to increase (at $t = t_t$) under a constant external supply that starts at $t = t_h$ (left). At the onset of the transition, the loss of plasma energy across the surface decreases simultaneously (right). In the interval $t_h < t < t_t$, the plasma is in the *L-mode*. It is in the *H-mode* after t_t.

has been called the 'H-mode' ('H' stands for 'high', or '*hoch*' in German). As a consequence, a previously observed state was named the 'L-mode' ('L' stands for 'low'). An appearance of a new state, the H-mode, has provided an attractive and challenging problem for us to understand. Since plasma parameters are improved, its understanding promotes a realization of controlled fusion. It is essential that this phenomenon illuminates key elements of plasmas transport.

A transition in transport gives rise to a change of the plasma profile. An abrupt change of loss flux implies that a change at the H-mode transition occurs very close to the edge. Consider the case where the change happens at radius r; then the response in outflux from the surface, a, needs a delay time,

$$\tau_d \sim \frac{(a-r)^2}{\chi} \tag{16.1}$$

where χ is the value before transition. The reduction of loss is so abrupt in experiments that the distance $a - r$ should be very small. Figure 16.2 shows the radial profile of plasma pressure near the edge. A steep pressure gradient evolves after the transition. This establishment of a steep pressure gradient is often called a 'transport barrier' formation.

The two distinct profiles in figure 16.2 are realized for a common heat flux from centre to edge. That is, a common heat flux is realized for different values of gradient. This means that heat flux must be a non-monotonic function of gradient. Figure 16.3(a) shows experimental observations on the plasma temperature near the surface as a function of heating power. (Total heating power is not exactly in proportion to, but is strongly correlated with, heat flux across the surface.) For a given flux, two states of plasma are realized. In order to express a gradient–flux relation, figure 16.3 (top) is transformed as figure 16.3

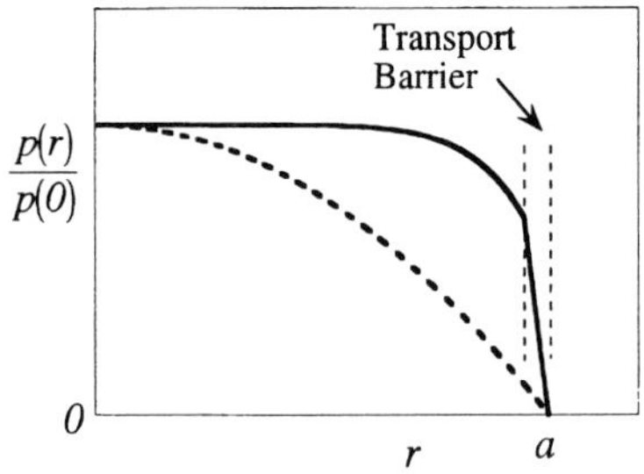

Figure 16.2. Two distinct pressure profiles before (dashed line) and after the transition (solid line).

(bottom) in which the axes are exchanged. Plasma temperature near the edge can be interpreted as the gradient at the edge.

This observation suggests that heat flux is a multivalued function of gradient. In general, two types of non-monotonic gradient–flux relation are illustrated in figure 16.4. The type of figure 16.4 (left) predicts that a heat flux could take a multiple value for fixed gradient. (Figure 'Z', or 'S', curve.) A *hysteresis* appears. In the case of figure 16.4 (right), the heat flux is not a monotonic function of gradient but is a single valued function. (Figure 'N' curve.) Both types of flux–gradient relation allow the existence of transition and a multiple structure of plasmas for a fixed external supply. The difference is seen in the time evolution of the plasma parameter at transition. In the case of figure 16.4 (left), a hard transition may occur. In contrast, a soft transition is predicted for the case of figure 16.4 (right).

16.2 Concept of Electric Field Bifurcation

The discovery of the H-mode transition has stimulated us to renew the picture of turbulent transport. If one writes a heat flux driven by plasma inhomogeneities as

$$q_r = -n\chi(\nabla p, \ldots)\nabla T \tag{16.2}$$

it could be a nonlinear function of gradients in plasmas, and may be a complex function of gradients. However, so long as the transport flux is expressed in terms of plasma parameters, one set of plasma parameters predicts one set of fluxes. (Figure 16.4 (right).) A multifold relation like figure 16.4 (left) could not be drawn. In other words, in order to construct a transport property which allows a hard type transition, the existence of a *hidden variable* is inevitable. Fluxes with different magnitudes appear for one set of plasma gradients, if there are additional variables that could control the transport.

As a candidate for 'hidden variable', the role of *field* is investigated [16.2].

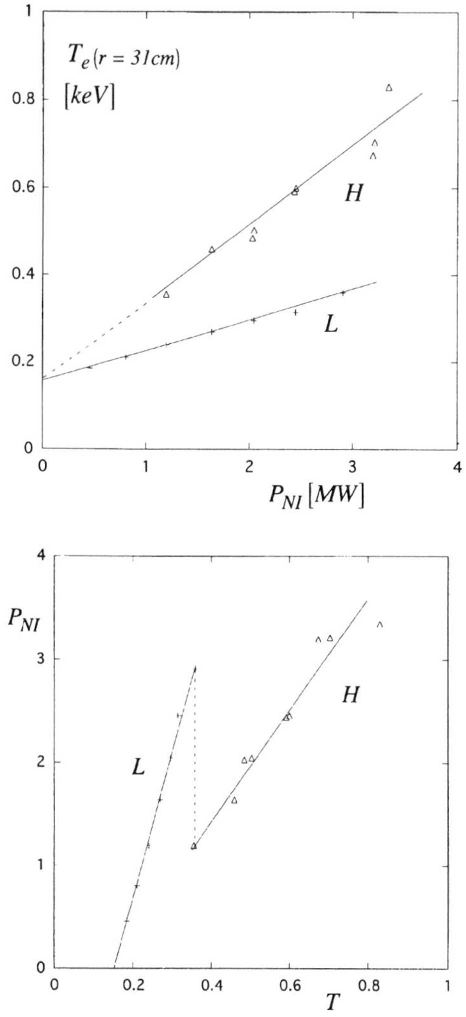

Figure 16.3. Experimental result of the two states of plasma as a function of heating power (top). The result of the ASDEX experiments [2.21]. The horizontal and vertical axes are exchanged on the bottom. Heat flux could be a multifold function of local plasma gradient.

A global structure in fields (such as a radial electric field) could be influential in plasma transport. A theoretical picture of transition is developed by the introduction of the concept of electric field bifurcation. Consider the situation

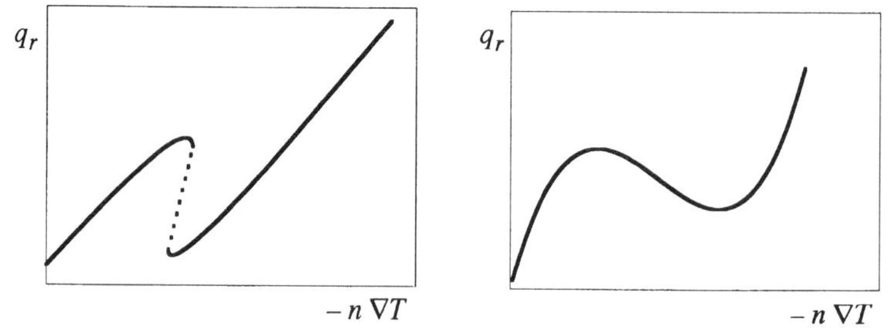

Figure 16.4. Examples of gradient–flux relation that allow multiple states.

where flux is dependent on the global structure of electric field E_r like

$$q_r = -n\chi(\nabla p, \ldots; E_r, \nabla E_r, \ldots)\nabla T. \tag{16.3}$$

The radial electric field is determined through another law, which is independent of equation (16.3). In a stationary state, a charge conservation law requires that particle fluxes of electrons and ions balance, i.e.,

$$\Gamma_e[\nabla n, \nabla p, \ldots; E_r, \nabla E_r, \ldots] = \sum_i Z_i \Gamma_i[\nabla n, \nabla p, \ldots; E_r, \nabla E_r, \ldots] \tag{16.4}$$

where Z_i is the charge number of the ion species. Equation (16.4) constrains the radial electric field profile. As is explained in the following chapters, a radial electric field structure has much more freedom than a pressure profile has. For instance, pressure is higher inside than outside in confined plasmas. Electric potential could be either higher or lower inside or outside. It is dramatic that multiple solutions of electric field structure are allowed to exist for given plasma parameters. Owing to the freedom of field, energy flux, equation (16.3), could be a multivalued function of gradient. A catastrophe structure [16.3] is formulated in the flux–gradient relation. Important roles of radial electric field in confinement are widely recognized [16.4]. In the following chapters, roles of global electric field in transport properties, the generation of field, bifurcation characteristics and self-organized dynamics are explained. This aspect of nonequilibrium transport is also an essential feature in plasmas.

REFERENCES

[16.1] Wagner F, Becker G, Behringer K, Campbell D *et al* 1982 *Phys. Rev. Lett.* **49** 1408

[16.2] Itoh S-I and Itoh K 1988 *Phys. Rev. Lett.* **60** 2276

[16.3] Thom R 1975 *Structural Stability and Morphogenesis* transl. D H Fowler (Reading: Benjamin)

[16.4] Itoh K and Itoh S-I 1996 *Plasma Phys. Control. Fusion* **38** 1

Chapter 17

Electric Field Effect on Collisional Transport

In this chapter, the influence of radial electric field E_r and its gradient on collisional transport is described. The effect of radial electric field on single particle orbits is discussed first. Whether an axial symmetry exists or not makes a difference in the effect of electric field. In a symmetric system, the influence of E_r appears as a second order effect. In contrast, it works as a first order effect if symmetry is broken. When the inhomogeneity of E_r is strong, a squeezing of the trapped particle orbit occurs. These facts are basis of effects of E_r on collisional transport.

17.1 Effect of Radial Electric Field on the Particle Orbit

Trapped particles exist due to inhomogeneity of magnetic field. Toroidicity gives rise to trapped particles (banana particles) which are localized near the side of low magnetic field (outside) of the torus (figure 3.6). Toroidal coils and helical coils generate another type of trapped particle: the ripple trapped particle. Discrete coils generate a toroidal magnetic field (figure 17.1(a)), and this field is not constant along the toroidal direction. Magnetic field is weaker between coils, and some particles are trapped within this local minimum of magnetic field. (A variation of magnetic field owing to discrete coils is noticeable in the outer side of a torus and is weak in the inner side of the torus.) This type of particle does not freely move in the toroidal direction, but its motion is restricted between coils. In a system with helical coils, magnetic field is weak between helical coils. Motion of some particles is localized between helical coils. Both of these particles are called ripple-trapped particles. Two typical examples of orbits of ripple-trapped particles are schematically illustrated in figure 17.1.

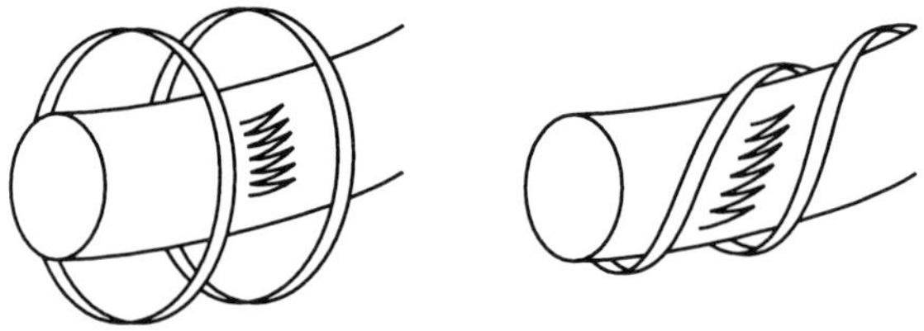

Figure 17.1. Schematic drawing of orbits of ripple-trapped particles. (Guiding centre motion is shown.) Toroidal-ripple-trapped particles (left) and helical-ripple-trapped particles (right).

17.1.1 Drift and Resonance

The trajectory of a trapped particle is strongly influenced by radial electric field. Its influence on ripple-trapped particles is most prominent. A banana orbit could also be modified, if E_r becomes strong enough. (In principle, cyclotron motion itself could change its character in an extreme case, but such a situation is not discussed here.)

The motion of the ripple-trapped particle is characterized by three representative elements, i.e., (i) gyro-motion, (ii) bounce motion of the guiding centre (i.e., banana orbit) and the (iii) drift of the centre of a banana. These elements constitute a deviation of the particle orbit from the magnetic surface. In normal situations a drift of the banana provides the largest contribution for deviation. The influence of radial electric field is noticeable if a potential change is a considerable fraction of a particle kinetic energy during its bounce motion (say banana or drift by ripples). Because a drift motion provides the largest excursion step in radius, this is most vulnerable to radial electric field. If the influence of electric field is so large that the drift orbit is strongly modified, we use the term '*resonance*'.

Toroidal-ripple-trapped Particles. A motion of ripple-trapped particles in toroidal ripples is limited by two adjacent coils (shown in figure 17.1 (left) for tokamaks). The motion of the banana centre is expressed in terms of a drift motion, $\boldsymbol{v}_G$, as

$$\boldsymbol{v}_G = v_D \hat{Z} + \frac{E_r}{B}\hat{\theta} \qquad (17.1)$$

where v_D is drift due to the toroidal field inhomogeneity,

$$v_D \approx W/eBR_M$$

W is particle energy, B is the main magnetic field and R_M is the scale-length of the toroidal field inhomogeneity. (For an analytic insight, a simple relation $R_M \approx R$ is employed unless specified.)

Equation (17.1) shows that drift motion of a ripple-trapped particle is a superposition of the vertical drift and poloidal drift. When the poloidal drift dominates over the vertical drift, a ripple-trapped particle becomes a banana particle. A trajectory in the poloidal cross-section (R, Z) of a particle which starts from an initial condition (R_1, Z_1, κ_1) (κ_1 being the pitch angle) is given by use of integrals of motion W and μ (magnetic moment) as

$$\Psi_0(R, Z) + \frac{R}{R_0} - 1 = 1 - \frac{W}{\mu B_0} \tag{17.2-1}$$

$$\Psi_0(R, Z) = \{\sqrt{\varepsilon_r(R, Z)} - \kappa_1^2\sqrt{\varepsilon_r(R_1, Z_1)}\}^2 - \kappa_1^4\varepsilon_r(R_1, Z_1) - \frac{e\phi}{\mu B_0}. \tag{17.2-2}$$

In this equation,

$$\kappa^2 = \frac{W - \mu B_0(R_0/R - \varepsilon_r) - e\phi}{2\varepsilon_r \mu B_0}$$

denotes a pitch angle of velocity with respect to the magnetic field line (not a magnetic gradient), B_0 the main magnetic field, ε_r the amplitude of the toroidal ripple, ϕ the static potential, e the charge (notation follows [17.1]).

This formula provides an estimate of the shift in a major radius direction as a ripple-trapped particle moves in the vertical direction. Also shown is a dispersion of this shift owing to a pitch-angle dependence. The value κ lies in between 0 and 1.

The magnitude of electric field for which orbits of ripple-trapped particles are considerably influenced is evaluated from equation (17.2). Consider a case where a radial electric field E_r exists in a radial region of thickness Δ (figure 17.2). If a poloidal rotation due to $\boldsymbol{E} \times \boldsymbol{B}$ motion is large enough,

$$|\Delta e E_r R / a W_\perp| \approx 1 \tag{17.3}$$

the particle moves in the poloidal direction by a distance of order a. The drift motion is strongly modified. (The toroidal ripple is usually weak inside the torus as is shown in [2.11]. The relation (17.3) gives an estimate of the condition for detrapping of ripple-trapped particles.)

A fraction of ripple-trapped particles is of the order of $\sqrt{2\varepsilon_r}$. Although a number of ripple-trapped particles is very small, their influence on transport could be important. This is because the drift of ripple-trapped particles is large for energetic particles and an effect may be prominently seen for cases with intense heating or with fusion products. In such cases, an interaction of particles with a wall would be localized in a limited region on the wall [17.2].

Helical-ripple-trapped particles. In helical systems, a similar description is available for a motion of helical-ripple trapped particles, by a replacement of ε_r by a helical ripple, ε_h. Leaving detailed arguments, which correspond to equation (17.2), to the literature [17.3, 17.4], a qualitative explanation is given here. The main difference is that ε_h could be as large as or greater than a

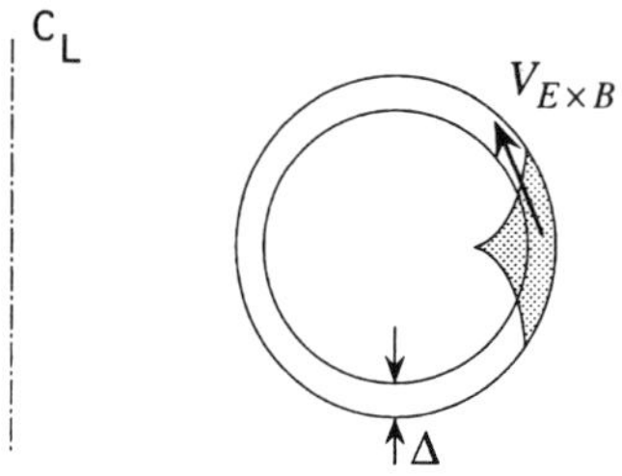

Figure 17.2. Trajectory of toroidal-ripple-trapped particles in the poloidal cross-section. With the effect of $\boldsymbol{E} \times \boldsymbol{B}$ drift (thick arrow). The hatched area shows where the toroidal ripple is strong.

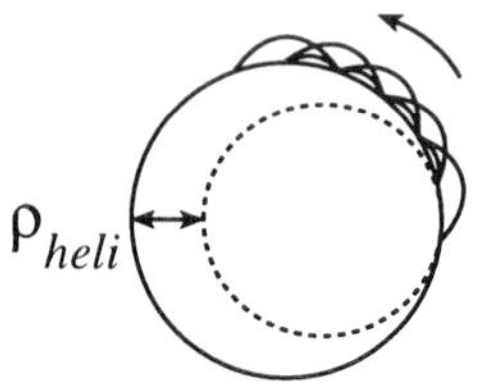

Figure 17.3. Projection of an orbit of a helical-ripple-trapped particle on the poloidal cross-section.

toroidicity $\varepsilon_t = r/R$ in helical devices. (Toroidicity ε_t is abbreviated by ε for simplicity). A considerable portion of particles is within this category of helical trapped particles, and could be essential in transport. A gradient of magnetic field, which helical ripple trapped particles feel, is around ε_h/r, and a curvature is directed in the radial direction. The drift velocity is given as

$$v_G \simeq v_D \hat{Z} + \left(\frac{W\varepsilon_h}{eBr} + \frac{E_r}{B} \right) \hat{\theta} \tag{17.4}$$

showing that a banana centre rotates in the poloidal direction with an angular frequency

$$\omega_{rot} \simeq r^{-1} \left(\frac{W\varepsilon_h}{eBr} + \frac{E_r}{B} \right) \tag{17.5}$$

as is schematically shown in figure 17.3. The deviation of a banana centre from an average magnetic surface, ρ_{heli}, is given by v_D/ω_{rot}, or

$$\rho_{heli} \simeq \frac{r^2}{\varepsilon_h R} \frac{1}{(1 + erE_r/\varepsilon_h W)}. \tag{17.6}$$

Two characteristics are seen in this simple result. First, the deviation ρ_{heli} is a function of geometrical factors (R, r, ε_h) and is independent of the strength of the magnetic field. As a result of this, ρ_{heli} does not scale with a gyro-radius or banana width, but can become of the order of a system size. Second, the ratio $erE_r/W\varepsilon_h$ plays an essential role in characterizing an orbit. If it satisfies a condition

$$\frac{erE_r}{\varepsilon_h W} \sim -1 \tag{17.7}$$

poloidal rotations due to the helical ripple and electric field, equation (17.5), cancel each other, and the shift ρ_{heli} becomes very large. It is called a *resonance* of a ripple-trapped particle. The resonance causes a variety of dynamic phenomena.

The resonance condition, equation (17.7), depends on the sign of the charge. If a charge is positive (i.e., ions), a resonance occurs when E_r is negative. In the opposite case ($E_r > 0$), it occurs for an electron orbit.

This sign dependence of the resonance conditions provides one mechanism for self-sustenance of the electric field. The former situation of ion resonance appears for a negative electric field. The negative radial electric field means that there is an excess of electrons, and that a mechanism of selective loss of ions exists. The loss of ions produces a negative radial electric field and gives rise to a resonance in ion orbits. The loss and electric field sustain each other. (The same argument is applied to electrons if $E_r > 0$.) Generation and sustenance of radial electric field is discussed later.

Banana orbit. A banana orbit itself is influenced by a radial electric field. Due to $\boldsymbol{E} \times \boldsymbol{B}$ motion, trapped particles (those trapped by toroidicity, not by ripples) drift in the toroidal direction with a velocity of E_r/B_p (B_p being the poloidal magnetic field, which is the same as B_θ). This toroidal velocity is associated with an angular frequency; a centrifugal force appears and affects a bounce motion. The parallel velocity and deviation of the guiding centre of a particle from the mean magnetic surface are given as [17.5]

$$u_\parallel^2 = u_{\parallel 0}^2 - \varepsilon(1 - \cos\theta)(u_{\perp 0}^2 + 2u_E^2) + u_g \delta_{dev}^2 \tag{17.8-1}$$

$$\delta_{dev} = (u_{\parallel 0} - u_\parallel)(1 - \varepsilon\Lambda_s \cos\theta) + 2\varepsilon(1 - \cos\theta)u_E \tag{17.8-2}$$

where $u = v/v_{th}$,

$$u_E = E_r/B_p v_{th} \tag{17.9-1}$$

and

$$u_g = \rho_p(\mathrm{d}E_r/\mathrm{d}r)/B_p v_{th} \tag{17.9-2}$$

are normalized velocities, δ_{dev} is a deviation normalized to poloidal gyro-radius, and a suffix 0 denotes poloidal location $\theta = 0$. In equation (17.8), Λ_s is the Shafranov Λ_s [2.11] and the poloidal magnetic field is given as $B_p = [1 + (r/R)\Lambda_s \cos\theta]B_{p0}$. It is related to the shift of magnetic surfaces in toroidal plasmas (section 3.2).

Note that u_g is defined including the sign of the charge. Influences of the radial electric field appears through normalized parameters u_E and u_g. Both normalized parameters have a dependence on particle mass; they are large for ions and small for electrons. The effect of radial electric field on banana orbits appears selectively in orbits of ions. This is in contrast to the influence on ripple-trapped particles, which effectively appear in orbits of both electrons and ions.

The influence of constant E_r is first explained. From relation (17.8), one sees that a boundary between trapped and transit particles is modified to

$$u_{\parallel 0}^2 = \varepsilon(1 - \cos\theta_b)(u_{\perp 0}^2 + u_E^2) \tag{17.10}$$

for a finite value of u_E, where θ_b denotes the poloidal location of the banana tip. The boundary is now given by the hyperbola shown in figure 17.4. The bounce frequency is increased to

$$\omega_b = \sqrt{\varepsilon/2}(v_{th}/qR)\sqrt{u_{\perp 0}^2 + u_E^2}. \tag{17.11}$$

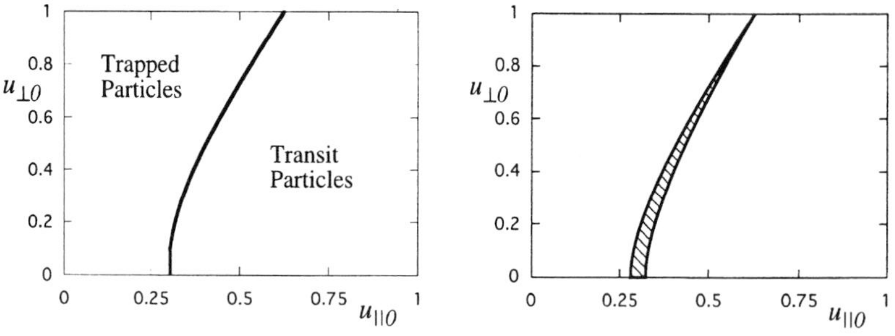

Figure 17.4. Boundary between banana particles and transit particles in velocity space of tokamaks. (Case of $\varepsilon(1 - \cos\theta_b) = 0.3$ and $u_E^2 = 0.3$.) Shaded area represents a possible loss cone region (right).

A note is made on coordinate transformation. One may imagine that a constant radial electric field does not affect a particle orbit; this idea is based on the fact that an electric field disappears in a frame which is moving with a constant velocity of $(\boldsymbol{E} \times \boldsymbol{B})B^{-2}$. However, in toroidal plasmas, this frame is rotating in the toroidal direction with a constant angular frequency. This is not a Galirei transformation. There appears a centrifugal force effect which modifies the banana orbit as (17.10) and (17.11). Since the $\boldsymbol{E} \times \boldsymbol{B}$ drift velocity is common to electrons and ions, the effect of centrifugal force is much weaker on electrons than on ions. Thus the magnitudes of parameters u_E and u_g are much smaller for electrons than for ions.

17.1.2 Orbit Squeezing and Drift Reversal by an Inhomogeneous Radial Electric Field

In the presence of a strong radial electric field such that a banana orbit is affected, gradient of the radial electric field plays an important role as well [17.5, 17.6]. A radial excursion δ_{dev} can be compressed or enlarged. The effective poloidal gyro-radius is given as [17.6]

$$\hat{\rho}_p = \rho_p / S \tag{17.12}$$

where ρ_p is a poloidal gyro-radius, and S is called a squeezing parameter, being defined as

$$S = 1 - u_g. \tag{17.13}$$

If absolute value of parameter S is greater than unity, i.e., $\mathrm{d}E_r/\mathrm{d}r$ is negative (or $\mathrm{d}E_r/\mathrm{d}r$ is positive and large, i.e., $u_g > 2$), the banana width is compressed (squeezed). Notice that there is also a *resonance* condition,

$$u_g \approx 1 \tag{17.14}$$

under which banana orbits expand greatly.

The precession velocity of a trapped particle in the toroidal direction is also affected. It has been given as [17.7]

$$v_{D.t} = -(1 + 2u_g)\frac{W_{\perp 0}}{ReB_p}. \tag{17.15}$$

Toroidal drift of trapped particles due to a bad curvature $v_{D.t}$ can be reduced. If the condition $1 + 2u_g < 0$, or

$$u_g < -\tfrac{1}{2} \tag{17.16}$$

is satisfied, trapped particles drift in the opposite direction (like in a good curvature). This situation is called *drift reversal*.

Note that effects of $\boldsymbol{E} \times \boldsymbol{B}$ drift, equations (17.10) and (17.11), are charge independent, while effects due to E_r' (equations (17.12) and (7.15)) have a charge dependence. Both affect plasma stability and transport. The gradient of E_r can influence a stability in the first order of $|\nabla E_r|$, while E_r itself affects it in the second order, E_r^2.

17.1.3 Loss Cone

A loss cone condition in a velocity distribution of particles is influenced by the radial electric field and its inhomogeneity.

Banana orbit. As a trapped-transit boundary is deformed by a radial electric field in a velocity space, a loss cone boundary is also modified as is shown in

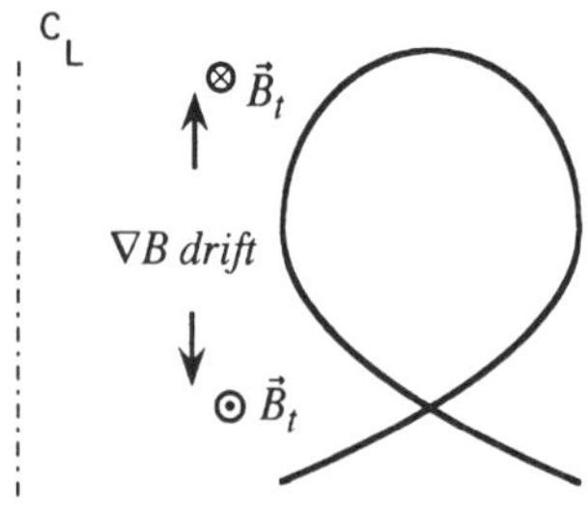

Figure 17.5. Poloidal cross-section of magnetic surface with a separatrix and a direction of ∇B drift of ions. Depending on the direction of toroidal field, the ∇B drift changes its direction to (or away from) an X-point of the separatrix.

figure 17.4 (right). This is the case where a poloidal limiter is located at $\theta = \theta_b$. A number of particles in the loss cone in a velocity space ($u_{\perp 0} = 0$) is reduced by a factor like

$$\exp(-u_E^2) \tag{17.17}$$

if the distribution is given by the Maxwellian.

In the presence of a separatrix of a magnetic surface (figure 17.5), an up–down symmetry does not hold. The influence of radial electric field on banana orbits has been studied. The boundary of the loss cone in velocity space is shifted to a higher energy region as the radial electric field becomes strong enough. [17.8] discusses an asymmetry with respect to a sign of E_r: the loss region first slightly increases as E_r becomes negative. The direction of ∇B drift depends on the direction of the toroidal magnetic field. Depending on whether $B_\zeta > 0$ or $B_\zeta < 0$, the vertical drift of ions due to ∇B is either in the direction of the X-point of the magnetic separatrix surface or away from it.

Ripple-trapped particles. Ripple-trapped particles easily reach a plasma surface due to their large deviation. For clarity of argument, we define a loss cone boundary by a trajectory touching a plasma surface. (Another definition is possible.) The boundary in radius–energy space is given in the literature [17.3, 17.4, 17.9]. A simplified expression for a case of a radially monotonic potential $\phi(r)$ is given as

$$W_+ < W < W_- \tag{17.18-1}$$

with

$$W_\pm = \frac{e\{\phi(a) - \phi(r)\}}{\varepsilon_h(a)(1 - r^2/a^2) + (a/R)(\pm 1 - r/a)}. \tag{17.18-2}$$

This expression is derived for particles which are deeply trapped in helical ripples. The region of the loss cone in phase space (i.e., in terms of particle energy and radial location) is shown in figure 17.6. When the initial condition

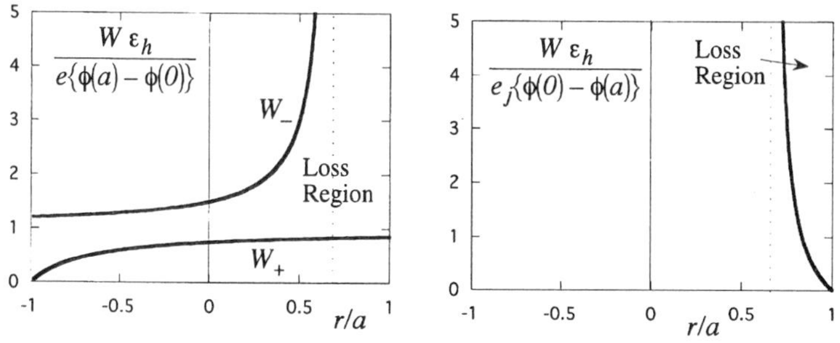

Figure 17.6. Loss cone region for deeply trapped particles in the presence of a helical ripple. Cases of $e_j\{\phi(a)-\phi(0)\} > 0$ and $e_j\{\phi(a)-\phi(0)\} < 0$ are shown. The potential profile is chosen as parabolic and $\varepsilon = \varepsilon_h/3$.

of energy and location of ripple-trapped particle is in the 'Loss region' of figure 17.6, the trajectory of the particle extends to the plasma surface. A case of $e_j\{\phi(a)-\phi(0)\} > 0$ (i.e., negative radial electric field for ions, or positive radial electric field for electrons) is shown in figure 17.6 (left). An opposite case, $e_j\{\phi(a)-\phi(0)\} < 0$ (i.e., positive radial electric field for ions, or negative radial electric field for electrons) is shown in figure 17.6 (right).

For a potential difference of the order of

$$\phi(a) - \phi(0) \approx \varepsilon_h(a)W/e_j \qquad (17.19)$$

a finite loss cone region spreads over the entire plasma column. A wide loss cone appears for ions if the radial electric field is negative. For electrons, this situation holds for a positive electric field.

17.2 Collisional Transport

A modification of the orbit due to radial electric field affects collisional transport. Deviation of the orbit from the magnetic surface, δ_{dev}, is affected by the radial electric field. In a dimensional argument, a collisional (neoclassical) transport coefficient has the dependence

$$D \sim \nu_{eff}\delta_{dev}^2$$

where ν_{eff} is the effective Coulomb collision frequency for relevant particles. (See section 4.2.) Care is necessary in defining collision frequency. The microscopic origin of Coulomb collisions is thermal fluctuations in a range of electron plasma frequency [2.3]. Thermal fluctuations, defined in the absence of inhomogeneity, are considered not to be modified by the electric field of present interest. However, the velocity distribution of relevant particles changes, as is shown in figures 17.4 and 17.6. Therefore the effective collision frequency is altered and its modification may be expressed in terms of average plasma parameters.

An assumption on the ordering in neoclassical transport theory is made as

$$\delta_{dev}/L \ll 1 \tag{17.20}$$

where L is a gradient scale-length. The argument to lead to ambipolarity of collisional transport relies on this assumption [17.10]. An example is that a like-particle collision does not contribute to particle flux. This does not necessarily hold for circumstances of confinement modes with strong radial electric field. (A new name for a new ordering would be useful in clarifying the physics.)

17.2.1 Neoclassical Transport Coefficient

Plasma fluxes (particle, momentum, current and heat) have been expressed in terms of the gradients (density, velocity, current and temperature) and toroidal electric field. Since parallel resistivity is proportional to electron–ion collision frequency, toroidal electric field is in proportion to collision frequency. Therefore all fluxes are proportional to collision frequency within the framework of neoclassical theory. Details are summarized in the literature [17.11, 17.12] and are not reproduced, but some expressions are given here. The flux is composed of two components; one is from an axisymmetric part of the distribution function, and the other is due to ripple-trapped particles.

Based on collisionality, a division of plasma parameter regimes is often made. If collision is rare and a trapped particle (banana particle) can circumnavigate one bounce motion (banana orbit),

$$\nu_j/\varepsilon < \omega_{b.j} \simeq \sqrt{\varepsilon} v_{th.j}/qR$$

the plasma is said to be in a 'banana regime'. (This regime is defined for ions and electrons separately.) If the plasma is collisional, and particles are subject to collisions within one poloidal motion,

$$\nu_j > \omega_{t.j} \simeq v_{th.j}/qR$$

it is called a collisional regime or a 'Pfirsch–Schlüter regime'. In intermediate collisionality,

$$\varepsilon^{3/2} v_{th.j}/qR < \nu_j < v_{th.j}/qR$$

the plasma is said to be in a 'plateau regime'. In the presence of a helical ripple, an additional classification is made. Helical-ripple-trapped particles have one-bounce motion without collision, if the condition

$$\nu_j/\varepsilon_h < \omega_{h.j} \simeq \sqrt{\varepsilon_h} v_{th.j} m/qR$$

is satisfied (m: toroidal pitch number of helical winding). This case is called a 'ripple-trapped regime'.

Axisymmetric Plasma. The influence of E_r on the axisymmetric component was found to be important if a parameter X defined as

$$X \equiv \rho_p E_r / T_i \; (= u_E) \tag{17.21}$$

becomes of order unity. For instance, a particle flux is affected as [17.13]

$$\Gamma^{NC} = -nD_p \left[\frac{n'}{n} + \gamma_j \frac{T'}{T} - \frac{e}{T}(E_r - B_\theta V_\parallel) \right] \exp(-X^2). \tag{17.22}$$

In a plateau regime, $\varepsilon^{3/2} v_{th.j}/qR < \nu_j < v_{th.j}/qR$, relations $\gamma_j = 3/2$ and

$$D_p = \frac{\sqrt{\pi}}{2} \frac{\varepsilon q \rho_i}{r} \frac{T}{eB}$$

hold for ions. A reduction by a factor of $\exp(-X^2)$ appears. This comes from the shift of the transition region (between trapped and transit particles) to a higher energy region, as is shown in figure 17.4. Fewer particles exist at the transition region for Maxwellian plasmas.

A similar reduction is seen in a calculation of bulk viscosity. This reduction could be important to understand experiments [17.14, 17.15]. There are three types of viscosity. One is the bulk (or parallel) viscosity which is denoted by μ_b. If a plasma moves in the direction of magnetic field variation, the flow is resisted by a force

$$-m_i n_i \mu_b \nabla_\parallel^2 V_\parallel.$$

This force is of the order of $m_i n_i \nu_i \varepsilon^2 q^{-2} V_\parallel$ and is very large. In other words, a poloidal flow, which is subject to a change of magnetic field strength, could be damped on a time scale of order of ν_i^{-1} in the absence of a radial electric field. The frictional force is

$$\langle \boldsymbol{B}_\theta \cdot \boldsymbol{\nabla} \cdot \boldsymbol{\Pi} \rangle = \frac{\sqrt{\pi}\varepsilon^2}{4r} m_i n_i v_{th} B (C_p V_\theta + C_t V_{p0}) \tag{17.23}$$

where V_θ is a poloidal velocity, $V_{p0} = -\rho_i v_{th.i} T_i'/2T_i$ is a drive due to the temperature gradient. (The relation with the poloidal velocity and radial electric field is explained in chapter 19.) Explicit forms of numerical integrals $C_p(X)$ and $C_t(X)$ are given in [17.14]. This viscous force has a local maximum with respect to V_θ: like a dynamic friction coefficient, it can be reduced at higher velocity. A useful formula in terms of the plasma dispersion function is given in [17.15].

Other viscosities, i.e., a gyro-viscosity [17.16] and a shear viscosity [17.17], are also subject to the effect of radial electric field. However, it is noted in [17.17] that collisional shear viscosity is less influenced by E_r, since mainly transit particles contribute to this viscosity.

The gradient of the radial electric field could also be important. It was shown that a parallel viscosity can be expressed [17.18]

$$\langle \boldsymbol{B} \cdot \nabla \cdot \boldsymbol{\Pi} \rangle = \frac{1}{S^{3/2}} \langle \boldsymbol{B} \cdot \nabla \cdot \boldsymbol{\Pi} \rangle_0 \tag{17.24}$$

where $\boldsymbol{\Pi}$ is a stress tensor, S is a squeezing parameter, and suffix 0 denotes an expression in the absence of inhomogeneous electric field. (For an explicit formula, see [17.18].)

Ripple-trapped particles. The nonaxisymmetric part of transport (ripple diffusion) is prominently influenced by radial electric field. For helically trapped particles, a deviation of orbit from the magnetic surface, δ_{dev}, is represented by ρ_{heli}. As is shown in equation (17.6), the deviation is strongly influenced by radial electric field. In the limit of large electric field, $|eE_r| \gg \varepsilon_h T/r$, the deviation is inversely proportional to radial electric field. A particle flux associated with helical-ripple-trapped particles is given in an integrated form over energy as

$$\Gamma_a^{NC} \simeq -\varepsilon^2 \varepsilon_h^{1/2} n v_D^2 \int_0^\infty \mathrm{d}W \frac{W^{5/2} e^{-W} \nu(W)}{\nu^2 + 1.5\sqrt{\varepsilon/\varepsilon_h}\,\omega_{rot}^2} \left(\frac{n'}{n} - \frac{eE_r}{T} + \frac{(W - 3/2)T'}{T} \right). \tag{17.25}$$

A similar result is obtained for energy and momentum fluxes with different weighting functions of W in the integrands [17.12]. It is seen that when the resonance condition equation (17.7) holds, i.e.,

$$\omega_{rot} \simeq 0$$

fluxes become large. Neoclassical flux is contributed from such particles that nearly satisfy the resonance condition. If E_r is very large, then neoclassical flux is reduced in proportion to ω_{rot}^{-2}. The dependence of fluxes on radial electric field is illustrated in figure 17.7. A qualitative difference is seen depending on the magnitude of the gradient. Let us look at the case $|T'/T| < |n'/n|$ for simplicity of argument. If the nonuniformity of density is weak, $|n'/n| \ll \varepsilon_h/r$, then the term ω_{rot}^2 is not modified by a radial electric field that is in the range of $|eE_r/T| \sim |n'/n|$. The dependence of flux on electric field is approximately expressed by a straight line in this range (figure 17.7 (left)). Contrary to this, when an inhomogeneity is strong, $|n'/n| \gg \varepsilon_h/r$, the term ω_{rot}^2 is modified much by a radial electric field of the order of $|eE_r/T| \sim |n'/n|$. The nonlinear dependence of flux on radial electric field becomes noticeable (figure 17.7 (right)).

A non-monotonic dependence of Γ on E_r is understood from a simple case of $T' = 0$ in equation (17.25). It is written as

$$\Gamma_a^{NC} = nD_h \left(-\frac{n'}{n} + \frac{eE_r}{T} \right) \tag{17.26}$$

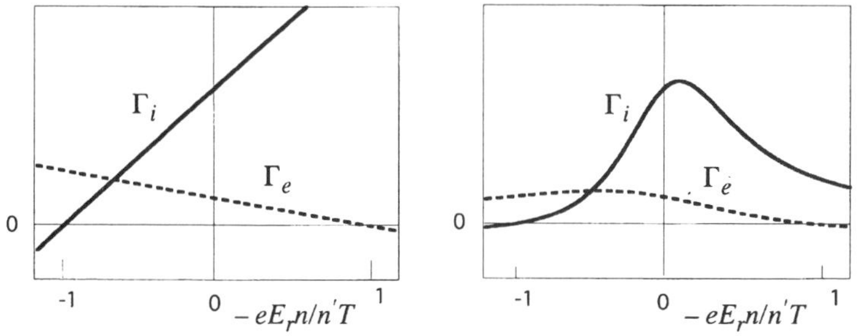

Figure 17.7. Neoclassical flux of helical-trapped particles. (Solid line: ions, dashed line: electrons.) In the weak gradient case, $|n'/n| \ll \varepsilon_h/r$ (left) and in the strong gradient case, $|n'/n| \gg \varepsilon_h/r$ (right).

$$D_h = \left[\varepsilon^2 \varepsilon_h^{1/2} v_D^2 \int_0^\infty \mathrm{d}W \frac{W^{5/2} e^{-W} \nu(W)}{\nu^2 + 1.5\sqrt{\varepsilon/\varepsilon_h}\omega_{rot}^2}\right]. \tag{17.27}$$

When the radial electric field is weak, $|eE_r| < \varepsilon_h T/r$, ω_{rot} is dominated by ∇B drift, as is given in equation (17.5). The integrand of equation (17.27) becomes approximately independent of E_r. Then D_h is independent of E_r, and the flux Γ_a^{NC} is a linear function of E_r. When the electric field is strong, $|eE_r| \gg \varepsilon_h T/r$ and $|E_r/rB| \gg (\varepsilon_h/\varepsilon)^{1/4}\nu$, one has $\omega_{rot} \sim E_r/rB$. Neglecting the ν^2 term in the denominator of equation (17.27), one has

$$D_h \sim \left[\varepsilon^{7/2} \varepsilon_h r^2 \int_0^\infty \mathrm{d}W\, W^{5/2} e^{-W} \nu(W)\right] \left(\frac{T}{erE_r}\right)^2. \tag{17.28}$$

The flux has an asymptotic dependence like

$$\Gamma_h^{NC} \propto \begin{array}{ll} \left(-\dfrac{n'}{n} + \dfrac{eE_r}{T}\right) & (|eE_r| < \varepsilon_h T/r) \\ \left(\dfrac{erE_r}{T}\right)^{-2} \left(-\dfrac{n'}{n} + \dfrac{eE_r}{T}\right) & (|eE_r| \gg \varepsilon_h T/r). \end{array} \tag{17.29}$$

Depending on the collision frequencies of electrons and ions, the magnitudes of electron and ion fluxes vary as is illustrated in figure 17.8.

The cross-field flux equation (17.29) of ions is not identical to that of electrons. (As is discussed in section 17.1, the effect appears in the first order of E_r when an axial symmetry of a toroidal plasma is violated.) Associated with this difference between ions and electrons, a radial current, $(e_i\Gamma_i - e\Gamma_e)$, appears in collisional transport. Its dependence on electric field is illustrated in figure 17.9.

The formula (17.25) has the advantage of giving us an insight into the physics mechanisms. However, deviation of a particle orbit could be comparable

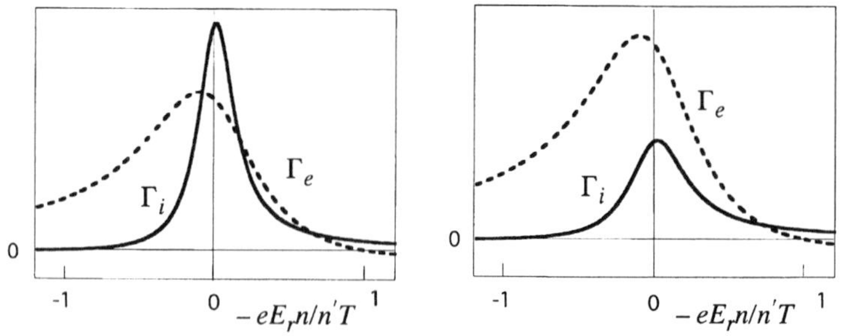

Figure 17.8. In the strong gradient case, $|n'n| > \varepsilon_h/r$, the relative magnitude of ion flux (solid line) and electron flux (dashed line) can change. A case of moderately high electron temperature, $T_e/T_i \sim (m_i/m_e)^{1/7}$ (left), and a case of high electron temperature, $T_e/T_i \gg (m_i/m_e)^{1/7}$ (right), are shown.

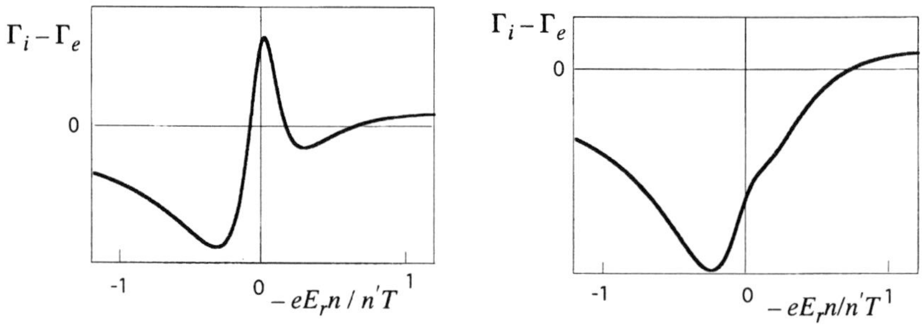

Figure 17.9. Radial current associated with the collisional transport of ripple-trapped particles. Cases of figure 17.8.

to a system size for ripple-trapped particles, and the analytic formula may not always work. The direct calculation of collisional flux based on an accurate orbit is sometimes used [17.19].

17.2.2 Loss Cone Loss (Direct Orbit Loss)

The boundary of the loss cone is important for an estimate of loss cone loss. For the case of banana particles, the number density of particles in the region of the loss cone boundary is reduced by a radial electric field, equation (17.17). The loss cone loss is estimated as

$$\Gamma^{lc} \propto n_i \nu_i \rho_p \exp(-X^2) \tag{17.30}$$

in the low collisionality limit [16.2]. Two aspects should be noted: (i) like-particle collisions (ion–ion in this case) cause a radial flux, and (ii) the flux is not necessarily proportional to the gradient. A more general formula was derived [17.14] as

$$\Gamma^{lc} \approx n_i \nu_i \rho_p \frac{\nu_*}{(\nu_* + X^4)^{1/2}} \exp[-(\nu_* + X^4)^{1/2}] \tag{17.31}$$

where $\nu_* = \nu_i/\omega_b$ is the collision frequency normalized to bounce frequency.

17.2.3 Bootstrap Current

Bootstrap current [2.4, 4.7] has attracted special attentions, because it shows a typical example of interference between various thermo-dynamical forces, and because it could be important in achieving fusion in toroidal plasmas (such as for a steady state sustenance of tokamak plasmas, or a possible cause of current-driven instability in helical systems).

The effect of E_r on the axisymmetric part of the bootstrap current (the usual one for tokamaks) appears of the order X^2. This is because ions and electrons move together of the linear order of X. Bootstrap current, and its counterpart, Ware pinch [4.6], work primarily on trapped particles, and they are of the order $\sqrt{\varepsilon}$. As is seen in figure 17.4, the number of trapped particles increases in the presence of a radial electric field. Its influences on bootstrap current and Ware pinch can appear if X becomes of order unity, through an increment of the trapped particle's population.

The effect of inhomogeneous radial electric field is subtle as well. As is shown in equation (17.24), magnitudes of all elements in neoclassical viscosity tensor for ions are reduced (or enhanced) by a factor $S^{-3/2}$. This does not necessarily mean an enhancement (or reduction) of plasma rotation and bootstrap current. Collisional friction forces are also reduced by the same order: owing to this ordering, the ratios between them, which determine the current and flow in the steady state, are not explicitly expressed by the squeezing parameter S. A change of effective collision frequency occurs, through which the ratio is affected [17.18].

Flux associated with ripple-trapped particles has a correction which is of linear order with respect to radial electric field. Drift motions of electron and ions can be different if they are in different collisionality regimes. This situation could happen if ions are in a ripple-trapped regime ($\nu_i/\varepsilon_h < \omega_{h.i} \simeq \sqrt{\varepsilon_h} v_{th.j} m/qR$) and electrons are not. In this case, a modification in frictional force between them appears linear in X. An example has been derived in [17.20], although it is complicated.

17.2.4 Impurities

Collisional transport is important for impurity ions due to their higher ionic charge. If impurity ions have heavier masses, a centrifugal force due to radial electric field is influential on a transport of heavier impurity. In a banana regime, impurity flow is given as [17.21]

$$v_{I,r} = \frac{m_i Z_I \nu_{ii}}{eB}\left[f_{trap}\frac{E_r}{B} + \frac{m_i}{reB}\left(\frac{E_r}{B}\right)^2 \left\{ (1+f_{trap})\frac{Z_i A_I}{Z_I A_i} - 1 \right\} \right] \tag{17.32}$$

where suffices I and i indicate impurities and main ions, respectively, Z is the charge and A is the mass number and f_{trap} is the ratio of trapped particles. ($f_{trap} \sim \sqrt{\varepsilon}$ holds in the absence of a strong electric field.) Relations $Z_i < Z_I$ and $A_i < A_I$ hold. The first term in the right-hand side denotes an inward pinch due to collisions with bulk ions, and the second term shows centrifugal force. Centrifugal force can be outward [17.21, 17.22]. Collisional transport theory predicts that impurities are expelled from the plasma, if the radial electric field is strong enough to satisfy the relation

$$\frac{1}{v_{th}}\left|\frac{E_r}{B}\right| > \frac{r}{\rho_i}\frac{f_{trap}}{(1+f_{trap})A_I/Z_I A_i - 1}. \tag{17.33}$$

This condition seems hardly to be realized except the case where A_I/Z_I is large. This case would occur when impurities preside not in a fully stripped state.

REFERENCES

[17.1] Itoh K, Sanuki H, Itoh S-I and Tani K 1991 *Nucl. Fusion* **31** 1405

[17.2] Nishitani T, Tobita K, Tani K *et al* 1993 *Plasma Physics and Controlled Nuclear Fusion Research 1992* vol 1 (Vienna: IAEA) p 351

Tobita K, Tani K, Nishitani T, Nagashima K and Kusama Y 1994 *Nucl. Fusion* **34** 1097

[17.3] Kovrizhnykh L M 1984 *Nucl. Fusion* **24** 851

[17.4] Mynick H E 1983 *Phys. Fluids* 26 1008

[17.5] Itoh S-I and Itoh K 1989 *Nucl. Fusion* **29** 1031

[17.6] Hazeltine R D 1989 *Phys. Fluids* B **1** 2031

[17.7] Itoh S-I, Itoh K, Ohkawa T and Ueda N 1989 *Plasma Physics and Controlled Nuclear Fusion Research 1988* vol 2 (Vienna: IAEA) p 23

[17.8] Miyamoto K 1994 *J. Plasma Fusion Res.* **70** 882

Chankin A V and McCracken G M 1993 *Nucl. Fusion* **33** 1459

[17.9] Itoh K, Sanuki H, Todoroki J *et al* 1991 *Phys. Fluids* B **3** 1294

[17.10] Kaufman A N 1958 *Phys. Fluids* **1** 252

[17.11] Hinton F L and Hazeltine R D 1976 *Rev. Mod. Phys.* **48** 239

Hirshman S P and Sigmar D J 1981 *Nucl. Fusion* **21** 1079

[17.12] Connor J W and Hastie R J 1971 *Phys. Fluids* **17** 114

Galeev A A and Sagdeev R Z 1977 *Review of Plasma Physics* vol 7, ed M A Leontovich (New York: Consultants Bureau) p 307

Potok P E, Politzer P A and Lidsky L M 1980 *Phys. Rev. Lett.* **45** 1328

Boozer A H and Kuo-Petravic G 1981 *Phys. Fluids* **24** 851

Hastings D E 1984 *Phys. Fluids* **27** 939

[17.13] Galeev A A and Sagdeev R Z 1967 *Zh. Eksp. Teor. Fiz.* **53** 348 (Engl. transl. 1968 *Sov. Phys.–JETP* **26** 233)
Stringer T E and Connor J W 1971 *Phys. Fluids* **14** 2177
Stix T H 1973 *Phys. Fluids* **16** 1260
Hazeltine R D and Ware A A 1978 *Plasma Phys.* **20** 673
[17.14] Shaing K C and Crume E Jr 1989 *Phys. Rev. Lett.* **63** 2369
[17.15] Stringer T E 1993 *Nucl. Fusion* **33** 1249
[17.16] Stacey W M 1989 *Plasma Phys. Control. Fusion* **31** 1468
Stacey W M 1993 *Phys. Fluids* B **5** 1413
Connor J W, Cowley S C, Hastie R J and Pan L R 1987 *Plasma Phys. Control. Fusion* **29** 931
[17.17] Hinton F L and Wong S K 1985 *Phys. Fluids* **28** 3082
[17.18] Shaing K C, Hsu C T and Hazeltine R D 1994 *Phys. Plasmas* **1** 3365
[17.19] Hirshman S P, Shaing K C, van Rij W I, Beasley C O and Crume E C Jr 1986 *Phys. Fluids* **29** 2951
Garabedian P R 1994 *Commun. Pure Appl. Math.* **47** 281
[17.20] Nakajima N and Okamoto M 1992 *J. Phys. Soc. Japan* **61** 833
[17.21] Ohkawa T 1994 *Comments Plasma Phys. Control. Fusion* **16** 1
[17.22] Yoshikawa S 1980 *Princeton University Research Report* PPPL-1710

Chapter 18

Electric Field Effect on Turbulent Transport

Anomalous transport has played a more important role than collisional transport in confinement of high temperature plasmas. The influence of radial electric field on anomalous transport (and vice versa) is an essential issue in understanding plasma structure and dynamics in a confinement device. We discuss here effects of electric field on anomalous transport which is caused by plasma fluctuations.

Study of anomalous transport has been developed by use of various methodologies. The best established one is a quasilinear theory, in which characteristics of microscopic perturbations are given by a linear response function of a plasma. Although this method has many limitations in explaining real plasmas, it is instructive to understand the relation of fluctuations and transport. Some typical results on this subject are revisited here.

Nonlinear theory for strong turbulence has been recognized to be inevitable for studying anomalous transport. Strong turbulence is characterized by the fact that the decorrelation rate of fluctuations is governed by nonlinear interactions (through $\tilde{\boldsymbol{E}} \times \boldsymbol{B}$ Doppler shift in electric turbulence), not by a linear growth rate. The inhomogeneous radial electric field is influential on nonlinear decorrelations. This problem will be reviewed next.

Based on this understanding, recent theoretical pictures for turbulent transport are discussed. As is well known, many complex transport phenomena have long been observed in plasmas (i.e., radial-pressure-gradient-driven toroidal current or particle pinch associated with energy flow, etc [2.4]). These processes of interference are discussed in a framework of interactions between thermodynamical fluxes and forces.

18.1 Linear Response

18.1.1 Quasilinear Theory

There are many instabilities in plasmas. The presence of an instability (frequency ω and wave number k) means that a perturbation electromagnetic field of the form $\exp(\mathrm{i}\boldsymbol{k}\cdot\boldsymbol{x} - \mathrm{i}\omega t)$ is emitted from plasma particles. In emitting a wave with the relevant momentum, plasma particles receive a force, and move in the radial direction. In the framework of quasilinear theory, each (k, ω) component is treated separately, and the linear response of the plasma is kept. The method for calculation is well established, and relations between gradients (density, momentum, temperature) and fluxes are derived, as in the case of collisional transport. Explicit formulae are given in the literature [18.1]. For instance, an ion flux is given as

$$\begin{pmatrix} \Gamma_i \\ P_{\zeta r}/m_i v_{thi} \\ q_i/T_i \end{pmatrix} = \mathbf{M}_i \begin{pmatrix} -\frac{n_i'}{n_i} + \frac{eE_r}{T_i} + \frac{T_i'}{T_i} - \frac{eB\omega}{k_\theta T_i} \\ -\frac{2V_\zeta'}{v_{th}} \\ -\frac{T_i'}{T_i} \end{pmatrix} \tag{18.1}$$

where $P_{\zeta r}$ is a radial flux of toroidal momentum, and q_i is a radial ion heat flux. Elements of matrix $\mathbf{M}$ are contributed from trapped particles and transit particles. The electric field mainly influences the contribution from trapped ions, through modifying ion number density and drift direction as discussed in chapter 17. For an illustration of the static electric field effect on the quasilinear flux, that from transit ions is shown as

$$M_{ij} = \frac{n}{B^2}\sum_k \int \frac{\mathrm{d}\omega}{2\pi}\left(\frac{\omega}{\overline{\overline{k}}_\parallel v_{th}}\right)^{i+j-2} \frac{\langle \tilde{E}_\theta(k,\omega)\tilde{E}_\theta(k,-\omega)\rangle}{|\overline{\overline{k}}_\parallel v_{th}|} \,\mathrm{Im}\, Z\left(\frac{\omega}{\overline{\overline{k}}_\parallel v_{th}}\right)\Lambda_0(b). \tag{18.2}$$

It is a second order term in fluctuating electric field. In equation (18.2), $\Lambda_0(b)$ denotes a finite-gyro-radius effect, $\Lambda_0(b) = I_0(b)\exp(-b)$, ($b = k_\perp^2\rho_i^2$ and $I_0(b)$ being the zeroth order modified Bessel function of the first kind), $\overline{\overline{k}}_\parallel$ is an effective parallel mode number which ions feel in the presence of a radial electric field,

$$\overline{\overline{k}}_\parallel^2 = k_\parallel^2 + (k_\theta\rho_i)^2(L_{n2}eE_r'T_i^{-1})^2$$

and $\mathrm{d}^2n/\mathrm{d}r^2 = -nL_{n2}^2$. (A discussion of this effective $k_\parallel$ is given in [18.2].) Equation (18.1) includes a mobility term (i.e., E_r term in the right column), suggesting that it would affect transport. It is noted that the nonuniform part of $\boldsymbol{E}\times\boldsymbol{B}$ rotation mainly contributes to mobility; a uniform part also appears in a Doppler shift of wave frequency ω, and they almost cancel each other.

In equation (18.1), the fluctuation amplitude is left undetermined. Further specification is necessary. The mixing length estimate, which is often employed as an assumption, relates the fluctuation level to the ratio of wavelength to gradient scale-length,

$$\tilde{n}/n \sim 1/kL_n \tag{18.3}$$

(or $\tilde{E}_\perp \sim T/eL_n$). A contribution of resonant particles to cause decorrelation, which appears in equation (18.2) as a form of Z-function, also appears in the expression for linear growth rate, γ_L. With the approximation equation (18.3), an estimate

$$D \sim \frac{\gamma_L}{k_\perp^2} \tag{18.4}$$

is given. This result is based on the picture where nonlinear damping rate, $Dk_\perp^2$, balances the linear growth rate. If one employs this formalism, a study of the effect of a radial electric field is reduced to the problem of analysing its effect on linear stability. This line of consideration is discussed in this subsection. A more fundamental argument on turbulence level is given in sections 18.2 and 18.3.

18.1.2 Linear Stability

Some mechanisms of linear stability, i.e., fluid-like response, Landau resonance and drift reversal, are shown here as examples.

An interchange mode is a typical example of plasma instability. In the presence of pressure gradient ∇p parallel to a 'gravity' g (either real gravity or a centrifugal force due to magnetic curvature) a fluid-like instability can occur with its linear growth rate $\gamma_L \sim \sqrt{g/L_p}$ as is discussed in chapter 6 ($L_p = -n/n'$ for $\nabla T = 0$). If a radial electric field, a component of which is in the direction of pressure gradient, exists, plasma moves with $\boldsymbol{E} \times \boldsymbol{B}$ velocity $V_{E\times B}$ in the direction perpendicular to the gradient. When the $\boldsymbol{E}$-field is radially inhomogeneous, the velocity $V_{E\times B}$ differs at a radial location (figure 18.1). The perturbed motion of the flute mode is in the direction of g (i.e., the x-direction in figure 18.1), and is deformed in the y-direction if the $\boldsymbol{E} \times \boldsymbol{B}$ velocity $V_{E\times B}$ depends on x (figure 18.1). As a result, the growth rate is modified to [18.3]

$$\gamma_L = \sqrt{\frac{g}{L_p} - \left\{\frac{V'_{E\times B}}{2kL_p}\right\}^2}. \tag{18.5}$$

An off-resonant type of stabilization is possible, if the condition

$$|V'_{E\times B}/2kL_p| = \gamma_{L0} \tag{18.6}$$

is satisfied, where γ_{L0} is the linear growth rate in the limit of $V'_{E\times B} = 0$. This order-of-magnitude estimate equation (18.6) is applicable for *linear stability* in a wide range, and is often employed in recent studies on improved confinement. This relation is similar to a well known stabilization mechanism by diamagnetic drift (or finite-gyro-radius effect),

$$\gamma_L = \sqrt{\gamma_{L0}^2 - \omega_*^2/4} \tag{18.7}$$

where ω_* is a drift frequency, and γ_{L0} is the linear growth rate in the limit of $\omega_* = 0$.

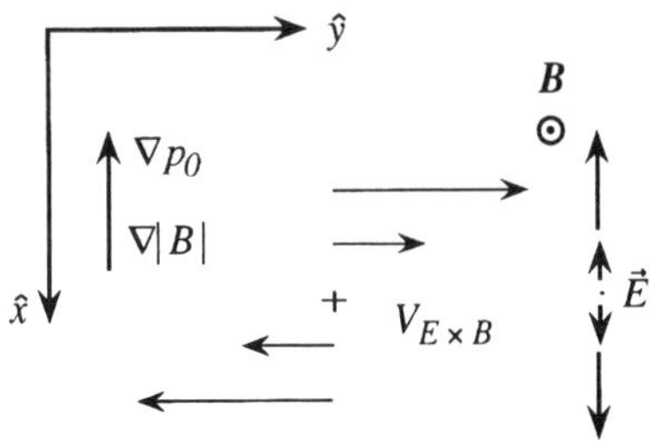

Figure 18.1. Schematic drawing of inhomogeneous plasma and electric field. A strong magnetic field is in the z-direction. Electric field and plasma inhomogeneity ∇p_0 are in the radial (x-) direction. $\boldsymbol{E} \times \boldsymbol{B}$ velocity is in the poloidal (y-) direction. Perturbation velocity of the flute mode $\tilde{V}$ is deformed by inhomogeneous $\boldsymbol{E} \times \boldsymbol{B}$ velocity due to the static electric field.

Another type of stabilization mechanism is seen in a wave–particle resonance. Landau damping is one of the main mechanisms for linear instabilities, and can also be important as a nonlinear mechanism. Ion motion is influenced by an inhomogeneous radial electric field. It has been reported that ion Landau damping is expected to occur for $\bar{\bar{k}}_{\parallel} v_{thi} \simeq \omega$, and the associated stability of kinetic modes takes place for [18.2]

$$\frac{L_{n2}\rho_i e E_r'}{T_i} \simeq \frac{\omega}{k_\theta v_{thi}} \tag{18.8}$$

even for a flute mode with $k_{\parallel} \simeq 0$.

Drift reversal due to an inhomogeneous electric field is also influential on stability. The toroidal drift velocity of trapped ions is modified by a factor $(1 + 2u_g)$. Equation (17.16) provides a condition for drift reversal of ions. If the condition $u_g < -\frac{1}{2}$ is satisfied, the sign of toroidal drift changes. Trapped particles drift as if the magnetic curvature were favourable. Resonance of trapped ions with waves has stabilizing influence [17.7]. Doppler shift is modified and the term ω_{Me} in equation (6.66) is replaced by $(1 + 2u_g)\omega_{Me}$. Owing to this modification, the growth rate equation (6.70) turns to be

$$\mathrm{Im}(\omega) \simeq (2\varepsilon)^{1/4}\sqrt{(1 + u_g)\omega_{Me}\omega_*}. \tag{18.9}$$

The trapped-ion mode is stabilized by drift reversal in the range of

$$u_g < -1. \tag{18.10}$$

18.2 Suppression of Turbulence

The formula (18.4) gives a first-step estimate, and equations (18.6), (18.8) and (18.10) provide a rough idea to understand the effect of radial electric field

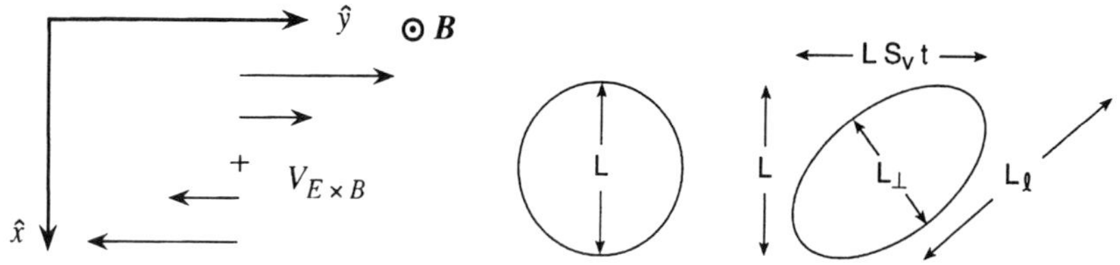

Figure 18.2. Flow in the y-direction, which is inhomogeneous in the x-direction, and stretching of a fluid element.

inhomogeneity. Understanding of turbulence itself is necessary. The turbulence level is suppressed by E_r', and is discussed in this section.

Figure 18.2 illustrates a sheared flow and stretching of a fluid element. Its mean velocity in the y-direction (poloidal direction), which has a shear in the x-direction (radial direction), is expressed as

$$V_y = S_v x \tag{18.11}$$

in local coordinates. Consider a deformation of an element which is circular at $t = 0$. A shift, in the y-direction, of the top of the element relative to the bottom is given as $LS_v t$ after an elapse of time of t, where L is the vertical size of the element. A circular element is stretched to an ellipse, in which the major axis has a length $L_1 \approx \sqrt{L^2 + (LS_v t)^2}$. Since area is preserved by this stretching, the length of the minor axis is given as

$$L_\perp = \frac{L}{\sqrt{1 + S_v^2 t^2}}. \tag{18.12}$$

This result shows that the perpendicular wavelength, $k_\perp^{-1}$, of any mode is compressed due to this shear flow. Owing to the reduction of $k_\perp^{-1}$, nonlinear damping is enhanced as is explained later. The deformation of the element is noticeable if $LS_v t$ is of the order of L, i.e., $t \sim 1/S_v$.

18.2.1 Decorrelation and Damping due to the Shear Flow

These general considerations also apply to plasma turbulence. The flow velocity shear is interpreted as

$$S_v = r\frac{\mathrm{d}}{\mathrm{d}r}\left(\frac{E_r}{Br}\right) \tag{18.13}$$

in a cylindrical geometry. Equation (18.13) is based on the picture where the flow is in the poloidal direction. If a flow is in the toroidal direction, a shear of flow is defined as $S_{v,t} = r\,\mathrm{d}(E_r/B_p)/\mathrm{d}r$. There is no essential difference in a theoretical framework between two cases. Let us consider a decay of a mode with wave number $k_{\perp 0}$. Noting a reciprocal relation between a correlation length L and the typical wave number $k_{\perp 0}$, an influence can be calculated. The stretching of a turbulent vortex, equation (18.12), indicates that the perpendicular wave number is effectively enhanced by a factor $(1 + S_v^2 t^2)$ [18.4–18.6]

$$k_{\perp eff}^2 = k_\perp^2 (1 + S_v^2 t^2). \tag{18.14}$$

As a consequence, the decorrelation rate, i.e., the nonlinear damping rate, which is defined by the relation

$$1/\tau_{corr} \approx D k_{\perp eff}^2 \tag{18.15}$$

becomes larger (for a given value of D) by this increment of perpendicular wave number.

The stretching of the turbulent vortex which has a decorrelation time τ_{corr} continues until the elapse time reaches τ_{corr}. Substituting the relation $t = \tau_{corr}$ into equation (18.14), the effective mode number $k_{\perp eff}$ (i.e., the inverse of decorrelation length) is estimated as

$$k_{\perp eff}^2 = k_{\perp 0}^2 (1 + S_v^2 \tau_{corr}^2).$$

Substituting this into equation (18.15), the equation that determines a consistent decorrelation rate $1/\tau_{corr}$ is derived as

$$\frac{1}{\tau_{corr}} = D k_{\perp 0}^2 (1 + S_v^2 \tau_{corr}^2). \tag{18.16}$$

Depending on the magnitude of flow shear, $|S_v|$, two limiting cases are derived from equation (18.16) as

$$\frac{1}{\tau_{corr}} = \begin{cases} D k_{\perp 0}^2 + S_v^2 D^{-1} k_{\perp 0}^{-2} & (|S_v| \ll D k_{\perp 0}^2) \\ (D k_{\perp 0}^2)^{1/3} S_v^{2/3} & (|S_v| \gg D k_{\perp 0}^2). \end{cases} \tag{18.17}$$

Figure 18.3 illustrates decorrelation rate $1/\tau_{corr}$ as a function of flow shear rate S_v. The limiting form of $\tau_{corr} \sim (D k_{\perp 0}^2)^{-1/3} S_v^{-2/3}$ is similar to the one which was derived from renormalization of turbulent electron parallel motion in a sheared magnetic field [18.7].

This result equation (18.17) shows that a decorrelation due to the shear flow is effective if the flow shear rate S_v reaches the level $D k_{\perp 0}^2$. If the diffusion coefficient D is constant, the relation $S_v \simeq D k_{\perp 0}^2$ is satisfied for the long wavelength mode even when S_v is small. That is, stabilization by the shear flow is more effective for longer wavelength modes. An increased effective wave number, equation (18.14) suggests a reduction of the turbulence level. If one employs the *ansatz* of the mixing length estimate, $\tilde{n}/n \sim 1/k_{eff} L_n$, the level of fluctuation may be reduced by a factor $k_{\perp 0}/k_{\perp eff}$ as

$$\frac{\langle \tilde{n}^2 \rangle}{\langle \tilde{n}^2 \rangle_{ref}} \approx \frac{1}{1 + S_v^2 \tau_{corr}^2} \tag{18.18}$$

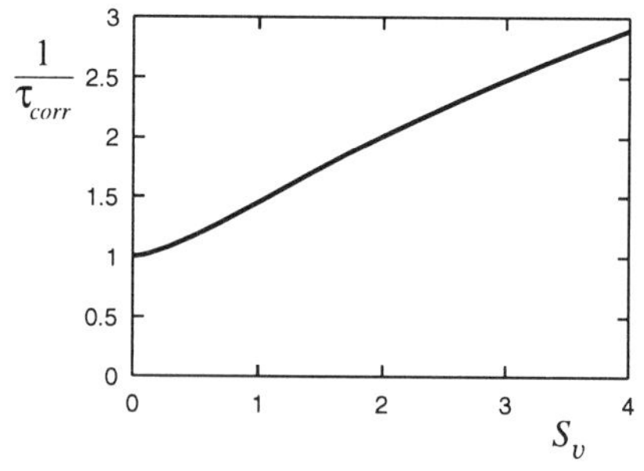

Figure 18.3. Correlation time τ_{corr} is shown as a function of flow shear rate S_v. (Both τ_{corr} and S_v are normalized to $Dk_{\perp 0}^2$.)

where the suffix *ref* indicates a reference case where $S_v = 0$ [18.4]. The Lorentzian correction appears [18.4, 18.6]. In the large shear limit, $\tau_{corr} \sim (Dk_{\perp 0}^2)^{-1/3} S_v^{-2/3}$, reduced fluctuations were discussed. If one employs the mixing length estimate, a similar reduction factor is expected for diffusivity.

These arguments on fluctuation level and transport, however, should not be taken as a conclusion. As was carefully discussed in [18.4], these analyses have been developed provided that basic properties of fluctuations (such as the wave number $k_{\perp 0}$ of a relevant mode and so on) are unchanged. These terms can also be functions of E_r and $\mathrm{d}E_r/\mathrm{d}r$. Effects of E_r and $\mathrm{d}E_r/\mathrm{d}r$, as a whole, can be simultaneously determined after the turbulence structure is properly solved. (Numerical simulations have also revealed a subtle balance between various nonlinear interactions [18.8, 18.9].)

18.2.2 Statistical Property Near Thermal Equilibrium

Decorrelation by sheared flow has been investigated intensively in neutral fluid dynamics, and the study is instructive for our understanding. Also in phase transition physics, suppression of fluctuations by sheared flow has been discussed. We revisit the study near thermal equilibrium. The form factor of fluctuations

$$S(\boldsymbol{k}) = \int \mathrm{d}\boldsymbol{r} \exp(\mathrm{i}\boldsymbol{k} \cdot \boldsymbol{r}) \langle \delta u(\boldsymbol{r} + \boldsymbol{r}_0) \delta u(\boldsymbol{r}_0) \rangle$$

(δu being a fluctuating quantity) has been calculated near thermal equilibrium. An analytic expression has been given [18.10]

$$S(\boldsymbol{k}) \simeq \frac{1}{\tau + k^2 + C|S_v k_y / L_k|^{2/5}} \tag{18.19}$$

where L_k is a typical value for the kinetic (Onsager) coefficient, and is a coefficient denoting damping in a Langevin equation. An inhomogeneous flow

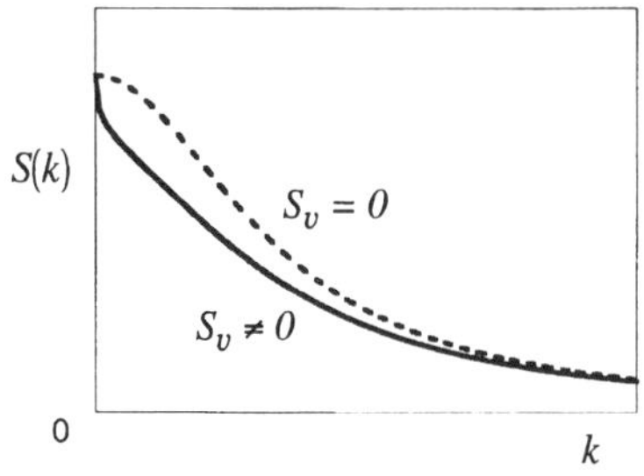

Figure 18.4. Form factor $S(k)$.

would efficiently suppress fluctuations of longer wavelength. The form factor is illustrated in figure 18.4

The suppression coefficient depends on the spatial size of fluctuations, as are shown in equations (18.17) for plasmas and in (18.19) for neutral fluids. This dependence on size is a characteristic feature of shear suppression. It is more effective for perturbations with larger spatial scales. Another example is seen in the study of bubble formation, which was analysed in neutral fluid dynamics [18.11]. Two forces work on a bubble in a sheared flow. Shearing force $\bar{\mu} S_v$ tends to stretch a bubble ($\bar{\mu}$ being a coefficient of fluid viscosity). Surface tension $\bar{\sigma}/L$ acts to restore spherical shape ($\bar{\sigma}$: the coefficient for surface tension, L: size of bubble). (See figure 18.5.) If the shearing force is greater than the surface tension, $\bar{\mu} S_v > \bar{\sigma}/L$, the bubble is destroyed. A critical size above which a bubble collapses

$$L_{cr} = \frac{\bar{\sigma}}{\bar{\mu} S_v}$$

was found.

On the other hand, bubble formation is a subcritical phenomena and there is a critical (minimum) size for nucleation, L_{min} [18.12]. Below this size, a bubble does not grow and disappears with a finite lifetime (even in the absence of sheared flow). This latter size is determined by a balance between the surface tension and the difference in free energy inside the bubble and fluid. Free energy associated with bubble formation is illustrated in figure 18.5 (right). If the velocity shear is strong enough that the condition

$$L_{cr} < L_{min}$$

is satisfied, then nucleation of the bubble is strongly suppressed. The condition $L_{cr} < L_{min}$ is rewritten as

$$S_v > \frac{\bar{\sigma}}{\bar{\mu} L_{min}}.$$

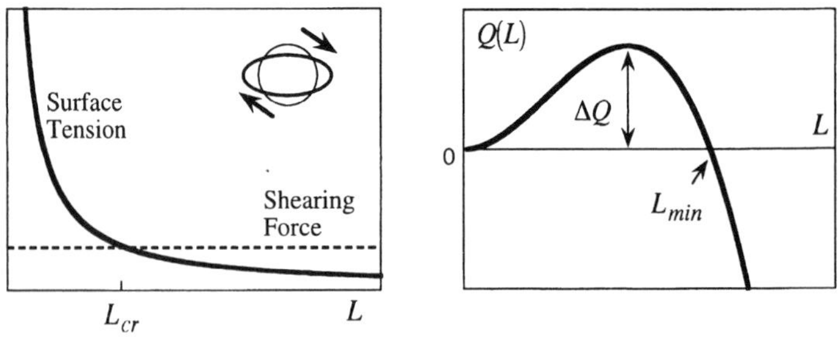

Figure 18.5. Shearing force and surface tension (per unit volume) against bubble size (left). Free energy of bubble formation as a function of bubble size (right). Activation energy ΔQ and minimum size L_{min} for nucleation are shown.

As the critical size for onset of nucleation L_{min} becomes larger, the suppression by shear flow appears at a lower level of flow shear rate.

18.3 Turbulent Transport

In this section, the study of nonlinear turbulence is discussed. As is illustrated in equation (18.1) and in the case of neoclassical transport, interactions between various thermodynamical forces (gradients) and fluxes appear in a magnetized plasma. The relation between gradients and fluxes may be written in matrix form. However, the expression of matrix form is not validated *a priori*. A form like equation (18.1) suggests a 'linear' relation between gradient and flux. On the contrary, as was shown in previous chapters, the elements of the matrix strongly depend on gradients. For instance, a factor like equation (18.18), which is also expected to appear in transport coefficients, includes inhomogeneity of global flow. The flux is given as a nonlinear function of various gradients; efforts to separate one element as a 'gradient' and to leave the others in the 'matrix element' retain some arbitrariness. Keeping this fact in mind, we shall distinguish the 'diagonal' transport coefficient and 'off-diagonal' terms in a heuristic manner.

18.3.1 Nonlinear Balance

In the study of the electric field effect on turbulent transport, one simplified method is a direct extension of the Kadomtsev formula. The increment of effective perpendicular wave number is taken into account, equation (18.14), and the mode growth rate is given as

$$\gamma_{NL} = \gamma_L - \frac{1}{\tau_{corr}} = \gamma_L - Dk_{\perp 0}^2(1 + S_v^2\tau_{corr}^2). \tag{18.20}$$

In a stationary state, $\gamma_{NL} = 0$, or $1/\tau_{corr} = \gamma_L$, one has

$$D = \frac{\gamma_L}{k_{\perp 0}^2} \frac{1}{(1 + S_v^2 \gamma_L^{-2})}. \tag{18.21}$$

This expression shows the effect of radial electric field in two ways. First, the linear growth rate can be reduced as is discussed in section 18.1. Second, the reduction of mixing length by sheared flow gives a further reduction of transport.

The result is simple, and many linear stability analyses could be substituted into it. However, the relevance of the result is not clear, as was the case of equation (18.4). Tests of a formula based on equation (18.4) have been performed on various unstable modes exhaustively, for cases where the E_r' effect is absent (see e.g., [4.10, 4.15]). The results were, however, far from satisfactory; the form (18.4) seems to be incomplete even in the absence of a radial electric field. Therefore the extension, equation (18.21), could be limited in its applicability to real plasmas.

If the mode growth is affected by nonlinear interactions, the transport coefficient could be different from a form like equation (18.4). This is really the case. An analysis on electron nonlinearity has shown that a drift wave in a sheared slab (which is known to be linearly stable [6.14]) can be unstable if the fluctuation amplitude is large enough [18.7].

In *nonlinear growth*, a stationary state is realized by the condition,

$$0 = \gamma_{NL} = \gamma_{ng} - \frac{1}{\tau_{corr}} \tag{18.22}$$

in which the suffix ng indicates growth by a nonlinear process. The nonlinear instability due to electron nonlinearity was shown to dominate over the linear one as is demonstrated in chapter 14. The balance, $\gamma_{ng} \approx 1/\tau_{corr}$, indicates the fact that both growth and damping are caused by nonlinear interactions. In other words, the nonlinear steady state turbulence is self-sustained.

18.3.2 Turbulent Transport Coefficient

The self-sustained turbulence is solved in an analytic manner for ballooning mode turbulence and interchange mode turbulence in chapters 9 and 10. For simplicity, the eigenvalue equation for interchange mode turbulence is shown as [18.13]

$$\frac{\mathrm{d}}{\mathrm{d}k} \frac{k_\perp^2}{\gamma + \lambda k_\perp^4} \frac{\mathrm{d}}{\mathrm{d}k} \left(\gamma + \chi k_\perp^2 + \omega_{E1} \frac{\mathrm{d}}{\mathrm{d}k} \right) \tilde{p} + \frac{G_0}{s^2} \tilde{p}$$

$$- \frac{1}{k_\theta^2 s^2} \left(\gamma + \mu k_\perp^2 + \omega_{E1} \frac{\mathrm{d}}{\mathrm{d}k} \right) k_\perp^2 \left(\gamma + \chi k_\perp^2 + \omega_{E1} \frac{\mathrm{d}}{\mathrm{d}k} \right) \tilde{p} = 0 \tag{18.23}$$

where G_0 is a combination of pressure gradient and bad curvature, s is a magnetic shear parameter and (μ, λ, χ) are *renormalized* viscosity, current diffusivity and thermal conductivity, respectively. In equation (18.23), normalization is employed for length and time with respect to plasma minor radius a and τ_{Ap}, respectively ($\tau_{Ap} = a/v_{Ap}$ and v_{Ap} is a poloidal Alfvén velocity).

$$\omega_{E1} = k_\theta \tau_{Ap} E_r' a/B \tag{18.24}$$

denotes a normalized sheared flow velocity due to an inhomogeneous radial electric field. The effect of E_r' is taken into account as perturbation up to the second order. If one neglects the inhomogeneous radial electric field, $\omega_{E1} \to 0$, equation (18.23) reduces to equation (9.17). The stationary state solution of equation (18.23) with ω_{E1}^2-corrections is obtained to give

$$\chi \sim \frac{1}{(1 + 0.5 G_0^{-1} \omega_{E1}^2)} \frac{G_0^{3/2}}{s^2} \left(\frac{c}{\omega_p}\right)^2 \frac{v_{Ap}}{a} \tag{18.25}$$

where ion and electron viscosities are of the same magnitude [18.13]. The typical correlation length of fluctuations has a pressure gradient dependence. Its inverse, the wave number, is given as

$$\langle k_\perp^2 \rangle \propto (1 + 0.5 G_0^{-1} \omega_{E1}^2) G_0^{-1}. \tag{18.26}$$

The saturation level of fluctuation

$$\frac{e\tilde{\phi}}{T} \sim \frac{1}{(1 + 0.5 G_0^{-1} \omega_{E1}^2)} \frac{G_0^{3/2}}{s^2} \left(\frac{c}{\omega_p}\right)^2 \frac{v_{Ap}}{a} \frac{eB}{T} \tag{18.27}$$

is determined simultaneously. Fluctuations are expressed in terms of structural parameters, G_0, s and ω_{E1}. These results show that the correlation length becomes shorter and the fluctuation level decreases in the presence of an inhomogeneous radial electric field shear effect. Contrary to the result of equation (18.14), the correlation length (and time as well) is consistently determined in equation (18.26). What is common and important is that the correlation length becomes longer and the fluctuation level higher as the pressure gradient increases. These features are typical characteristics of transport processes in nonequilibrium matter. (It is in contrast to collisional diffusion, which is generated by thermal fluctuations, i.e., Coulomb collision.)

The same analysis is extended to tokamak plasma transport due to ballooning mode turbulence. The result is given as

$$\chi \sim \frac{1}{(1 + h_1 \omega_{E1}^2 + h_2 \omega_{E2}^2)} F(s, \alpha) \alpha^{3/2} \left(\frac{c}{\omega_p}\right)^2 \frac{v_A}{qR} \tag{18.28}$$

where $\alpha = -q^2 R \beta'$, and numerical coefficients (h_1, h_2) are functions of s and α (see [18.14] for the details.) ω_{E2} denotes the curvature of the radial electric field. In the absence of radial electric field shear, equation (18.28) is

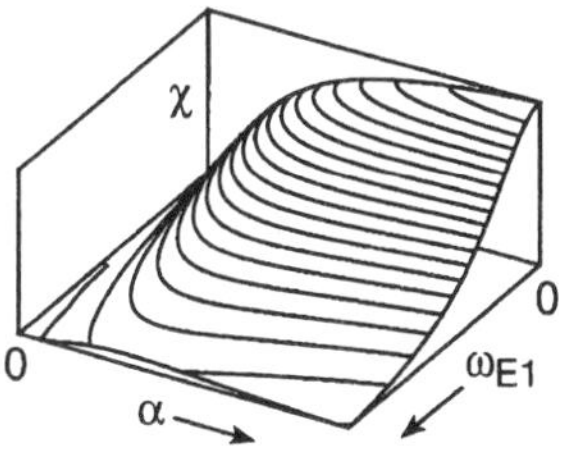

Figure 18.6. Turbulent driven transport coefficient is shown as a function of the pressure gradient α and the gradient of the radial electric field ω_{E1}.

reduced to equation (10.36). Figure 18.6 illustrates the dependence of transport coefficient on pressure gradient and gradient of radial electric field. Ion viscosity also follows the formula equation (18.28). A characteristic scale-length of fluctuations is also shown to decrease by a similar factor if the electric field shear is increased.

The result equation (18.28) (or equation (18.25)) indicates that electric field inhomogeneity, as a whole, suppresses anomalous transport. The efficiency for suppression (a coefficient of inhomogeneity parameters) critically depends on pressure gradient and other plasma parameters. Both the gradient and curvature of electric field is effective in reducing transport coefficients. The formulae (18.25) and (18.28) show that the difference appears in the form of a configurational parameter and that pressure gradient plays a main and common role in generating the anomalous transport.

REFERENCES

[18.1] Balescu R 1991 *Phys. Fluids* B **3** 564
Hazeltine R D, Mahajan S M and Hitchcock D A 1981 *Phys. Fluids* **24** 1164
Shaing K C 1988 *Phys. Fluids* **31** 2249
Itoh S-I 1992 *Phys. Fluids* B **4** 796

[18.2] Sanuki H 1984 *Phys. Fluids* **27** 2500

[18.3] See, e.g., Lehnert B 1966 *Phys. Fluids* **9** 1367

[18.4] Shaing K C, Lee G S, Carreras B A, Houlberg W A and Crume E C Jr 1989 *Plasma Physics and Controlled Nuclear Fusion Research 1988* vol 2 (Vienna: IAEA) p 13
Shaing K C, Crume E C Jr and Houlberg W A 1990 *Phys. Fluids* B **2** 1492

[18.5] Biglari H, Diamond P H and Terry P W 1990 *Phys. Fluids* B **2** 1

[18.6] Zhang Y Z and Mahajan S M 1992 *Phys. Fluids* B **4** 1385

[18.7] Molvig K, Hirshman S P, Rechester A B, White R B *et al* 1981 *Plasma Physics and Controlled Nuclear Fusion Research 1980* vol 1 (Vienna: IAEA) p 73

[18.8] Hasegawa A and Wakatani M 1987 *Phys. Rev. Lett.* **59** 1581

Wakatani M, Watanabe K, Sugama H and Hasegawa A 1992 *Phys. Fluids* B **4** 1754

Carreras B A, Lynch V, Garcia L and Diamond P H 1993 *Phys. Fluids* B **5** 1491

[18.9] For a recent review, see, e.g., Staebler G M 1998 *Plasma Phys. Control. Fusion* **40** 569

[18.10] Onuki A and Kawasaki K 1979 *Ann. Phys., NY* **121** 456

[18.11] Taylor G I 1934 *Proc. R. Soc.* A **146** 501

[18.12] Lord Rayleigh 1914 *Phil. Mag.* **34** 94

For recent review, see, e.g., Mel'nikov V I 1991 *Phys. Rep.* **209** 1

[18.13] Itoh K, Itoh S-I, Fukuyama A, Sanuki H and Yagi M 1994 *Plasma Phys. Control. Fusion* **36** 123

[18.14] Itoh S-I, Itoh K, Fukuyama A and Yagi M 1994 *Phys. Rev. Lett.* **72** 1200

Chapter 19

Generation of Radial Electric Field

19.1 Evolution of Radial Electric Field

19.1.1 Charge Conservation

A radial electric field is generated by a radial current. If ions are lost faster than electrons, then the radial electric field becomes more negative. Poisson's equation which is averaged on a magnetic surface can be written as

$$\varepsilon_0 \frac{\partial}{\partial t} E_r = -J_r^{tot} \tag{19.1-1}$$

where J_r^{tot} is a total current in the radial direction being the sum of dc current, J_r^{dc}, and polarization drift current, J_r^p. The latter is in proportion to $\partial E_r / \partial t$, and equation (19.1-1) is usually rewritten as

$$\varepsilon_0 \varepsilon_\perp \frac{\partial}{\partial t} E_r = -J_r^{dc} \tag{19.1-2}$$

where $\varepsilon_\perp$ is a perpendicular dielectric constant. It should also be noticed that equation (19.1) could be a nonlocal equation to determine the structure. As is seen in the previous sections, a particle loss such as an orbit loss is not a local quantity, but is nonlocally determined by the global structure of the radial electric field inside a plasma. Equation (19.1) must be considered as an equation that determines the global structure simultaneously with inclusion of a long range interaction.

19.1.2 Flow, Friction and Radial Current

$\boldsymbol{E} \times \boldsymbol{B}$ *Flow*. We first illustrate that a radial current $J_r[E_r]$ is related to the friction of a plasma flow on a magnetic surface. The radial ion current is given by the equation of motion of ions. In order to study the response against radial electric field, a limit of $\nabla p \to 0$ is shown. The dc component of radial current is discussed first, and the polarization drift is discussed next. The force balance equation is given as

$$n_i e(\boldsymbol{E} + \boldsymbol{V}_i \times \boldsymbol{B}) + \boldsymbol{f}_i = 0 \tag{19.2}$$

where $\boldsymbol{f}_i$ is a force, and includes the friction and inertial effects. Let us write this force in symbolic form as

$$\boldsymbol{f}_i = -m_i n_i \begin{pmatrix} \nu_r V_r \\ \nu_\theta V_\theta \\ \nu_\zeta V_\zeta \end{pmatrix}. \tag{19.3}$$

Generally speaking, 'friction coefficients' $(\nu_r, \nu_\theta, \nu_\zeta)$ could include operators. For instance, an (anomalous) ion viscosity drives a force like $f_{viscous} = m_i n_i \nabla_\perp \cdot \mu_i \nabla_\perp V$. The coefficients ν_r, ν_θ and ν_ζ are not identical but anisotropic in toroidal plasmas.

When the friction force is not strong, $\nu_r, \nu_\theta, \nu_\zeta \ll eB/m_i$, a Taylor expansion with respect to $\nu m_i/eB$ is performed, and a solution $\boldsymbol{V}_i$ of equation (19.2) is expressed in an expansion parameter series $\nu m_i/eB$. (We also take the limit where $(B_\theta/B_\zeta)^2$ is a smallness parameter.) In the lowest order, the flow is expressed in terms of $\boldsymbol{E} \times \boldsymbol{B}$ flow on the magnetic surface. Two patterns of incompressible flow are known in an axisymmetric torus: the toroidal flow, $\boldsymbol{V} = (0, 0, V_\zeta)$, and the poloidal flow with parallel flow, $\boldsymbol{V} = (0, V_\theta, 0) - (0, 2\varepsilon, 2q)V_\theta \cos\theta$ (figure 19.1.) The zeroth order force balance, $\boldsymbol{E} + \boldsymbol{V}_i \times \boldsymbol{B} = \boldsymbol{0}$, provides the relation

$$V_\theta - \left(\frac{B_\theta}{B_\zeta}\right) V_\zeta = \frac{-E_r}{B}. \tag{19.4}$$

Poloidal and toroidal velocities are determined, together with radial flow velocity, in the first order equation with respect to $\nu m_i/eB$. The poloidal and toroidal components of equation (19.2), $-n_i e V_r B + f_\theta = 0$ and $n_i e V_r B_\theta + f_\zeta =$

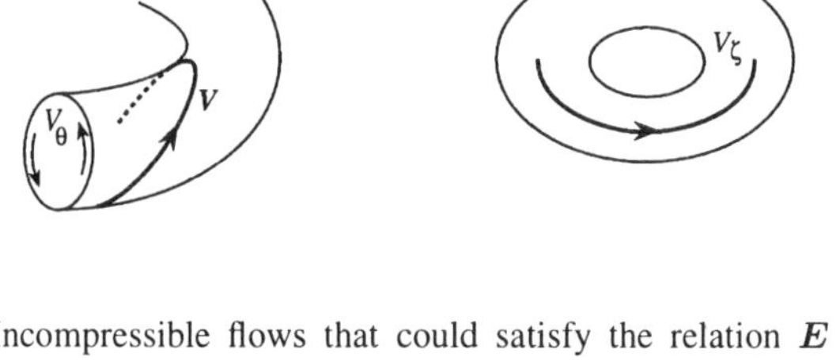

Figure 19.1. Incompressible flows that could satisfy the relation $\boldsymbol{E} + \boldsymbol{V}_i \times \boldsymbol{B} = 0$: (a) poloidal flow with parallel flow, $\boldsymbol{V} = (0, V_\theta, 0) - (0, 2\varepsilon, 2q)V_\theta \cos\theta$, and (b) toroidal flow, $\boldsymbol{V} = (0, 0, V_\zeta)$.

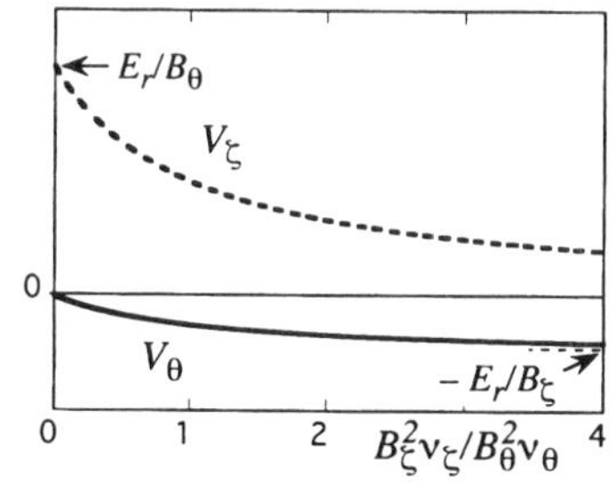

Figure 19.2. $\boldsymbol{E} \times \boldsymbol{B}$ flow velocity on a magnetic surface (at the poloidal angle $\theta = \pi/2$) as a function of the ratio of damping rate ν_ζ/ν_θ.

0, respectively, with equation (19.4) provide the solution

$$\boldsymbol{V} = \frac{E_r}{B} \frac{B_\zeta^2 \nu_\zeta}{B_\theta^2 \nu_\theta + B_\zeta^2 \nu_\zeta} \begin{pmatrix} \dfrac{m_i \nu_\theta}{e B_\zeta} \\ -1 + 2\varepsilon \cos\theta \\ \dfrac{B_\theta \nu_\theta}{B_\zeta \nu_\zeta} + 2q \cos\theta \end{pmatrix}. \tag{19.5}$$

The toroidal flow and the poloidal flow are shown in figure 19.2 as a function of the ratio of the damping rate ν_ζ/ν_θ. As the toroidal damping coefficient ν_ζ increases, the toroidal velocity V_ζ decreases and the $\boldsymbol{E} \times \boldsymbol{B}$ flow approaches a poloidal rotation.

Radial flux. The result equation (19.5) shows that a radial current $en_i V_r$ is generated by friction. If the toroidal friction is weak, $\nu_\zeta B_\zeta^2 \ll \nu_\theta B_\theta^2$, $\boldsymbol{E} \times \boldsymbol{B}$ flow is mainly in the toroidal direction, $\boldsymbol{V} = (0, 0, V_\zeta)$, and the radial current is in proportion to ν_ζ.

$$en_i V_r = \frac{m_i n_i \nu_\zeta}{B_\theta^2} E_r. \tag{19.6-1}$$

On the other hand, when the toroidal friction is strong, $\nu_\zeta B_\zeta^2 \gg \nu_\theta B_\theta^2$, $\boldsymbol{E} \times \boldsymbol{B}$ flow causes a poloidal flow, $\boldsymbol{V} = (0, V_\theta, 0) - (0, 2\varepsilon, 2q) V_\theta \cos\theta$, and the radial current is in proportion to ν_θ. In this limit, the radial ion current is given as

$$en_i V_r = \frac{m_i n_i \nu_\theta}{B_\zeta^2} E_r. \tag{19.6-2}$$

Damping rate by friction and viscosity. An effective damping coefficient ν_θ is expressed in terms of the local damping process and shear viscosity. Let us consider the term ν_θ which is composed of a scalar component and an operator such as a radial derivative. A local damping rate ν_{local}, that is expressed in terms of local plasma parameters, includes a transit time magnetic pumping equation (17.23). The friction due to ion viscosity (shear viscosity) is expressed by an operator of the second order radial derivative. The dissipation,

$$-\boldsymbol{f}\cdot\boldsymbol{V} = \nu_\theta V_\theta^2$$

is composed of a poloidal flow component, $V_\theta(\nu_{local} - \boldsymbol{\nabla}_\perp \cdot \mu_i \boldsymbol{\nabla}_\perp)V_\theta$, and of the return flow component, $V_\parallel(\nu_{local} - \boldsymbol{\nabla}_\perp \cdot \mu_i \boldsymbol{\nabla}_\perp)V_\parallel$. Because the return flow along the field line, $V_\parallel = V_\theta 2q\cos\theta$, has a poloidal dependence, the latter dissipation is averaged over the magnetic surface, and is given as $2q^2 V_\theta(\nu_{local} - \boldsymbol{\nabla}_\perp \cdot \mu_i \boldsymbol{\nabla}_\perp)V_\theta$. The dissipation is enhanced by the return flow and is explicitly calculated for the poloidal flow (figure 19.1(a)) as

$$-\boldsymbol{f}\cdot\boldsymbol{V} = (1+2q^2)V_\theta(\nu_{local} - \boldsymbol{\nabla}_\perp \cdot \mu_i \boldsymbol{\nabla}_\perp)V_\theta. \tag{19.7}$$

The friction coefficient ν_θ is given in terms of local damping rate and ion viscosity as

$$\nu_\theta = (1+2q^2)(\nu_{local} - \boldsymbol{\nabla}_\perp \cdot \mu_i \boldsymbol{\nabla}_\perp). \tag{19.8}$$

In the limit $\nu_\zeta B_\zeta^2 \gg \nu_\theta B_\theta^2$, $\boldsymbol{E}\times\boldsymbol{B}$ flow causes poloidal rotation, and there appears an enhancement factor in toroidal plasmas of $(1+2q^2)$.

A possible nonlinear relation between radial electric field and radial current is caused by the nonlinear dependence of the coefficient ν_θ on radial electric field.

Polarization drift. When a radial electric field changes in time, a polarization drift occurs. The origin of this drift is from an acceleration of $\boldsymbol{E}\times\boldsymbol{B}$ velocity. A simple way to consider the effect of this inertial force on a flow is to employ the replacement

$$\nu_{local} \to \nu_{local} + \frac{\partial}{\partial t}. \tag{19.9}$$

The correction to radial drift is obtained from equations (19.6). When toroidal friction is weak, $\nu_\zeta B_\zeta^2 \ll \nu_\theta B_\theta^2$, and flow in the toroidal direction appears, the polarization drift is given as

$$e n_i V_r^p = \frac{m_i n_i}{B_\theta^2}\left(\frac{\partial}{\partial t}E_r\right). \tag{19.10-1}$$

The superscript p indicates polarization drift. In another limit, if toroidal friction is strong, $\nu_\zeta B_\zeta^2 \gg \nu_\theta B_\theta^2$, $\boldsymbol{E}\times\boldsymbol{B}$ flow causes poloidal flow, and the polarization drift is given as

$$e n_i V_r^p = \frac{m_i n_i}{B_\zeta^2}(1+2q^2)\left(\frac{\partial}{\partial t}E_r\right). \tag{19.10-2}$$

It is noted that polarization drift is in proportion to the mass of the particles. Therefore, the polarization drift selectively work on ions and drives a radial current as

$$J_r^p = n_i e_i V_r^p. \tag{19.11}$$

It is a convention to express a polarization current in terms of a perpendicular dielectric constant as

$$J_r^p = \varepsilon_0(\varepsilon_\perp - 1)\frac{\partial}{\partial t}E_r. \tag{19.12}$$

From equations (19.10) and (19.11), one sees that $\varepsilon_\perp$ is of the order of c^2/v_A^2 and crucially depends on the flow pattern as

$$\varepsilon_\perp - 1 = \frac{c^2}{v_A^2}\frac{q^2}{\varepsilon^2} \qquad \text{(toroidal flow)} \tag{19.13-1}$$

$$\varepsilon_\perp - 1 = \frac{c^2}{v_A^2}(1 + 2q^2) \qquad \text{(poloidal flow).} \tag{19.13-2}$$

19.1.3 Various Elements of Radial Current

There are many processes which cause radial current. If one writes this current equation in terms of a radial electric field for singly charged ions, one has

$$\frac{\varepsilon_0\varepsilon_\perp}{e}\frac{\partial}{\partial t}E_r = \Gamma_{e-i}^{anom} - \Gamma_i^{lc} - \Gamma_i^{bv} - \Gamma_i^{v\nabla v} - \Gamma_i^{NC} + \Gamma_e^{NC} - \Gamma_i^{cx}. \tag{19.14}$$

Terms on the right-hand side represent the following processes [16.4]. (i) Γ_{e-i}^{anom} is a contribution from a difference in bipolar parts of an anomalous cross-field flux (i.e., the excess flux of electrons relative to that of ions). (ii) Γ_i^{lc} is that from loss cone loss of ions. (iii) Γ_i^{bv} is due to a bulk viscosity coupled to a magnetic field inhomogeneity (i.e., transit time magnetic pumping). (iv) $\Gamma_i^{v\nabla v}$ is a Reynolds stress in a global flow, which includes the one due to toroidicity. (v) Γ_i^{NC} and Γ_e^{NC} are contributions from collisional fluxes (e.g., a ripple diffusion or a contribution of gyro-viscosity). (vi) Γ_i^{cx} is the ion loss owing to a charge exchange process. It should be noted that externally driven rf waves can be included in the term of Γ_{e-i}^{anom}.

Each of these terms has a dependence on radial electric field. In order to provide a perspective, we choose some characteristic terms.

19.2 Collisional Transport

19.2.1 Axisymmetric Plasmas

The dependence of radial current on radial electric field is discussed in chapter 17. Qualitative dependences of these terms are shown as

$$\Gamma_i^{lc} \sim f_{lc}\nu_i\rho_p n_i \exp(-X^2) \tag{19.15}$$

$$\Gamma_i^{bv} = \frac{\varepsilon^2 n_i T_i}{erB}(X + X_0)\,\mathrm{Im}\,Z(X + i\nu_{**}) \tag{19.16}$$

where $Z(X + i\nu_{**})$ is the plasma dispersion function [4.19], $\nu_{**} = \nu_i qR/v_{thi}$ and

$$X_0 = -\rho_p(n'/n + \gamma_{NC}T_i'/T_i) + V_{i\|}/v_{thi}.$$

The numerical coefficient f_{lc} is a function of geometrical factors such as ε. Formula (19.16), which is available for both banana and plateau regimes, is given in [17.15]. In a more collisional case, equation (19.16) has a limiting form

$$\Gamma_i^{bv} \sim \varepsilon^2 \nu_i \rho_p n_i \frac{X + X_0}{\nu_{**}^2 + X^2}. \tag{19.17}$$

The form of radial flux due to charge exchange with neutral particles is given as [17.5, 18.4, 19.1]

$$\Gamma_i^{cx} = -n_0\langle\sigma_{cx}v\rangle n_i \rho_p (X_0 + X - qV_\theta/\varepsilon v_{thi}) \tag{19.18}$$

where n_0 indicates the density of neutral particles and $\langle\sigma_{cx}v\rangle$ is a charge exchange cross-section. In the case where plasma is rotating only in the toroidal direction (i.e., $V_\theta = 0$), equation (19.18) reduces to a form being in proportion to (X_0+X). Figure 19.3 illustrates an example of radial current owing to bulk viscosity as a function of radial electric field. Nonlinear dependence of radial current on radial electric field is clearly observed.

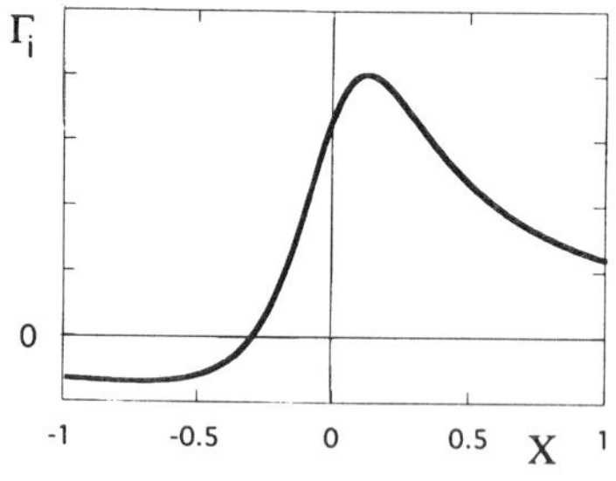

Figure 19.3. Example of radial current and the dependence on radial electric field, $X = e\rho_p E_r/T_i$. The contribution of bulk viscosity in tokamaks, for which electrons have only a small contribution, is taken into account.

19.2.2 Asymmetric Systems

The collisional transport in an asymmetric plasma is discussed in section 17.2.1. Ripple-trapped particles carry a radial current as

$$J_r^{NC} = e_i\Gamma_{a.i}^{NC} - e\Gamma_{a.e}^{NC}. \tag{19.19}$$

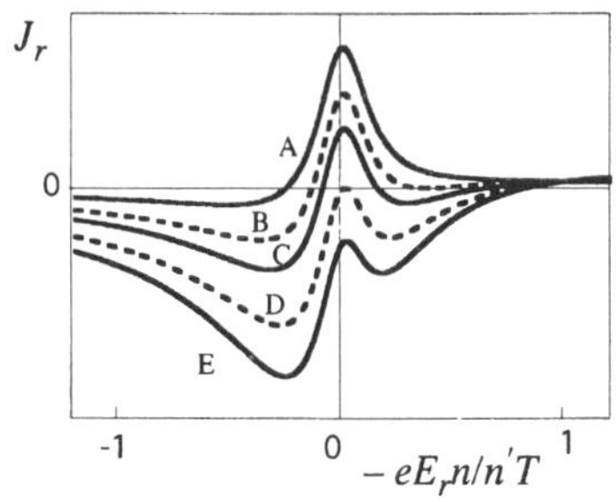

Figure 19.4. Example of radial current and the dependence on radial electric field. A current owing to ripple-trapped particles is shown based on figure 17.8. In cases A–E, ion flux is fixed, and electron temperature is increased from A to E.

An explicit form of $\Gamma_{a.j}^{NC}$ is given in equation (17.25). As is explained in section 17.2.1, $\Gamma_{a.j}^{NC}$ has a nonlinear dependence on E_r. As a result of this, the radial current from ripple-trapped particles has a complicated dependence on radial electric field. In the limit of large electric field, the direction of radial current is the same as that of radial electric field. However, in an intermediate strength of electric field, the direction of current could be opposite to the radial electric field. An example is illustrated in figure 19.4. In the region of weak radial electric field, the sign of radial current varies, depending on collisionalities of electrons and ions.

19.3 Anomalous Transport

Ambipolarity of turbulent driven flux is some times assumed *a priori*. However, this is *not true*. If turbulence is driven locally, i.e., waves being emitted by electrons are absorbed by ions on the same magnetic surface, momentum balance on each magnetic surface holds: this leads to ambipolarity. However, fluctuations can transmit momentum across magnetic surfaces (e.g., a convective damping of waves). Then an exchange of momentum between different magnetic surfaces takes place [4.11, 19.2], and a shear viscosity appears. Near the plasma edge, the convection of waves can be more important. These contributions give rise to a nonambipolar (bipolar) component in anomalous transport. Based on these considerations, a model (zero-dimensional expression) was proposed as

$$\Gamma_{e-i}^{anom} \sim \hat{D}[X, X'](\lambda_p - X) \tag{19.20}$$

where $\lambda_p = -\rho_p(\nabla n/n + \alpha_0 \nabla T/T)$ is the normalized gradient [16.2] (α_0 is the numerical coefficient). In the presence of strong inhomogeneity of pressure (i.e., the situation which is relevant to the H-mode), anomalous transport could play an important role for electric field generation.

19.3.1 Influence of Ion Viscosity

Contributions of anomalous transport to generation of a radial electric field have been investigated in accordance with the progress of the theory of turbulence. Phenomena like bootstrap current or rotation drive by a pressure gradient are analysed by the study of the off-diagonal elements of the transport matrix in nonlinear theory. By the renormalization of the Lagrangean derivative, the matrix formula is obtained. For instance, one-point renormalization gives the torque, the driving force for current and heating in the form

$$-\boldsymbol{\nabla}\cdot\boldsymbol{D}\begin{pmatrix}\nabla_r E_r'\\ \nabla_r J_\parallel\\ \nabla_r p_0\end{pmatrix} \tag{19.21}$$

where $J_\parallel$ is a parallel current, and the elements of matrix $\boldsymbol{D}$ are given by fluctuations and decorrelation rates (chapter 9). (Notice that vorticity is proportional to E_r'.)

Ion viscosity has an essential role in determining plasma flow and radial current. As is discussed in equation (19.5), the direction of flow is determined by competition between a local drag ν_{local} of poloidal flow and a damping rate ν_ζ of toroidal rotation. For the latter, ion viscosity is important. Let us study the simplified case where toroidal damping is dominated by shear viscosity, $-\nu_\zeta \simeq q^2\varepsilon^{-2}\nabla_\perp\mu_i\nabla_\perp$. If ion viscosity is small,

$$|\mu_i\nabla_\perp^2| \ll \varepsilon^2\nu_{local} \tag{19.22}$$

then the relation $\nu_\zeta B_\zeta^2 \ll \nu_\theta B_\theta^2$ is satisfied. The flow is in the toroidal direction, and the radial current is given by (19.6-1) as

$$en_i V_r = -\frac{m_i n_i}{B_\theta^2}\boldsymbol{\nabla}_\perp\cdot\mu_i\boldsymbol{\nabla}_\perp E_r. \tag{19.23-1}$$

In another limit, $|\mu_i\nabla_\perp^2| \gg \varepsilon^2\nu_{local}$, the relation $\nu_\zeta B_\zeta^2 \ll \nu_\theta B_\theta^2$ is satisfied. The flow is in the poloidal direction and the radial current is given by equation (19.6-2), i.e.,

$$en_i V_r = \frac{m_i n_i(1+2q^2)}{B_\zeta^2}(-\nu_{local} + \boldsymbol{\nabla}_\perp\cdot\mu_i\boldsymbol{\nabla}_\perp)E_r. \tag{19.23-2}$$

In both cases, ion current $e\Gamma_i$ includes the viscous term $\boldsymbol{\nabla}_\perp\cdot\mu_i\boldsymbol{\nabla}_\perp E_r$, and the dynamic equation (19.1) turns out to be a nonlinear diffusion equation. In the case where poloidal flow is driven, $|\mu_i\nabla_\perp^2| \gg \varepsilon^2\nu_{local}$, the equation for radial electric field has the form

$$\varepsilon_0\varepsilon_\perp\frac{\partial}{\partial t}E_r = (e\Gamma_e - e_i\Gamma_i)_{local} + \frac{m_i n_i}{B_\zeta^2}(1+2q^2)\boldsymbol{\nabla}_\perp\cdot\mu_i\boldsymbol{\nabla}_\perp E_r \tag{19.24}$$

where the first term in the right-hand side represents a radial current which is determined by the local value of radial electric field. (Compared to the case of $|\mu_i \nabla_\perp^2| \ll \varepsilon^2 \nu_{local}$, the coefficient of the second term in equation (19.24) is smaller by the factor $2\varepsilon^2$.) It is noted that the radial current from electron viscosity is not kept in equation (19.24). This is because the radial current due to viscosity is in proportion to particle mass, and is small for electrons for cases of $\mu_e \sim \mu_i$.

19.3.2 Generation and Damping of Flow

Microscopic fluctuations in a plasma can cause a global flow and a global (static) radial electric field. Generation of a global electric field by fluctuations is symbolically written in equation (19.20). In this section, mechanisms to generate a static and global flow from small scale fluctuations are discussed. Anomalous transport, that influences evolution of the radial electric field through ion viscosity, can cause selective loss of ions or electrons at the same time, and drives a radial electric field and associated flow. A generation of flow by fluctuations is expected when a mirror symmetry does not hold. The radial electric field and associated poloidal flow, which are generated by plasma fluctuations, have been studied intensively in relation to the H-mode physics. As an example, one can consider an inverse cascade process in plasma turbulence. Owing to nonlinear interactions, the inverse cascade takes place, and microscopic fluctuations accumulate to a longer wavelength region and give rise to a global structure [14.2]. There is a possibility that this inverse cascade process also generates global and static radial electric field.

If one writes an equation of poloidal momentum P_θ in the absence of radial current, it contains terms

$$\frac{\partial}{\partial t}\overline{P_\theta} = -\frac{\partial}{\partial \gamma}\overline{\langle \tilde{V}_r \tilde{P}_\theta \rangle} - \frac{2T}{R}\overline{\cos\theta \frac{\partial}{\partial \theta} n} \qquad (19.25)$$

where the overbar indicates an average on the magnetic surface.

The first term in the right-hand side of equation (19.25) is understood as a convection of momentum. If a perturbation of momentum $\tilde{P}_\theta$ is in phase with a radial velocity fluctuation $\tilde{V}_r$, there appears a net radial flux of poloidal momentum. An example of a flow pattern in fluctuations, which has a correlation between $\tilde{P}_\theta$ and $\tilde{V}_r$, is shown in figure 19.5. In this case, $\tilde{P}_\theta$ takes positive (negative) value if $\tilde{V}_r$ is positive (negative). The average $\langle \tilde{V}_r \tilde{P}_\theta \rangle$ does not cancel out, but gives a finite value. The divergence of this flux appears as a local force.

The evaluation of the term $\langle \tilde{V}_r \tilde{P}_\theta \rangle$ has been performed by calculation of turbulent-driven flux. The quasilinear response of the momentum transport is explained in chapter 4. Here, heuristic arguments on nonlinear terms are first explained. One example of an evaluation of a nonlinear term is discussed.

Argument on a symmetry. The phase relation between $\tilde{P}_\theta$ and $\tilde{V}_r$ is as important as that between $\tilde{n}$ and $\tilde{V}_r$ for determination of particle flux.

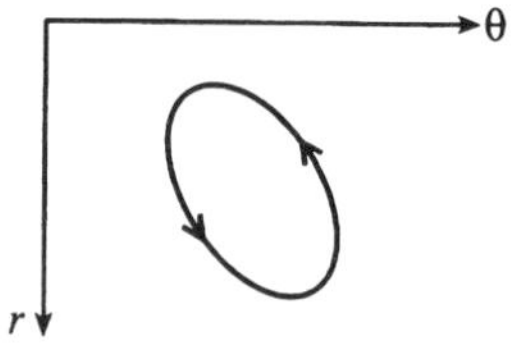

Figure 19.5. Correlation between $\tilde{P}_\theta$ and $\tilde{V}_r$ makes a net momentum flow $\langle \tilde{P}_\theta \tilde{V}_r \rangle$.

One heuristic argument is as follows [19.3]. (i) This term is a quadratic of fluctuation level if the correlation is independent of the level. (ii) If fluctuations are not propagating in plasma, there is no preferential momentum in the absence of a symmetry breaking $V'_{E\times B}$; therefore it may be proportional to $V'_{E\times B}/\gamma_0$ (γ_0 being a typical growth rate of the mode). Based on these dimensional considerations, one has an analytic estimate for the divergence of turbulent-driven flux (i.e., torque) as

$$\frac{\partial^2}{\partial r^2}\overline{\langle \tilde{V}_r \tilde{P}_\theta \rangle} \propto -\left(\frac{p_0}{\ell^2}\right)\left|\frac{\tilde{n}}{n}\right|^2 \frac{1}{\gamma_0} V'_{E\times B}. \tag{19.26}$$

Coefficient p_0/ℓ^2, ℓ being a scale-length (which should be determined self-consistently), is put to identify the dimension of a force. Based on a drift wave model, the coefficient p_0/ℓ^2 is discussed in [19.3]. A dimensional dependence like equation (19.26) may have applicability if the mode characteristics are properly chosen. An alternative expression is discussed for toroidal plasmas [19.4]. This type of study is widely developed in the literature [19.5].

The second term of the right-hand side of equation (19.25) remains finite if plasma parameters (static as well as fluctuating quantities) depend on poloidal angle. An up–down difference of these is expected in a toroidal plasma and could also be important for anomalous transport. An expression was derived [19.6]

$$\frac{2T}{R}\overline{\cos\theta \frac{\partial}{\partial \theta} n} \simeq \frac{q^2}{\varepsilon}\frac{1}{n}\overline{\cos\theta \frac{\partial}{\partial \gamma}\langle \tilde{V}_r \tilde{n} \rangle}\, \overline{P_\theta}. \tag{19.27}$$

This term is also proportional to average poloidal momentum. If the $\cos\theta$ component of radial flow $\langle \tilde{V}_r \tilde{n} \rangle$ exists, then the average $\overline{\cos\theta (\partial/\partial r)\langle \tilde{V}_r \tilde{n} \rangle}$ remains finite. In such a case, there arises a force which is proportional to $\overline{P_\theta}$. If there is no poloidal rotation, $\overline{P_\theta} = 0$, the torque in the right-hand side of equation (19.27) vanishes.

Both the terms (19.26) and (19.27) are in proportion to global flow V_θ or its derivative V'_θ. When one considers the case where the coefficient of $\overline{P_\theta}$ in equation (19.27) is much more slowly varying in space than $\overline{P_\theta}$ itself, the radial derivative of force equation (19.27) may be written as $(\partial/\partial r)(\overline{(2T/R)\cos\theta(\partial/\partial\theta)n}) \simeq ((q^2/\varepsilon)(1/n)\overline{\cos\theta(\partial/\partial r)\langle\tilde{V}_r\tilde{n}\rangle})(\partial/\partial r)\overline{P_\theta}$. Then the spatial derivative of equation (19.25), $(\partial^2/\partial t\partial r)\overline{P_\theta} = -(\partial^2/\partial r^2)\langle\tilde{V}_r\tilde{P}_\theta\rangle - (\partial/\partial r)(\overline{(2T/R)\cos\theta(\partial/\partial\theta)n})$, is symbolically written

$$\frac{\partial}{\partial t}P'_\theta = A_1 V'_\theta + A_2 P'_\theta.$$

where A_1 and A_2 are coefficients that include turbulent fluctuations. The mechanism contained in equation (19.25) can enhance the global flow and its structure, if the coefficient $n_i^{-1}A_1+A_2$ is positive and if there is a seed for a shear of global flow [19.7]. This feature resembles the situation of a dynamo problem. For instance, α- and β-dynamos are proportional to average magnetic field and to current, respectively. A seed component of the global scale is required. A connection to the physics of solar dynamics [19.8] has been pointed out.

Spontaneous torque. An estimate for anomalous torque must be developed under the same theory that is used in the analysis of diagonal terms in transport flux. The theory of self-sustained turbulence was extended to estimate equation (19.21) as [19.9]

$$P_{\perp r} = -m_i n_i \mu \left\{ \nabla V_{E\times B} + \frac{M_{12}}{M_{11}} v_{Ap} \nabla^2 \beta \right\} \tag{19.28}$$

with $\mu \simeq \chi$ and $M_{12}/M_{11} \sim (4\varepsilon)^{-1}F^{-2}(\alpha, s)\sqrt{m_i/m_e}c\omega_p^{-1}$. Radial inhomogeneity of plasma pressure causes a radial flow of momentum. Model equations have been proposed as

$$\Gamma_{e-i}^{anom} \sim \hat{D}[X, X'](\lambda_p - X) + \hat{\mu}_\perp[X, X']\hat{\nabla}_\perp^2 X \tag{19.29}$$

where $\lambda_p = -\rho_p(\nabla n/n + \alpha_0 \nabla T/T)$.

A density gradient can generate a poloidal flow and radial electric field. This nature has been confirmed by numerical simulations. It was shown that potential energy has the nature to cause an inverse cascade, if parallel electron motion is impeded [18.8]. Numerical simulation has shown that as the fluctuations develop, a zonal flow is generated [18.8, 19.6].

19.3.3 Anomalous Bootstrap Current and Dynamo

Anomalous bootstrap current is known to be small in toroidal plasmas. This is because wave vectors of fluctuations are mainly in the poloidal direction, and little momentum in the parallel direction is exchanged by electrons [18.1, 19.10]. The importance of the dynamo term for structural formation in the L–H-transition has also been pointed out [19.11]. The interaction of current generation and flow generation was discussed. The mechanism is found to be effective in large scale dynamics (like celestial plasmas) and its application to a laboratory plasma remains a future problem [19.12].

REFERENCES

[19.1] Ohkawa T and Hinton F L 1987 *Plasma Physics and Controlled Nuclear Fusion Research 1986* vol 2 (Vienna: IAEA) p 221

[19.2] Berk H L and Molvig K 1983 *Phys. Fluids* **26** 1385

[19.3] Diamond P H, Liang Y-M, Carreras B A and Terry P W 1994 *Phys. Rev. Lett.* **72** 2565

[19.4] Zhang Y Z and Mahajan S M 1995 *Phys. Plasmas* **2** 4236

[19.5] Rozhanskii V and Tendler M 1992 *Phys. Fluids* B **4** 1877
Sugama H and Horton C W 1995 *Plasma Phys. Control. Fusion* **37** 345
Dnestrovskij A Yu, Parail V V and Vojtsenkhhovich I A 1993 *Plasma Physics and Controlled Nuclear Fusion Research 1992* vol 2 (Vienna: IAEA) p 371

[19.6] Drake J F, Finn J M, Guzdar P *et al* 1992 *Phys. Fluids* B **4** 488
Drake J F, Antonsen T M, Finn J M *et al* 1993 *Plasma Physics and Controlled Nuclear Fusion Research 1992* vol 2 (Vienna: IAEA) p 115

[19.7] Stringer T E 1969 *Phys. Rev. Lett.* **22** 1770

[19.8] See, e.g., Zeldovich Ya B, Ruzmaikin A A and Sokoloff D D 1983 *Magnetic Fields in Astrophysics* (New York: Gordon and Breach)
Ruediger G 1989 *Differential Rotation and Stellar Convection* (New York: Gordon and Breach)

[19.9] Itoh K, Itoh S-I, Fukuyama A and Yagi M 1996 *J. Phys. Soc. Japan* **65** 760

[19.10] Itoh S-I and Itoh K 1988 *Phys. Lett.* **127A** 267

[19.11] Yoshizawa A 1991 *Phys. Fluids* B **3** 2723

[19.12] Yoshizawa A 1990 *Phys. Fluids* B **2** 1589
Yoshizawa A and Yokoi N 1993 *Astrophys. J.* **407** 540

Chapter 20

Electric Field Bifurcation

The equation that governs dynamics of a radial electric field predicts a bifurcation of the solution E_r. Since plasma transport coefficients are dependent on electric field structure, electric field bifurcation simultaneously causes bifurcations in the plasma transport. This bifurcation has been discussed as a main working hypothesis for understanding the structural transition in confined plasmas.

20.1 Hard and Soft Bifurcation

20.1.1 Bifurcation and Cusp

The transport coefficient, i.e., the ratio of flux to gradient, is discussed based on the concept of a contraction of variables. Coefficient $\chi[\nabla p, \ldots : E_r, E_r', \ldots]$ is a function of various gradients and electric field. In stationary states, an electric field is an implicit function of plasma parameters through the charge neutrality equation

$$\Gamma_e[E_r(r); \nabla p, \ldots] = Z_i \Gamma_i[E_r(r); \nabla p, \ldots]. \tag{20.1}$$

This coupling induces bifurcations in transport coefficients as well. The transport coefficient, $\chi[\nabla p, \ldots : E_r, E_r', \ldots]$, is usually considered to be a smooth function of variables as is shown in chapters 17 and 18. Nevertheless, a bifurcation in the radial electric field at a particular set of plasma parameters gives rise to a discontinuity of this coefficient at this set of plasma parameters. Plasma flux could be subject to a bifurcation at a certain value of gradient.

Theories of electric field bifurcation in plasmas are classified into two, depending on a characteristic relation between gradient and flux. Figure 20.1 schematically illustrates examples of two cases, showing the relations between the particle and heat fluxes as a function of gradient ∇n and/or ∇T. Radial electric field and fluctuation level against gradients are also plotted. Figure 20.1(a) shows a *hard transition* model. In this case, the flux (and fluctuation level etc as well) takes multiple values at certain values of fixed gradients. Figure 20.1(b) presents those for a *soft transition* model. In the latter

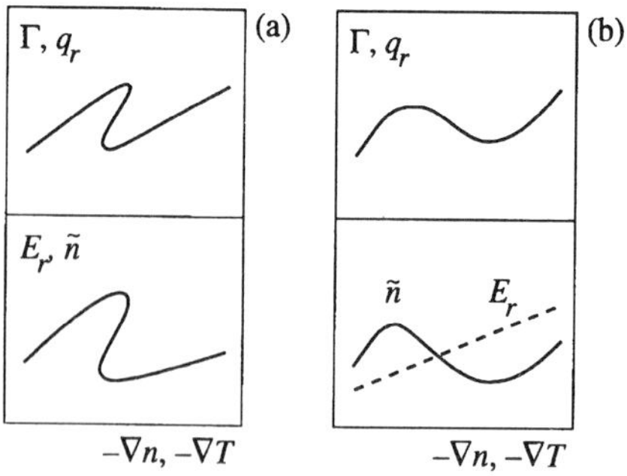

Figure 20.1. Hard transition (a) and soft transition (b).

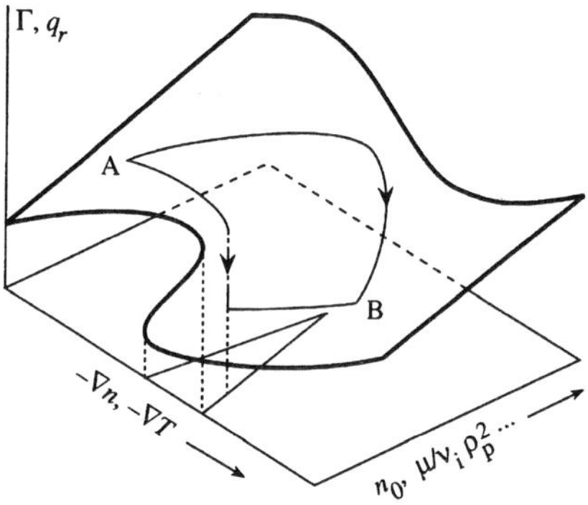

Figure 20.2. Cusp of a hard transition.

case, the flux can be a decreasing function of gradients, but is a single value function. If a system has the nature of a hard transition, a soft transition is also available by introduction of a change of parameters. For the case of a hard transition, a cusp type bifurcation is obtained as is illustrated in figure 20.2. Gradients $-\nabla T$ and $-\nabla n$ play the role of *order parameter*, and those such as neutral density are *control parameters*. A change from a state with large flux to a state with smaller flux ('A' to 'B' in figure 20.2) could occur either by crossing the transition points, or by following a smooth change as is illustrated.

20.1.2 Asymmetric Systems

The electric field bifurcation is predicted in asymmetric systems. As is shown in figure 19.4, the E_r–J_r relation is a nonlinear and nonmonotonic function.

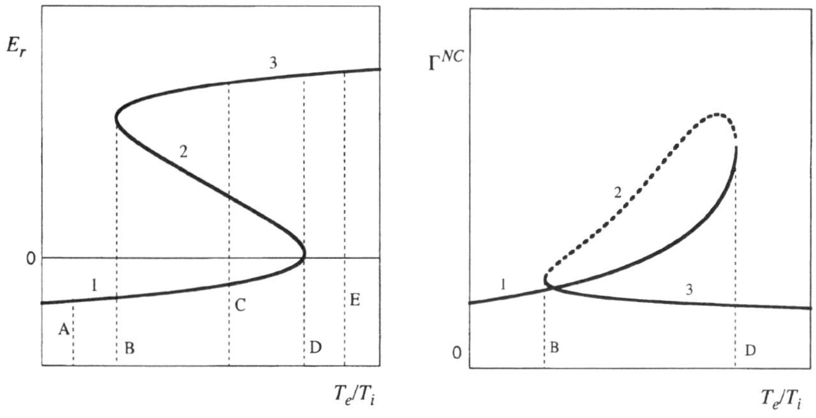

Figure 20.3. Solution of equation (20.1) as a function of the ratio of temperatures (left). In this case, ripple-trapped particles dominate the radial current. Symbols A–E correspond to those in figure 19.4. Neoclassical particle flux as a function of the ratio of the temperature (right). Line 1 is the branch of the negative radial electric field, and line 3 is the one with positive electric field. Line 2 is the intermediate solution.

Owing to this nonlinearity, multiple solutions of equation (20.1) exist. In the case of line A of figure 19.4, equation (20.1) has one solution of negative E_r. In the case B, a bifurcation takes place, and three solutions exist for the case of C. At D, a transition occurs again. In the case E, only one positive solution of E_r exists. As a summary of these results, the solution of equation (20.1) is illustrated in figure 20.3 (left). In this case, the ratio of temperatures plays a role in controlling the bifurcation.

A bifurcation of electric field is associated with that of plasma flux. Neoclassical flux, which is evaluated by a solution of the radial electric field of equation (20.1), is shown in figure 20.3 (right). At critical conditions (B or D in figure 20.3 (right)), a sudden jump of flux takes place. As is shown in equation (17.29), particle flux becomes smaller if the absolute value of radial electric field is larger. In an intermediate branch, flux could be larger. As is shown later, the intermediate branch is unstable, so that a jump between branches 1 and 3 is expected.

20.1.3 Axisymmetric Systems

A bifurcation in axisymmetric systems has attracted attention related to the H-mode transition. The Itoh–Itoh model and Shaing model with respect to the H-mode transition belong to the class of hard transitions [16.2, 17.14]. Those in [19.3, 19.5, 20.1, 20.2] belong to the class of soft transitions. The physics of

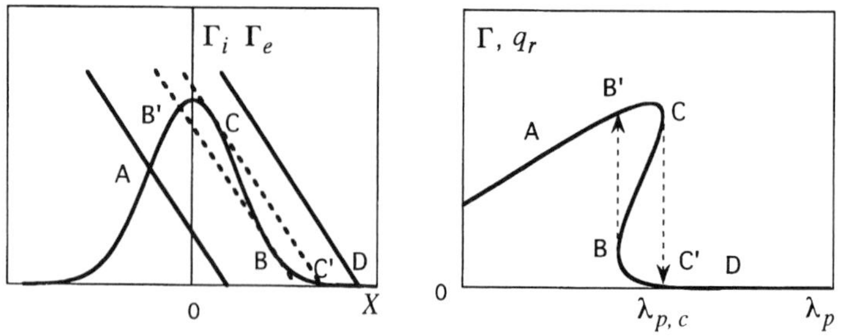

Figure 20.4. Example of the hard transition. Balance between the ion orbit loss and the bipolar electron flux as a function of electric field $X = e\rho_p E_r/T_i$ (left) and the gradient–flux relation that satisfies the charge neutrality condition (right).

transition is explained in the following. Bifurcations of hard transition type have been found by a combination of equations (19.14)–(19.20). A first example is the analysis of the balance between Γ_{e-i}^{anom} and Γ_i^{lc} [16.2] and another is that between Γ_i^{lc} and Γ_i^{bv} [17.14].

Figure 20.4 illustrates these examples of bifurcation. Figure 20.4 (left) illustrates an ion orbit loss flux, equation (19.15) (Gaussian shape curve), and an electron particle flux, equation (19.20) (straight lines), as a function of radial electric field, $X = e\rho_p E_r/T_i$. A crossing point of two lines gives a self-consistent flux that is a solution of charge neutral condition equation (20.1). When a plasma gradient is weak, two lines have a cross point at A with a large flux. If the gradient increases, the solution moves from A to C through B′. At critical condition C, a transition from C to C′ occurs. The solution further moves to a point D as the plasma gradient increases. When the gradient starts to decrease, the solution moves as D → C′ → B. The back-transition takes place from B to B′. The gradient–flux relation, which satisfies equation (20.1), is summarized in figure 20.4 (right).

The other example is shown in figure 20.5. The ion flux via orbit loss (solid line), and that by bulk viscosity equation (19.16) (dashed line), are drawn as a function of radial electric field. When a bipolar component is absent for electrons, the charge neutrality condition is simplified as $\Gamma_i = 0$. The crossing points of two lines in figure 20.5 represent the solution of equation $\Gamma_i = 0$. Multiple solutions are possible as is illustrated in figure 20.5. As plasma parameters gradually change, a sudden jump takes place in the radial electric field.

It is seen from these figures that the radial electric field shows a hard transition in tokamaks at critical plasma parameters. The plasma fluxes jump at the critical point as well.

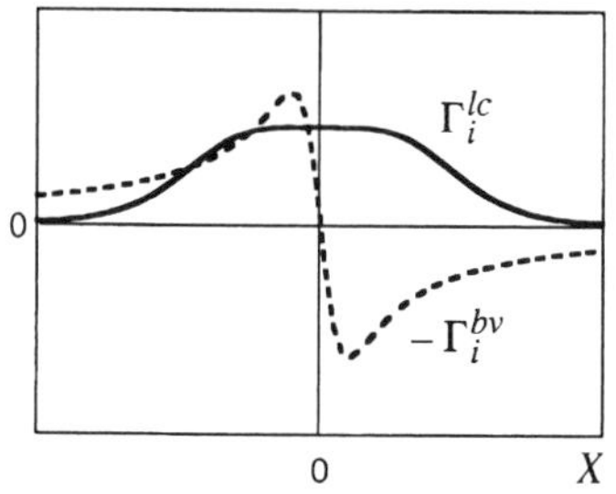

Figure 20.5. Another example of hard transition. The balance between ion orbit loss (solid line) and contribution of ion bulk viscosity (dashed line) is shown as a function of $X = e\rho_p E_r / T_i$. (Based on [17.14].)

20.2 Hysteresis and Limit Cycle

A hard transition model predicts a fast transition at critical points. This system has hysteresis and can generate a limit cycle oscillation.

Figure 20.6 illustrates loss rate as a function of a plasma parameter in the case of a hard transition type. This curve is seen to belong to a family of *backward* bifurcation (which usually appears in a subcritical turbulence). There is a hysteresis, as is shown by a dotted line. The central branch is thermodynamically unstable, so that a limit cycle oscillation is possible. Let us consider the situation where a critical condition is satisfied under a gradual change of global parameters. As is shown later in chapter 22, the limit cycle oscillation appears suddenly at the critical condition.

A hysteresis also exists in the case of a soft bifurcation. This belongs to a class of normal bifurcations called pitch-fork bifurcations. A limit cycle oscillation can also be available in this type. The limit cycle appears slowly in accordance with a gradual change of global parameters. This is a point that

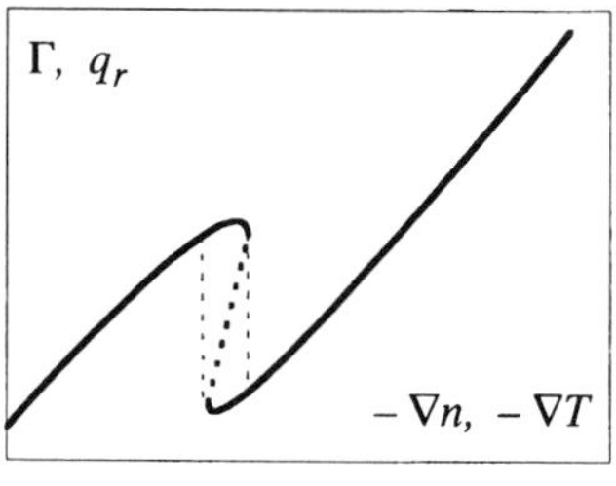

Figure 20.6. Hysteresis in a relation between gradient and loss.

discriminates the relevance of models.

A hard transition can in principle happen sequentially. In such a case, a chaotic behaviour or an intermittence of bursts can appear. These problems have been studied intensively in statistical physics (e.g., [20.3]), but not much work has been done in plasma transport theory.

20.3 Time Scale of Fast Transition

20.3.1 Dielectric Constant

The dielectric constant $\varepsilon_\perp$ determines how fast the stationary state equation (20.1) is realized. In a magnetized plasma, the dielectric constant takes a large value. This is because polarization drift of ions screens the charge separation. As radial electric field starts to increase, so does the cross-field $\boldsymbol{E} \times \boldsymbol{B}$ velocity. The acceleration in $\boldsymbol{E} \times \boldsymbol{B}$ velocity causes a polarization drift of ions. This polarization drift is in the radial direction, and total drift velocity in slab geometry is given as

$$\boldsymbol{V}_\perp = \frac{\boldsymbol{E} \times \boldsymbol{B}}{B^2} + \frac{m_i}{eB^2} \frac{\partial}{\partial t} \boldsymbol{E}.$$

This polarization ion drift is directed so as to reduce the change of radial electric field. In a slab plasma, the dielectric constant is given by

$$\varepsilon_\perp^s = 1 + c^2 v_A^{-2} \approx c^2 v_A^{-2}. \tag{20.2}$$

A superscript s indicates slab geometry.

In toroidal plasmas, this coefficient can be enhanced [20.4, 20.5]. As is discussed in relation to the Pfirsch–Schlüter current [2.3] or to damping of poloidal flow, equation (19.5), a flow in the poloidal direction can be coupled to a toroidal flow. This is because plasma flow is nearly incompressible, $\boldsymbol{\nabla} \cdot \boldsymbol{V} \approx 0$. If the flow is constrained to the magnetic surface, as in the neoclassical theory, i.e., $V_r = 0$, a divergence of poloidal flow $\boldsymbol{\nabla}_\perp \cdot \boldsymbol{V}_p$ must be compensated by a parallel divergence $\boldsymbol{\nabla}_\parallel \cdot \boldsymbol{V}_\parallel$. As is discussed in section 19.1.2 the poloidal flow V_p drives the parallel flow,

$$V_\parallel \sim (2q\cos\theta) V_p \tag{20.3}$$

(see figure 19.1a). In this case, the polarization drift is enhanced as is shown in equation (19.10-2). This is interpreted as the ion inertia being effectively increased. As a result of this, the dielectric constant is multiplied by an enhancement factor M_{eff},

$$\varepsilon_\perp^{tor} = M_{eff} \varepsilon_\perp^s \tag{20.4-1}$$

$$M_{eff} \simeq 1 + 2q^2. \tag{20.4-2}$$

In general, M_{eff} alters for different collisionality and is a function of radial electric field. See, e.g., [20.4, 20.5] for details of neoclassical theory.

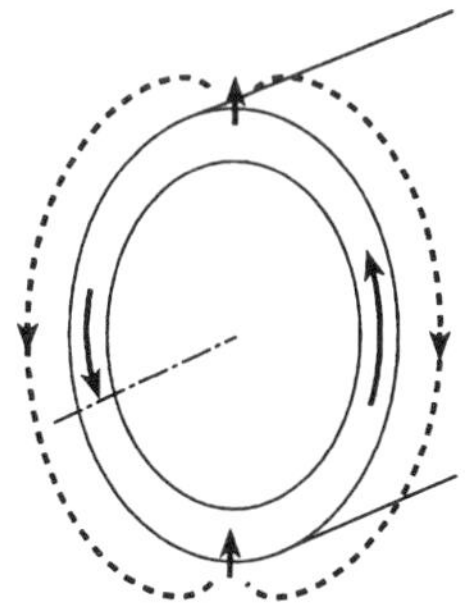

Figure 20.7. The radial flow may compensate the divergence of poloidal flow.

This neoclassical enhancement factor M_{eff} depends on the flow pattern in toroidal plasmas. The factor may be different from equation (20.4-2). Two possible cases are noted. The factor M_{eff} can be larger than the formula of equation (20.4-2). If the damping for toroidal flow is weaker, $\nu_\zeta B_\zeta^2 \ll \nu_\theta B_\theta^2$, $\boldsymbol{E} \times \boldsymbol{B}$ flow is directed in the toroidal direction, not in the poloidal direction, equation (19.5). If this pattern of flow is realized, the inertial effect is much enhanced and the polarization current becomes larger as is given by equation (19.10-1). The dielectric constant increases as

$$M_{eff} \simeq q^2 \varepsilon^{-2}. \tag{20.5}$$

The factor M_{eff} could also be smaller than equation (20.4-2) in some cases. The plasma flow across a magnetic surface possibly exists in a rapid transition like the L–H-transition (figure 20.7). The divergence of poloidal flow can be compensated in a different way, namely, $\partial V_r/\partial r \approx -r^{-1}\partial V_\theta/\partial\theta$. If a thin layer of width ℓ near the edge is considered, a radial velocity of the order

$$V_r \sim (\ell/R) V_\theta \sin\theta \tag{20.6}$$

is sufficient to balance the divergence of poloidal flow. The $\sin\theta$ dependence of radial flow is generated by a poloidal inhomogeneity of static potential. As is discussed in chapter 21, a radial flow with $\sin\theta$ dependence could be possible. The analysis of the dielectric constant in toroidal plasmas with inhomogeneous radial flow remains for future analysis.

20.3.2 Time Constant at Bifurcation

Based on the arguments on dielectric constant, a typical time scale of a jump of radial electric field is estimated. During a transition from one solution to the other, the radial electric field takes a value which is in between two stable solutions. The excess current during this interval is calculated from

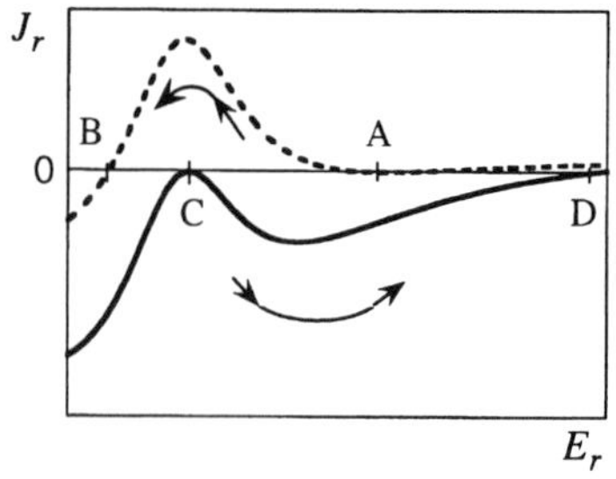

Figure 20.8. Relation between the electric field and current, that induces the transition. Transitions from A to B and C to D are shown. The solid and dashed lines correspond to D and B in figure 19.4, respectively.

equation (19.14). If one studies the case of poloidal rotation, $\nu_\zeta B_\zeta^2 \gg \nu_\theta B_\theta^2$, a typical value is evaluated through the component equation (19.16). The radial current is estimated to be of the order of

$$J_r \sim \pm\varepsilon^2 \frac{v_{thi}}{qR} \frac{m_i n_i}{B_p^2} E_r \tag{20.7}$$

in the limit of weak radial electric field. (The sign depends on the direction of bifurcation: the 'plus' sign is for the case from a larger E_r to a smaller E_r, and vice versa. See figure 20.8.) By substituting equations (20.2), (20.4-1) and (20.7) into equation (19.1), one has the relation

$$\frac{\partial}{\partial t} E_r \sim \pm \frac{1}{M_{eff}} \frac{v_A^2}{\varepsilon_0 c^2} \left(\varepsilon^2 \frac{v_{thi}}{qR} \frac{m_i n_i}{B_p^2} \right) E_r. \tag{20.8}$$

One finds that the radial electric field changes in a time scale of the order of

$$\tau_{tr} \sim \frac{1}{q^2} \frac{qR}{v_{thi}} M_{eff}. \tag{20.9}$$

If one employs a neoclassical dielectric constant, equation (20.4-2), $M_{eff} \simeq 1 + 2q^2$, equation (20.9) becomes

$$\tau_{tr} \sim 2(1 + q^{-2}) \frac{qR}{v_{thi}}. \tag{20.10}$$

A typical time for a transition of radial electric field is estimated to be an ion-transit time, qR/v_{thi}.

20.4 Model Equations

A model system is constructed to include essential elements, i.e., nonlinearity, bifurcation, spatial diffusion and dissipation. In order to understand the dynamics of a problem, a nonlinear diffusion equation which contains a bifurcation is necessary. This type of model is studied first. Then a point model is considered. A local approximation model (point model) with simple and tractable equations provides fruitful physics, when a proper choice of scale-length is made. The Lorenz model for a neutral fluid is one of the most famous model equations of this kind.

20.4.1 Extended Ginzburg–Landau Type Equations

Equations that govern an evolution of profiles $\{n(r), T(r), E_r(r)\}$ are obtained based on a concept of time-scale separation. A characteristic time for a change of turbulence, which is given as $\mathrm{O}(\alpha^{-1/2}\tau_{Ap})$, is considered to be much faster than that of plasma parameters. As plasma parameters and their profiles change, fluctuations develop so fast that stationary relations for turbulent transport, such as equations (18.25) and (18.28), are immediately established. By this time-scale separation, transport equations of plasma parameters and electric field are derived as a closed set of equations. They are given in symbolic form by use of a nonlinear transport matrix as

$$\frac{\partial}{\partial t}n(r) = \nabla \cdot D[X, X'; n(r), T(r), \ldots]\nabla n(r) + S_p \tag{20.11}$$

$$\frac{\partial}{\partial t}T(r) = \nabla \cdot \chi[X, X'; n(r), T(r), \ldots]\nabla T(r) + P \tag{20.12}$$

$$\upsilon\frac{\partial}{\partial t}X(r) = \nabla \cdot \mu[X, X'; n(r), T(r), \ldots]\nabla X(r) + N[X, X'; n(r), T(r), \ldots] \tag{20.13}$$

where $X = e\rho_p E_r/T_i$, S_p is the particle source and P is heating power. (In general, terms like 'pinch' can also be included in equations (20.11) and (20.13).) A smallness parameter υ indicates the fact that the time scale in equation (20.13), determined by flux and viscosity, could be shorter than those in transport processes that determine temperature and density. The nonlinear term N in equation (20.13) indicates source terms of radial electric field in equation (19.14), which allow a bifurcation of hard transition type. Transport coefficients in the forms

$$D = D[X, X'; n(r), T(r), \ldots] \tag{20.14-1}$$

$$\mu = \mu[X, X'; n(r), T(r), \ldots] \tag{20.14-2}$$

$$\chi = \chi[X, X'; n(r), T(r), \ldots] \tag{20.14-3}$$

are derived from a nonlinear turbulence theory. This set of equations belongs to a type of extended-time-dependent Ginzburg–Landau equation (E-TDGL equation), and has been used in the study of self-organized dynamics [20.6]. Details are explained in chapter 22. (The term 'extended' indicates that the set of equations (20.11)–(20.13) includes diffusion terms.)

20.4.2 Zero-Dimensional Model

The set of equations (20.11)–(20.13) includes information on both time and space. By a replacement of the operator ∇ with a numeric, the set of equations is simplified as a one-point model for local variables. The E-TDGL type equation is reduced to

$$\frac{\partial}{\partial t}n = -\frac{1}{\tau_{p,\ell}[n, T, X]}n + S_p \tag{20.15}$$

$$\frac{\partial}{\partial t}T = -\frac{1}{\tau_{E,\ell}[n, T, X]}T + P \tag{20.16}$$

$$v\frac{\partial}{\partial t}X(r) = -\frac{1}{\tau_{v,\ell}[n, T, X]}X + N[n, T, X]. \tag{20.17}$$

The suffix ℓ indicates that the loss time is evaluated within a layer ℓ of interest. A similar formulation, including a hard transition, was obtained from equations (20.11)–(20.13).

Another type of bifurcation (e.g., soft type) is also modelled in a local dynamical model. A set of equations for plasma pressure, electric field inhomogeneity and the fluctuation level was obtained [19.3]. In this model, time scales of changes in fluctuations and in global parameters are considered to be of the same order as those of dynamics of the electric field. Coupled equations are proposed

$$\dot{A} = \gamma_0 A - \alpha_1 A^2 - \alpha_2 U A \tag{20.18}$$

$$\dot{U} = -\mu U + \alpha_3 A U \tag{20.19}$$

$$\dot{G} = -\alpha_5 G - \alpha_4 G A + P \tag{20.20}$$

where A denotes fluctuation level, $A \sim |\tilde{n}/n|^2$, U flow shear, $U \sim |V_E'|^2$, G pressure gradient and P heating power, respectively. γ_0 indicates linear growth rate, and μ represents collisional damping. (This particular form was derived by dimensional arguments, and is not necessarily unique. Similar, but not identical, zero-dimensional dynamical models are given by other groups (say, [19.5]). The set of equations (20.18)–(20.20) is shown as a representative one of this kind.) This type of dynamical model has two types of steady state, i.e.,

$$A = \frac{\gamma_0}{\alpha_1} \qquad U = 0 \tag{20.21-1}$$

$$A = \frac{\mu}{\alpha_3} \qquad U = \left(\gamma_0 - \frac{\alpha_1 \mu}{\alpha_3}\right)\frac{1}{\alpha_2}. \tag{20.21-2}$$

These two branches smoothly merge at the condition $\gamma_0\alpha_3 = \mu\alpha_1$. A soft transition (pitch-fork bifurcation) is predicted. These types of model would be useful in analysing the phenomenon of soft transition.

REFERENCES

[20.1] Ohkawa T, Hinton F L, Liu C S and Lee Y C 1983 *Phys. Rev. Lett.* **51** 2102
[20.2] Hinton F L 1991 *Phys. Fluids* B **3** 696
[20.3] See, e.g., Tominaga H and Mori H 1994 *Prog. Theor. Phys.* **91** 1081
[20.4] Shaing K C, Hazeltine R D and Sanuki H 1992 *Phys. Fluids* B **4** 404
[20.5] Hirshman S P 1978 *Nucl. Fusion* **18** 917
Hirshman S P 1978 *Phys. Fluids* **21** 1295
Hinton F L and Robertson J A 1984 *Phys. Fluids* **27** 1243
Hugill J 1994 *Plasma Phys. Control. Fusion* **36** B173
Wobig H and Kisslinger J 1995 *Plasma Phys. Control. Fusion* **37** 893
Rozhansky Y and Tendler M 1996 *Reviews of Plasma Physics* vol 19 ed B B Kadomtsev (New York: Consultants Bureau) p 147
[20.6] Itoh S-I, Itoh K, Fukuyama A and Miura Y 1991 *Phys. Rev. Lett.* **67** 2485

Chapter 21

Interface—Spatial Structure of Electric Field

Bifurcation of a radial electric field causes steep gradients of electric field and plasma parameters. In this chapter, the radial structure of electric field is illustrated. A short description is also given of a possible variation of plasma parameters and electric potential in the poloidal direction.

21.1 Interface of Domains

The spatial structure of the electric field has vital importance, because inhomogeneity of the radial electric field is an important mechanism for reducing a turbulent transport. On the basis of the set of equations (20.11)–(20.13), the spatial structure has been solved. An analytic insight is discussed here and a detailed transport simulation is discussed later.

21.1.1 Interface

The investigation of a spatial interface at a discontinuity in a medium which has multiple states (such as the L- and H-phases in this case) is a key issue. A simplified model shown in figure 21.1 is employed, and a bifurcation with a hard transition is considered. Parameter g is a representative plasma parameter that controls the bifurcation. A charge neutrality condition equation (20.1), $J_r[E_r; g] = 0$, yields a relation $E_r[g]$, which is a multiple value function. (In the analysis [20.6], a parameter is introduced as $g = D/L_n \nu_i \rho_p$. In the case of figure 20.3, a parameter $g = T_e/T_i$ controls the transition.) Radial variation of the plasma parameter induces a change of controlling parameter g (i.e., the abscissa of figure 21.1). In a range of parameter $g_1 < g < g_2$, multiple solutions exist. The parameter g becomes a function of plasma radius, when plasma parameters are inhomogeneous. A possible solution for radial electric field (and transport coefficients as well) varies in space owing to multiple solutions of

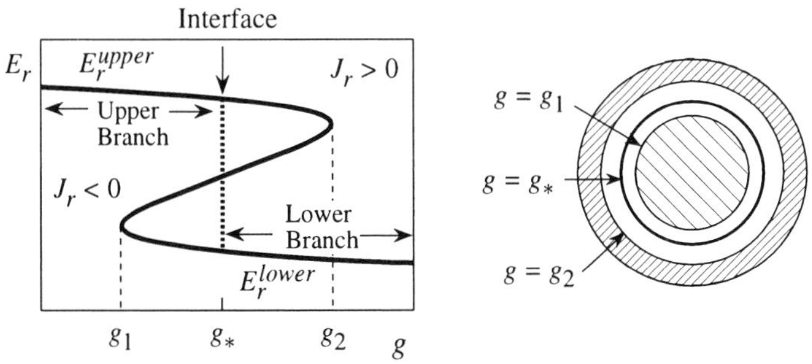

Figure 21.1. Interface between two branches (left). The relation $E_r[g]$, which is the solution of charge neutrality condition $J_r[E_r; g] = 0$, is shown by the solid line. Upper branch and lower branch merge across an interface (at the position $g = g_*$). The field jumps at the interface. The parameter g is a function of the plasma minor radius. The surfaces, where $g = g_1$, $g = g_*$ and $g = g_2$ hold, respectively, are illustrated on the right.

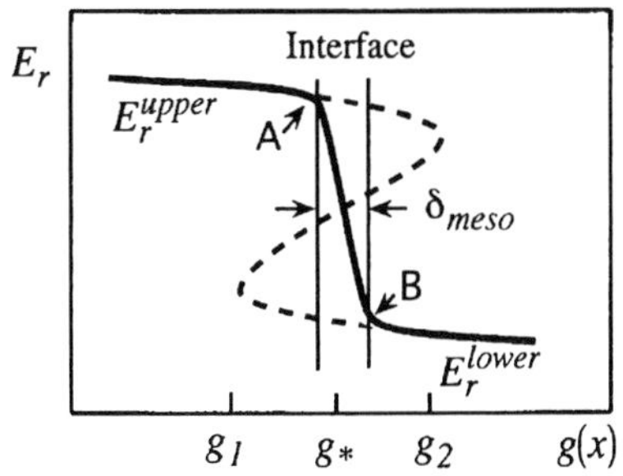

Figure 21.2. Internal structure of interface. It has a finite thickness, δ_{meso}. A meso-phase appears in between two branches. A and B indicate boundaries between an interface and upper and lower branches, respectively.

equation (20.1) as is illustrated in the figure. At each plasma location, the electric field takes one value at one time; there should appear an *interface*, across which different branches of the solution touch. Across an interface, the radial electric field varies from one branch to the other. A strong inhomogeneity of radial electric field is expected at the interface.

The interface itself has an internal structure. Its width is small but finite. Figure 21.2 shows a schematic structure of the interface. Viscosity makes the interface a zone of finite thickness. Two branches are connected by a layer of finite width following the E-TDGL equations in equation (20.13). In this layer, the radial electric field takes an intermediate value between the upper branch and the lower one. Therefore it is called a *meso-phase*. The location and internal structure of the interface are discussed in this section.

21.1.2 Maxwell's Construction

There has been analysis of an interface, motivated by study of an electric field in nonaxisymmetric systems [21.1, 21.2], and the concept of an ambipolaron [21.1] was proposed. The interface layer in neoclassical theory was found to be thermodynamically stable: this layer can drift in the radial direction until it reaches a stationary state, and the stationary solution is shown to be unique for given plasma parameters. Under the condition that the spatial variation of the plasma profile is small compared to a layer width and that it is considered constant in time, the location of the interface was discussed [21.2]. At the interface, a parameter g (the parameter in figure 21.1 controlling electric field and transport) satisfies the equation $g = g_*$, and g_* is given by the relation

$$\Delta\Phi(g_*) = 0 \tag{21.1}$$

where

$$\Delta\Phi(g) \equiv \int_{E_r^{lower}}^{E_r^{upper}} J_r^{local}[E_r; g]\, \mathrm{d}E_r \tag{21.2}$$

and E_r^{upper} and E_r^{lower} are solutions $E_r[g]$ on upper and lower branches of figure 21.1, respectively. The superscript *local* indicates that a viscous term is not included in this radial current (i.e., the term N in equation (20.13)). Schematic forms of the integrand of equation (21.2) and $\Delta\Phi(g)$ are illustrated in figure 21.3. The relation equation (21.1) is a counterpart of *Maxwell's construction* (Maxwell's rule of equal areas) at a phase transition in thermodynamics [21.3].

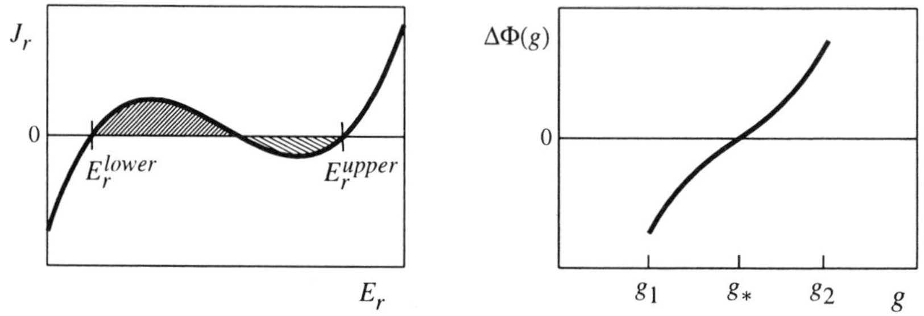

Figure 21.3. The relation between electric field and local current (left), and integral function $\Delta\Phi(g)$ (right). If hatched areas in the left panel are equal, then $\Delta\Phi(g) = 0$ holds.

The rule of equal area for the location of an interface is derived from equation (20.13). In the stationary state, equation (20.13) takes the form

$$\nabla_{\perp} \cdot \mu_{eff} \nabla_{\perp} E_r(r) - J_r^{local} = 0. \tag{21.3}$$

This relation is rewritten as

$$\nabla \cdot \mu_{eff} \nabla E_r(r) - \frac{\mathrm{d}}{\mathrm{d}E_r} F[E_r] = 0 \tag{21.4}$$

where a potential function for current is introduced as

$$F[E_r] = \int_{E_r^{lower}}^{E_r} \mathrm{d}E_r J^{local}[E_r]. \tag{21.5}$$

Function F satisfies the relation $F[E_r^{lower}] = 0$ at $E_r = E_r^{lower}$. The function F and $\Delta\Phi$ are related, and an identity

$$\Delta\Phi = F[E_r^{upper}]$$

holds. We here assume that effective viscosity is very small so that the second derivative term in equation (21.4) is effective only at the interface. For an analytic insight, μ_{eff} is assumed constant. Under this assumption, equation (21.4) is integrated as

$$\frac{1}{2}\mu_{eff}\{\nabla E_r(r)\}^2 - F[E_r] = \text{ constant.} \tag{21.6}$$

The constant in the right-hand side of equation (21.6) is given by a boundary condition. The gradient $|\nabla E_r|$ is large only inside the interface, and is small on branches E_r^{lower} and E_r^{upper}. Solution $E_r(r)$ smoothly connects to two single valued outside branches, and the gradient ∇E_r is approximated to vanish at $E_r = E_r^{lower}$ (the connection at the point B in figure 21.2). Under these constraints, the constant in equation (21.6) is to be close to zero. One has

$$\tfrac{1}{2}\mu_{eff}\{\nabla E_r(r)\}^2 = F[E_r]. \tag{21.7}$$

Integral relation (21.7) is shown in figure 21.4 for various values of $\Delta\Phi$.

Another condition for an interface is that a solution smoothly connects to an upper branch (point at A in figure 21.2). That is, ∇E_r is approximated to vanish at $E_r = E_r^{upper}$ as well. From this constraint, we have the condition

$$F[E_r^{upper}] = 0 \tag{21.8}$$

which must be satisfied at the ridge of the interface, point A in figure 21.2. This equation is identical to equation (21.1).

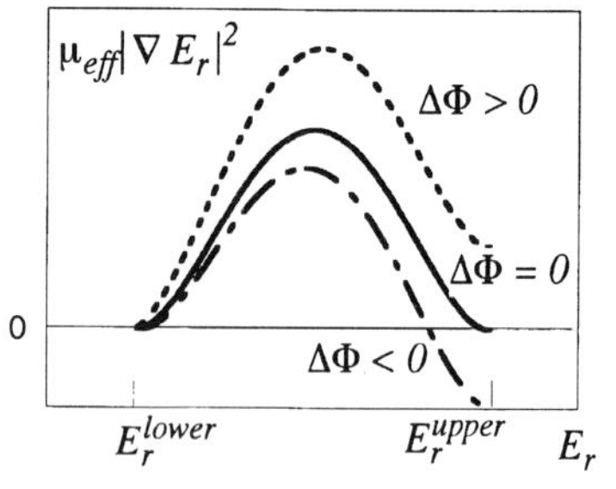

Figure 21.4. Solutions of ∇E_r as a function of E_r. When $\Delta\Phi = 0$ holds, the boundary conditions are satisfied.

21.1.3 Mesophase

An interface is not a strict surface, but has an internal structure with a finite width. There appears a region called a meso-phase. In the meso-phase, viscosity plays a dominant role in determining profile of the radial electric field, which is described by equation (21.7). Equation (21.7) provides the gradient of the radial electric field as

$$|\nabla E_r(r)| = \sqrt{\frac{2}{\mu_{eff}} F[E_r]}. \tag{21.9}$$

The solution of the radial electric field in a meso-phase is formally given from equation (21.9) as

$$\int_{E_r^{lower}}^{E_r} \sqrt{\frac{\mu_{eff}}{2F[E_r]}}\, \mathrm{d}E_r = \pm(r - r_{lower}) \tag{21.10}$$

where r_{lower} denotes a radius where the solution in the meso-phase merges to a single valued solution $E_r = E_r^{lower}$ (point B in figure 21.2). The sign $\pm$ depends on the radial locations of upper branch, interface and lower branch. (If the minor radius of point A in figure 21.2 is smaller than that of point B, $r_A < r_B$, sign $-$ is chosen. On the other hand, if $r_A > r_B$ holds, sign $+$ is chosen.)

The result (21.9) or (21.10) allows us to estimate the thickness of the meso-phase. The thickness of the meso-phase, δ_{meso}, is estimated as

$$\delta_{meso}^{-1} \simeq |E_r^{-1} \nabla E_r|_{max} = \left| E_r^{-1} \sqrt{\frac{2}{\mu_{eff}} F[E_r]} \right|_{max}. \tag{21.11}$$

Note that the effective viscosity for the electric field is given by the ion viscosity μ_i. In the case of a weak bulk viscosity, $|\mu_i \nabla_\perp^2| \gg \varepsilon^2 \nu_{local}$, the plasma rotates in the poloidal direction. The effective viscosity for the electric field is given in equation (19.23-2) and the relation between μ_{eff} and μ_i is given as

$$\mu_{eff} = (1 + 2q^2)\frac{m_i n_i}{B^2}\mu_i. \tag{21.12}$$

The maximum value of F in equation (21.11) is calculated by use of the formula of excess current given in equation (20.7). An order of magnitude estimate is given as $F_{max} \simeq \varepsilon^2 (v_{thi}/qR) m_i n_i B_p^{-2} E_r^2$. Combining this estimate with equation (21.12), we have

$$\frac{2F_{max}}{\mu_{eff}} \simeq \frac{1}{\mu_i}\frac{v_{thi}}{qR}E_r^2. \tag{21.13}$$

By substituting equation (21.13) into equation (21.11), one has an estimate for the thickness of the meso-phase as

$$\delta_{meso} \simeq \sqrt{\frac{\mu_i R}{v_{thi}}} = \sqrt{\left(\frac{\nu_{**}}{q}\right)\frac{\mu_i}{\nu_i}}. \tag{21.14}$$

The thickness of the meso-phase is in proportion to $\sqrt{\mu_i}$. Substitution of equations (20.7) and (21.12) into equation (21.3) gives the same result.

It is noted that transport coefficient μ_i is dependent on (a decreasing function of) radial electric field gradient, i.e., the sharpness of spatial structure. This dependence can cause a self-sustaining of a steep gradient. If viscosity is reduced, then the interface becomes thinner as equation (21.14). Although equation (21.14) is an analytic estimation based on an assumption of constant μ_i, the relation between δ_{meso} and μ_i, i.e., δ_{meso} is an increasing function of μ_i, holds in general cases. Consequently, the viscosity is further reduced when the thickness of the meso-phase decreases. There is a nonlinear link between turbulent transport and the structure of the radial electric field. This nonlinear system thus 'self-organizes' a thin transport barrier through interactions of electric field and transport coefficient.

The picture of the interface is extended to the study of the transport barrier of the H-mode. In this case, spatial variations of plasma parameters, such as density and temperatures, could become as steep as those of the transport coefficient and electric field, and the result is dependent on boundary conditions as well [20.6]. Figure 21.5 illustrates the spatial distribution of transport coefficient D near the plasma edge in the L- and H-phase. A meso-phase is formed and observed in the H-mode case.

Ion orbit loss has a nonlocal influence on the radial electric field structure, and the thickness of the layer is evaluated as

$$\delta \sim \sqrt{\hat{\rho}_p^2 + \mu_\perp/\nu_*} \tag{21.15}$$

where $\hat{\rho}_p$ is the squeezed poloidal gyro-radius, equation (17.12), and $\mu_\perp$ is the shear viscosity of ions. An application to the externally biased limiter case is presented in [17.15], where an explicit plasma parameter dependence is incorporated. The analysis of the H-mode plasma is also performed [21.4].

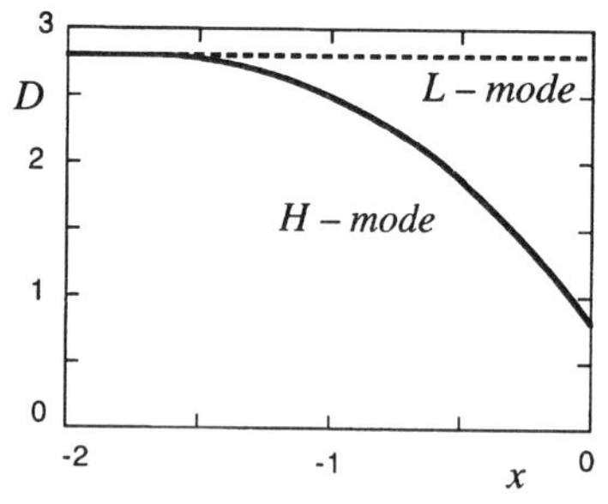

Figure 21.5. Example of the meso-phase which appears in the H-mode plasma. (Quoted from [20.6].)

21.2 Poloidal Structure and Shock Formation

A fast rotation associated with a radial electric field causes changes in the poloidal structures of plasma parameters. As is discussed in section 19.1, radial electric field E_r gives rise to two patterns of flow, i.e., a toroidal rotation $\boldsymbol{V} = (0, 0, V_\zeta)$ or a poloidal rotation coupled with a parallel flow $\boldsymbol{V} = (0, V_\theta, 0) - (0, \varepsilon, q)2V_\theta \cos\theta$.

21.2.1 Effect of Toroidal Flow

The poloidal variation of plasma parameters could be caused by a toroidal flow $\boldsymbol{V} = (0, 0, V_\zeta)$. In toroidal plasmas, the influence of centrifugal force associated with a toroidal rotation has been discussed in the framework of neoclassical theory. Due to the centrifugal force, the population of ions is high in the low field side of a torus, and the electrostatic potential there can be higher than in the high field side. The inhomogeneous part of the static potential has been calculated as [17.17]

$$\frac{e\tilde{\phi}_0}{T_e} = \frac{\varepsilon V_\zeta^2}{c_s^2} \cos\theta \tag{21.16}$$

where the suffix 0 indicates a stationary state. This electrostatic potential gives rise to an in–out asymmetry of density and an up–down asymmetry in $\boldsymbol{E} \times \boldsymbol{B}$ velocity. For a range of radial electric field, $X \equiv e\rho_p E_r/T_i \sim 1$, the toroidal velocity reaches the ion thermal velocity, $V_\zeta \sim v_{thi}$, if a flow pattern of the form $\boldsymbol{V} = (0, 0, V_\zeta)$ is realized. (See equation (19.5).) In such a case, the poloidal inhomogeneity of the static potential satisfies the relation $|e\tilde{\phi}_0/T_e| \sim \varepsilon$. The poloidal variation remains of the order of ε.

21.2.2 Effect of Poloidal Flow

The poloidal rotation coupled with a parallel flow, $\boldsymbol{V} = (0, V_\theta, 0) - (0, \varepsilon, q)2V_\theta \cos\theta$, is generated by a radial electric field if toroidal friction is bigger than poloidal friction, $\nu_\zeta B_\zeta^2 \gg \nu_\theta B_\theta^2$. A fast poloidal rotation associated with a radial electric field bifurcation can cause a poloidal shock in tokamaks. Theoretical study on shock formation was first performed with the analysis of fast flows of the order of sound velocity [21.5]. Because experimental plasma parameters in the 1970s were far from the condition which is necessary for shock formation, the progress in shock formation study has awaited the development of H-mode physics [20.4, 21.6].

A poloidal flow compresses itself as a fluid element moves from the outside of the torus to the inside. After [21.6], the equation of motion is given in the form of a Burgers equation with an external force,

$$\frac{\partial}{\partial t}V_\theta + V_\theta \frac{\partial}{r\partial\theta}V_\theta - \frac{\mu}{2}\frac{\partial^2}{r^2\partial\theta^2}V_\theta = -\frac{F_p}{2}\sin\theta \tag{21.17}$$

where F_p stands for an effective force due to the toroidicity, and μ is related to the bulk viscosity μ_b. This is in the limit where the terms Γ_i^{bv} and $\Gamma_i^{v\nabla v}$ of equation (19.14) are kept explicitly and $\sin\theta$ components of other terms are rewritten in the form of an effective external force F_p. The term of dispersion ($\partial^3/\partial\theta^3$, the third derivative with respect to poloidal angle) is small for parameters of interest, and is omitted here. Equation (21.17) describes a simple diffusion equation, if force and flow are small. As is shown in equation (19.14), a large electric field and high poloidal velocity are possible if terms on the right-hand side of equation (19.14) become significant.

As the rotation torque, F_p in equation (21.17), becomes large, a fast plasma rotation with a poloidal Mach number (X) of order unity is possible. In such a situation, poloidal profiles of velocity and density are not uniform. An example is shown in figure 21.6. The poloidal distribution of electrostatic potential is plotted in figure 21.6. In this case, the electrostatic potential is higher on the inside of the torus ($\theta \sim \pi$) than on the outside ($\theta \sim 0$). It is noted that a shock structure appears inside the torus, where the static potential is subject to a strong poloidal variation. In figure 21.6, a sharp change is shown by a vertical dotted line. The poloidal variation of electrostatic potential on the magnetic surface is of the order of

$$|\tilde{\phi}_0| \sim \sqrt{\varepsilon}T_e/e \tag{21.18}$$

for the range of radial electric field, $X \equiv e\rho_p E_r/T_i \sim 1$. The poloidal variation reaches the order of $\sqrt{\varepsilon}$. A flow with a form $\boldsymbol{V} = (0, V_\theta, 0) - (0, \varepsilon, q)2V_\theta \cos\theta$ causes a more prominent poloidal variation of static potential than the toroidal flow.

The solution in figure 21.6 implies a weak shock formation in the poloidal direction. As a result of this, there appears an up–down asymmetry of density and/or poloidal flow. The influence on radial flow is also important. A radial

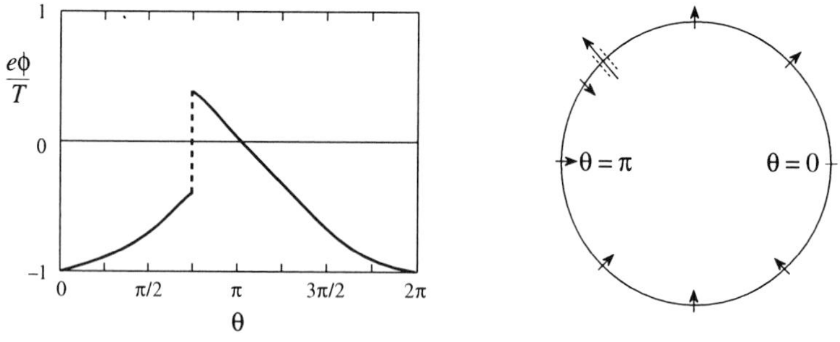

Figure 21.6. Shock formation in the poloidal direction. The poloidal angle dependence of electrostatic potential on one magnetic surface is shown (left). (Near the condition $X \equiv e\rho_p E_r/T_i \sim 1$, quoted from [20.4].) Radial component of $\boldsymbol{E} \times \boldsymbol{B}$ velocity is illustrated in poloidal cross-section (right). Near the location of poloidal shock, V_r can become large.

flow which has a parity of $\sin\theta$ appears. When a poloidal variation of static potential satisfies the condition (21.18), the poloidal electric field may reach the level of $E_\theta = -\nabla_\theta \tilde{\phi}_0 \sim \sqrt{\varepsilon} T_e/e\ell_p$, where ℓ_p is the poloidal length of shock, across which a sharp variation of potential takes place. Radial velocity, $V_r \sim E_\theta/B$, could be of the order of $V_r \sim \sqrt{\varepsilon} T_e/e\ell_p B$ near the position of poloidal shock. Comparing this radial velocity with the poloidal velocity at the condition $X \equiv e\rho_p E_r/T_i \sim 1$, $V_\theta = E_r/B \sim v_{thi}\varepsilon/q$, one has the relation $V_r = (\sqrt{\varepsilon}\rho_p/\ell_p)V_\theta$. (In discussing this relation, an assumption $T_e \sim T_i$ is made.)

This shock formation and up–down asymmetry of the electric field are reported to play a part in the L–H-transition physics. For example, the up–down asymmetry has an influence on radial fluxes through gyro-viscosity [17.16]. The poloidal electric field caused by a nonuniform ion loss is also studied in [21.7]. As is discussed in section 20.3.1, a modification of the flow pattern is influential on a transition time scale. The radial flow associated with poloidal shock, the magnitude of which is estimated as $V_r = (\sqrt{\varepsilon}\rho_p/\ell_p)V_\theta$, could also be important. This value easily reaches (or exceeds) the level of equation (20.6). The neoclassical amplification of dielectric constant can be reduced. This shortens the transition time from the L-state to H-state or vice versa.

Finally, an influence on MHD equilibrium itself is noted. It has been shown that, if poloidal flow is large enough, the MHD equilibrium equation (Grad–Shafranov equation) changes its character from an elliptic equation to a hyperbolic one [21.8]. The condition for this is expressed for low β plasma as

$$\frac{V_\theta^2}{v_A^2} \sim \frac{\gamma\mu_0 p}{B^2}$$

where γ is the specific heat ratio here. An order of magnitude estimate of this criterion is given as $V_\theta \sim c_s$. This value is much higher than that for the electric field bifurcation, $V_\theta \sim v_{thi}\varepsilon/q$. In the range of electric field for which a bifurcation theory in relation to the H-mode is discussed, a change of MHD equilibrium equation is not expected to occur.

REFERENCES

[21.1] Hastings D E, Hazeltine R D and Morrison P J 1986 *Phys. Fluids* **29** 69

[21.2] Yahagi E, Itoh K and Wakatani M 1988 *Plasma Phys. Control. Fusion* **30** 1009
Shaing K C 1984 *Phys. Fluids* **27** 1567

[21.3] Haken H 1976 *Synergetics* (Berlin: Springer) section 9.3
See also Landau L D and Lifshitz E M 1987 *Fluid Mechanics* 2nd edn, transl. J B Sykes *et al* (Oxford: Pergamon) section 102

[21.4] Ida K, Hidekuma S, Kojima M *et al* 1992 *Phys. Fluids* B **4** 2552

[21.5] Hazeltine R D, Lee E P and Rosenbluth M N 1971 *Phys. Fluids* **14** 361
Greene J M, Johnson J L, Weimer K E and Winsor N K 1971 *Phys. Fluids* **14** 1258

[21.6] Taniuti T, Moriguchi H, Ishii Y, Watanabe K and Wakatani M 1992 *J. Phys. Soc. Japan* **61** 568

[21.7] Tendler M, Daybelge U and Rozhansky R 1993 *Plasma Physics and Controlled Nuclear Fusion Research 1992* vol 2 (Vienna: IAEA) p 243
Xiao H, Hazeltine R D and Valanju P M 1994 *Phys. Plasmas* **1** 3641

[21.8] Zehrfeld H P and Green B J 1972 *Nucl. Fusion* **12** 529
Hameiri E 1983 *Phys. Fluids* **26** 230
Kerner W and Tokuda S 1987 *Z. Naturf.* a **42** 1154
Zelazny R, Stankiewicz R, Ikowski A Ga and Potempski S 1993 *Plasma Phys. Control. Fusion* **35** 1215

Chapter 22

Self-Organized Dynamics

22.1 Limit Cycles and ELMs

Evolution equations for plasma parameters have been analysed, and limit cycle oscillations were investigated. The characteristics of the solution and the relationship to edge-localized modes (ELMs) [22.1] are discussed.

22.1.1 Dithering ELMs

The dynamical equations of E-TDGL type, equations (20.11)–(20.13), predict the existence of a limit cycle oscillation, which is a sequence of transitions and back transitions [20.6, 22.2]. In this analysis, the parameter g, which controls the transport, was chosen as

$$g \equiv \frac{\rho_p}{L_n} \frac{\hat{D}_L}{\nu_i \rho_p^2} \tag{22.1}$$

and the nonlinear term N in equation (20.13) is modelled by a cubic equation of $X = e\rho_p E_r / T_i$. This model is employed based on assumptions that L-to-H-transition occurs if the gradient is sharp and collisionality is low, i.e., $\nu_* < \nu_c \sim 1$ and $\lambda_p = \lambda_{pc} \sim \nu_* \rho_p^2 / \hat{D}_L$ [17.5]. The schematic dependence is illustrated in figure 22.1. The loss rate as a function of control parameter is shown in figure 22.1 (left). When the control parameter g is governed by the development of density, a hysteresis relation between density and loss rate is drawn and is given in figure 22.1 (right).

The dynamical equations are solved for density and electric field under a constant supply. The supply is shown by a dashed line in figure 22.1 (right). The H-phase and L-phase are realized one by one; in each stage the radial profile of the transport coefficient takes the one for the L-mode or for the H-mode in figure 21.5. The confinement property of the plasma evolves as follows. In the H-phase (the lower branch of figure 22.1), confinement is good. The loss rate is smaller than the source rate, so that an increment in edge density occurs

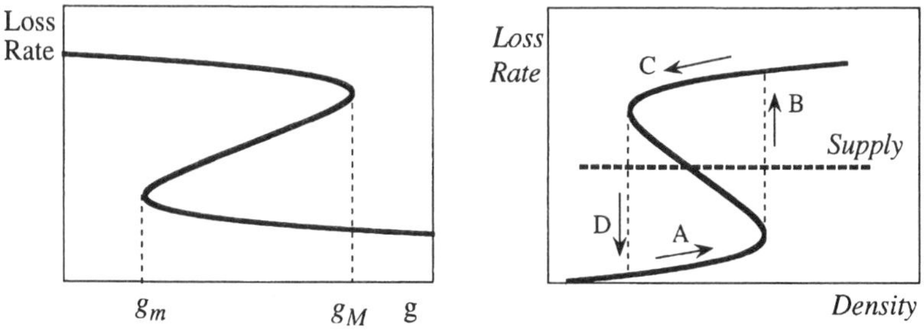

Figure 22.1. Schematic dependence of loss rate on control parameter (left). It is interpreted in terms of local density for a point model (right).

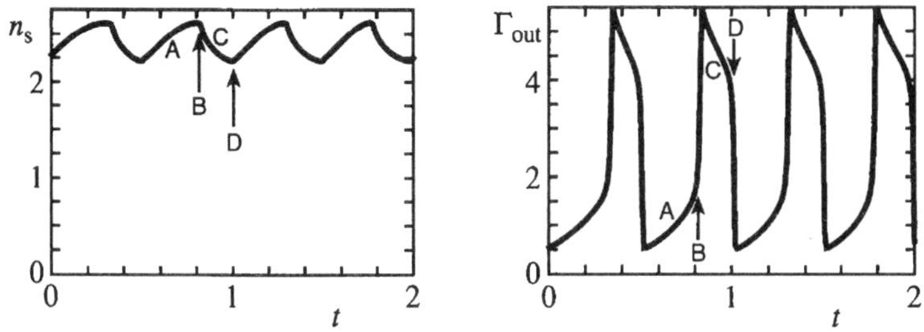

Figure 22.2. Self-generated transitions and limit cycle oscillation. Solution of equations (20.11)–(20.13).

(path A in figure 22.1 (right)). Owing to this improvement, the condition g_m in figure 22.1 (left) is reached. Then a sudden increment in the loss rate takes place (path B), causing a jump of outflux. On the upper branch, the loss rate exceeds the supply. A rapid decay of density in the edge pedestal follows (path C). If the parameter g increases, the condition g_M in figure 22.1 is realized. Then a transition to the H-phase takes place (path D). These processes repeat themselves. Under a constant supply from the core plasma, a limit cycle oscillation occurs. This is a self-organized oscillation generated by a hysteresis and hard transition characteristics in plasma transport.

Figure 22.2 illustrates a limit cycle oscillation seen in edge density and flux out of the plasma. This limit cycle oscillation is possible near the threshold condition of the L–H-transition, which is evaluated in a stationary state [22.2]. Small and frequent ELMs are observed experimentally in the vicinity of threshold power for the L–H-transition, and are called dithering ELMs. This limit cycle oscillation constitutes a model of this kind of ELM.

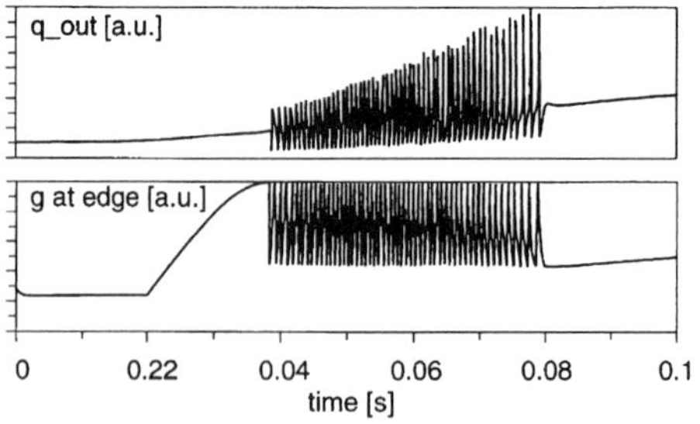

Figure 22.3. Slow change of external parameters and an abrupt start of self-generated oscillations between L- and H-states. (Quoted from [22.3].)

The appearance of the limit cycle oscillation in the system of equation (20.11)–(20.13) is studied under a slow evolution of global plasma parameter [22.3]. The result is shown in figure 22.3. Heat flow from the core is increased gradually in this analysis. It is shown that, as plasma parameters reach the threshold condition, the first burst appears, followed by periodic bursts. Finally self-regulated oscillation disappears as the heating power becomes larger. A stationary H-mode is established. It is observed from the figure that the height of the first burst is almost the same as the following one; and the last one is similar to the preceding one. The oscillation amplitude jumps abruptly from zero to finite value and from finite amplitude to zero. This feature is characteristic of a hard type of bifurcation which contains a hysteresis. The shape of each burst may be compared. When the average plasma flux is close to that of the L-mode, each period is composed of a short and sharp H-state (and a long and round L-phase). In contrast, the shape of each period is characterized by a short and sharp L-phase, at the end. As heating power becomes larger, then probability of staying in the L-state becomes shorter [22.2].

These oscillations can also be found in the zero-dimensional models. If a model takes into account slow transitions (such as [19.3, 19.5]) then a gradual appearance of oscillations, rather than a sudden bursting, is predicted at the boundary between a stationary state and an oscillating state.

Multiple hysteresis can appear in the gradient–flux relation. When more than two nonlinear terms are kept in equation (19.14), multiple hysteresis is predicted to exist. An example is analysed in [22.4]. When effects of ion orbit loss, bulk viscosity and anomalous transport are taken into account, two hystereses are predicted as is shown in figure 22.4. Two types of limit cycle oscillation $E \to E' \to D \to D' \to E$ and $B \to B' \to C \to C' \to B$ are possible. It is found that limit cycles occur sequentially as is shown in figure 22.4(b). This type of self-organized oscillation corresponds to compound dithers.

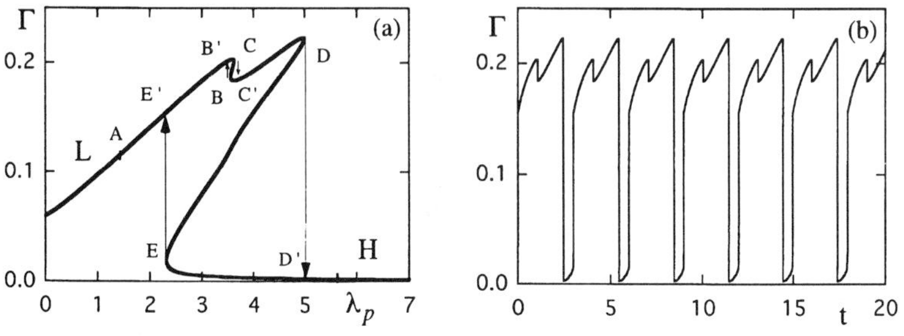

Figure 22.4. Multiple hystereses (a) and compound dithers (b). (Quoted from [22.4].)

22.1.2 Other Limiting Phenomena

Other types of ELM have also been found. When the pressure gradient at an edge becomes too high, an ELM burst usually occurs. In the high heating limit, a stationary H-mode no longer exists, and periodic and giant ELMs are observed. They are called type-I ELMs. A condition for this burst to occur has been compared to stability analyses based on ideal MHD theory [22.1, 22.5]. With a steep pressure gradient at the edge, ideal MHD modes with high mode numbers are predicted to become unstable, and their relation with ELMs has been analysed (see, e.g., [16.1]). Although ideal MHD modes are candidates for ELMs, the linear theory has a difficulty in explaining ELMs: this is because magnetic perturbations are observed to appear very abruptly in ELMs, and the fast change of growth rate is beyond that calculated by a linear MHD theory [2.24]. The discreteness in the height of ELMs' bursts at the onset of occurrence is not explained. A theory of a transport catastrophe and subsequent periodic bursts

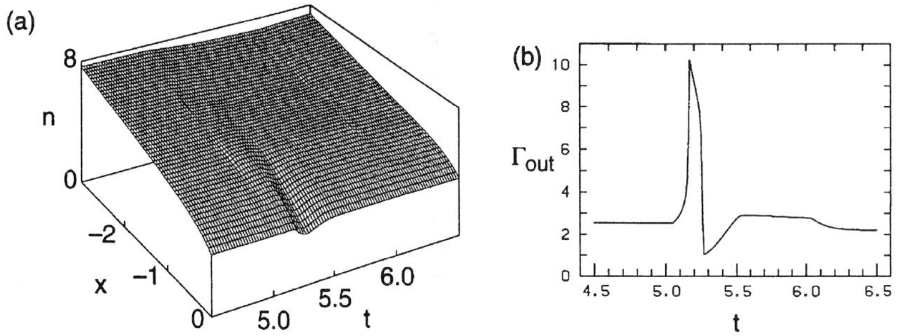

Figure 22.5. Density profile development and associated burst as a transient response of transport barrier against a large pulse from inside.

has been proposed [22.6]. This is now being applied to the physics of ELMs [22.7]. A catastrophe is predicted to occur at a critical pressure; the period is expected to vary roughly inversely proportionally to heating power. This is in line with the observations, and the cause of type-I ELMs could be a beta-limiting phenomenon. The Kelvin–Helmholtz instability could also be the origin of the destruction of the H-mode, and investigations have been initiated [22.8].

The structure of the transport barrier in figure 21.5 is stable against perturbation. If an event happens, the barrier is destroyed. However, if a violent instability and transport catastrophe are terminated, then a transport barrier is re-established again [22.2]. The temporal evolution of edge plasma density and outflux associated with the transient response of the transport barrier is shown in figure 22.5

REFERENCES

[22.1] Keilhacker M, Becker G, Behringer K *et al* 1984 *Plasma Phys. Control. Fusion* **26** 49
Gohil P, Mahdavi M A, Lao L, Burrell K H *et al* 1988 *Phys. Rev. Lett.* **61** 1603
[22.2] Itoh S-I, Itoh K and Fukuyama A 1993 *Nucl. Fusion* **33** 1445
[22.3] Zohm H 1994 *Phys. Rev. Lett.* **72** 222
[22.4] Toda S, Itoh S-I, Yagi M *et al* 1996 *Plasma Phys. Control. Fusion* **38** 1337
[22.5] Burrell K H, Allen S L, Bramson G, Brooks N H *et al* 1989 *Plasma Physics and Controlled Nuclear Fusion Research 1988* vol 1 (Vienna: IAEA) p 193
Doyle E J, Groebner R J, Burrell K H *et al* 1991 *Phys. Fluids* B **3** 2300
[22.6] Itoh K, Itoh S-I and Fukuyama A 1995 *Plasma Phys. Control. Fusion* **37** 707
Haas F A and Thyagaraja A 1995 *Plasma Phys. Control. Fusion* **37** 415
[22.7] Itoh S-I, Itoh K, Fukuyama A and Yagi M 1996 *Phys. Rev. Lett.* **76** 922
[22.8] Itoh S-I and Itoh K 1987 IPP III/126, Max-Planck-Institute für Plasmaphysik
Chiuch T, Terry P W, Diamond P H and Seddack J E 1986 *Phys. Fluids* **29** 231
Sugama H and Wakatani M 1993 *Phys. Plasmas* B **3** 1110

Chapter 23

Concluding Remarks

και ουδεν μενει—and nothing lasts
(Heraclitus)

In this monograph, turbulent transport and structural formation in magnetically confined and inhomogeneous plasmas is discussed.

One of the characteristic features of transport phenomena in confined plasmas is that the plasma inhomogeneity is the order parameter that governs the transport. The fluctuations are self-sustained: they can be driven through subcritical excitation, being independent of the linear stability of the confined plasma. When the system is in a nonequilibrium state but close to the thermal equilibrium, the transport is governed by collisional processes. Collisional diffusion describes the cross-field transport that is driven by the thermal fluctuations. It is independent of the inhomogeneity of the plasma. When turbulent transport is the dominant process, the system is in a far-nonequilibrium state. Turbulent fluctuations are strongly affected by the plasma inhomogeneity. As a result of the theory described in chapters 9 and 10, the power index to the gradient parameter is obtained. The transition from collisional transport to turbulent transport is shown. This transition condition, which is given in terms of the pressure gradient, specifies the boundary between the near-thermal equilibrium and the far-nonequilibrium. The global structure, the released energy and the fluctuations are self-sustaining in the structural formation in plasmas. In a nonuniform plasma far from thermal equilibrium, the thermal diffusivity is an increasing function of the gradient. It is also shown that there is an interference between the gradients of the velocity and of the pressure, and that these are related to the transport of the momentum and energy. The gradient of the pressure could cause the flux of the momentum. The symmetry violation due to the fluctuations allows the mixing of fluxes.

The other essential element in the structural formation of plasmas is the role of the fields, that induce the long range interactions. As an example, the influence of the radial electric field is discussed in detail. Owing to the coupling between the plasma inhomogeneity and the radial electric field structure, bifurcations and structural transitions appear in the plasma. The strong inhomogeneity of the radial electric field can modify the turbulence and turbulent transport. Owing to the coupling with the radial electric field structure, the relation between the pressure gradient and the energy flux changes. The flux can be a non-monotonic function of the gradient and become a multifold function. The radial structure of the interface, across which the polarity of the radial electric field changes (from positive to negative, or from negative to positive), is discussed. One example of a structural transition is the transition between the L-mode and H-mode. Plasmas could establish a steep gradient, which is not restricted by the global size of the plasma. This self-sustaining of the plasma gradient becomes possible by the coupling with the structure of the fields. The impact of the change of the poloidal magnetic field is also studied. Various structural formations are available: models of some of the typical structures are developed.

In this monograph, the plasma is described within the framework of the fluid dynamics, and one particular class of fluctuations is considered. These choices lead to a highly simplified treatment. Encyclopaedic description of the phenomena is not within the scope of this text. There are other kinds of fluctuation; the wave–particle interactions provide additional freedom for the dynamics. At the price of this incompleteness, we try here to put forward the point of views that the plasma structure, fluctuations and turbulent transport are regulating each other, and that the structural formation and structural transition of plasmas are a typical example of the physics of the far-nonequilibrium systems.

The structure and dynamics of plasmas are, in reality, more varied in their appearance than is described in this monograph. For instance, an abrupt crash of plasma inhomogeneity has been observed. Even though the transport-like process is one of the governing mechanisms for the evolution in plasmas, much faster phenomena also control the plasma dynamics. Also for such catastrophic types of event, a number of theories have been developed. The methodology in this monograph has been extended into such a direction, and a turbulence–turbulence transition has been identified. Hysteresis is obtained in the relation between the gradient and flux. The catastrophic enhancement of turbulence propagates in the plasma as an avalanche. In addition to such dynamic events, nonlocal transport problems have also attracted attention.

The study of confined plasma will have an impact for the present and future research of physics. The structural formation in high temperature plasmas is a challenging branch of modern physics, and it is a key to understanding basic phenomena that occur in our present cosmos and that have occurred in the early stage of our universe as well as that will occur in the final stage of our universe.

Chapter 24

Annex: Structural Formation in Tokamak Plasmas and Various States of Confined Plasmas

The global structure of a plasma $(n(r), \boldsymbol{V}(r), T(r))$ with electromagnetic field $(E_r, E_\zeta; B_\theta, B_\zeta)$ is governed by transport equations,

$$\frac{\partial}{\partial t} n + \boldsymbol{\nabla} \cdot \boldsymbol{\Gamma} = S_p \tag{24.1-1}$$

$$m_i n_i \frac{\mathrm{d}}{\mathrm{d}t} \boldsymbol{V} = \boldsymbol{J} \times \boldsymbol{B} - \nabla p - \boldsymbol{\nabla} \cdot \boldsymbol{\Pi} + \boldsymbol{F}_{ext} \tag{24.1-2}$$

$$\frac{m_e}{ne^2} \frac{\mathrm{d}}{\mathrm{d}t} J_\zeta + \frac{1}{\sigma_c} J_\zeta - \boldsymbol{\nabla} \cdot \lambda \boldsymbol{\nabla} J_\zeta = E_\zeta + E_{drive} \tag{24.1-3}$$

$$\frac{\mathrm{d}}{\mathrm{d}t} nT + \frac{3}{2} T \boldsymbol{\nabla} \cdot \boldsymbol{\Gamma} = \boldsymbol{\nabla} \cdot q + P_{in} \tag{24.1-4}$$

where S_p is the particle source, $\boldsymbol{F}_{ext}$ is the external force, E_{drive} is the drive for the toroidal current and P_{in} is the local heating power.

Through the studies of plasma turbulence and turbulent transport, cross-field plasma fluxes $\{\boldsymbol{\Gamma}, \boldsymbol{\Pi}, \boldsymbol{q}\}$ (i.e., particle flux, momentum flux and energy flux) are expressed in terms of global plasma parameters and gradients. An example is

$$\begin{pmatrix} \Pi_{\theta r} \\ \Pi_{\zeta r} \\ E_\zeta \\ q_r \end{pmatrix} = -M\nabla \begin{pmatrix} V_\theta \\ V_\zeta \\ J_\zeta \\ p \end{pmatrix}$$

as discussed in preceding chapters.

As is schematically drawn in figure 24.1, transport properties are functionals of plasma profile (gradients) and external magnetic field. The plasma profile, in turn, is determined by the balance between the transport process and external supplies of particles, momentum and energy. This nonlinear link between gradients and transport coefficients is an intrinsic feature of far-nonequilibrium states. It is shown that the link is an origin of structural formations.

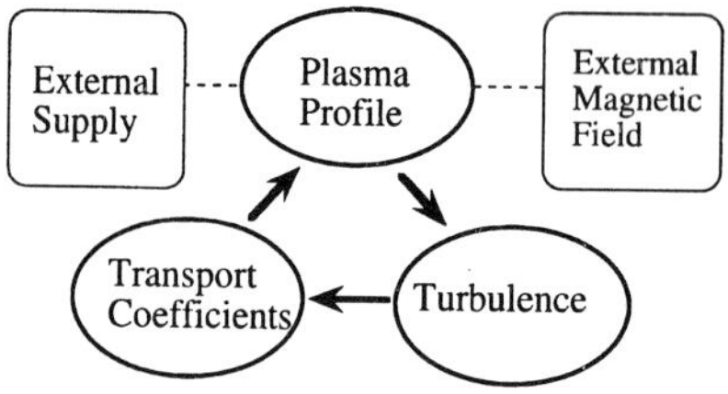

Figure 24.1. Inter-relation between plasma profile, turbulence and transport properties under external magnetic field and supplies.

The expressions of plasma fluxes, which are derived from a stationary solution of turbulence, allow us to investigate the structure of nonequilibrium plasmas, which are sustained under external supplies. In this chapter, we discuss the physics of structural formation in confined plasmas, based on a study of transport equations. The study of the plasma profile by use of the transport code has a long history [24.1]. In transport simulation, a basic assumption is employed: the time scales are separated between the dynamics of turbulence and the evolution of global plasma profiles. During the slow evolution of global plasma parameters, the turbulence state is considered to be in a stationary state. That is, turbulence relaxation is fast enough, and turbulence level and turbulence-driven fluxes are expressed in terms of plasma profiles and are not influenced by the time derivative of global plasma parameters. This approximation allows a reduction of variables, and equation (24.1) is closed within global parameters. With this simplification, the stationary structures and the evolution of plasma profiles are investigated.

In this chapter, examples of self-organized structures are shown. First, the mutual relation of pressure gradient and thermal conductivity is studied, and the resilience of the plasma profile is discussed. Second, the interaction of pressure and current profiles is illustrated. The formation of an internal transport barrier is shown. Third, the induction of radial plasma flow by an external momentum source is discussed. Fourth, the interaction of a radial electric field and turbulent transport is analysed, and a demonstration of transport barrier formation at the edge is made. It is illustrated that plasma profiles are subject to various transitions, owing to the nonlinear nature of turbulent transport processes. At the end, some discussion on experimental observations is presented.

24.1 Transport Simulation of Energy and Current

As an example, the formation of a plasma pressure profile is explained first. Equations for plasma pressure and internal magnetic field are solved. The structural formation and a soft transition are illustrated. (The influence of radial electric field, i.e., plasma flow, is explained in the next section.) A one-dimensional transport simulation of plasma is performed in a model of cylindrical tokamaks. Tokamak plasmas with circular cross-sections are considered. The coordinates (r, θ, ζ) are approximated by cylindrical coordinates. The transport equation is solved in the radial direction. Transport codes have been developed, and it is now a routine exercise if an expression of plasma fluxes is explicitly formulated [24.1].

24.1.1 Transport Equations

The evolution of plasma pressure and current is studied for a *given density profile*. The development of density profile $n(r)$ is not solved as is discussed later. In this section, equations (24.1-3) and (24.1-4) are solved. With this simplification, the self-organized dynamics of plasma temperature and current profiles is explained.

To solve equations (24.1-3) and (24.1-4), transport properties for energy and current must be specified.

Ohm's Law. In the left-hand side of Ohm's law, equation (24.1-3), three terms appear as mechanisms that impede a free electron motion along magnetic field lines. Within these three terms, i.e., electron inertia, collisional resistivity and current diffusivity, the collisional resistivity has a dominant contribution to the transport process of global plasma current in tokamaks. The current diffusion is effective for microscopic structures as is discussed in preceding chapters. However, it is not so effective for the evolution of global current profile $J_\zeta(r)$. If the relation $\lambda_N \ll 1/\sigma_c$ holds in equation (24.1-3), plasma current profile $J_\zeta(r)$ is not affected by anomalous transport. The relation $\lambda_N \ll 1/\sigma_c$ is rewritten as

$$(\mu_0 v_A / qR)\sigma_c \ll (a\omega_p/c)^4 \alpha^{-3/2}$$

This condition is usually satisfied in laboratory experiments.

There are various methods to drive a toroidal current by an external supply. In this section, we consider the case where the current is driven under the mechanism of bootstrap current, and symbolically write

$$E_{drive} = \eta_{NC} J_{BS} \tag{24.2}$$

where η_{NC} is used to estimate the parallel resistivity $1/\sigma_c$ based on neoclassical theory and J_{BS} is the bootstrap current density. The neoclassical formula (i.e., the result of collisional transport theory) of η_{NC} and J_{BS} are given, e.g., in [4.7, 24.2]. It is given as

$$J_{BS} \sim -\sqrt{r/R} B_p^{-1} \nabla p. \tag{24.3}$$

(See also equation (4.32-3).)

Based on these considerations, Ohm's law (24.1-3) is simplified as

$$E_\zeta = \eta_{NC}(J_\zeta - J_{BS}). \tag{24.4}$$

Ohm's law and the magnetic diffusion equation include the neoclassical resistivity and the bootstrap current, which is driven by a pressure gradient.

Heat flux. The flux–gradient relation for plasma energy is simply written as

$$q_{r,j} = -n_j \chi_j \nabla T_j \qquad (j = e, i). \tag{24.5}$$

Transport equation. With these forms, equations (24.4) and (24.5), and Maxwell's equation, equations (24.1-3) and (24.1-4) are rewritten as one-dimensional transport equations:

$$\frac{\partial}{\partial t}\left(\frac{3}{2}n_e T_e\right) = -\frac{1}{r}\frac{\partial}{\partial r}r\frac{3}{2}n_e\chi_e\frac{\partial T_e}{\partial r} + P_{OH} + P_{ie} + P_{He} \tag{24.6}$$

$$\frac{\partial}{\partial t}\left(\frac{3}{2}n_i T_i\right) = -\frac{1}{r}\frac{\partial}{\partial r}r\frac{3}{2}n_i\chi_i\frac{\partial T_i}{\partial r} - P_{ie} + P_{Hi} \tag{24.7}$$

$$\frac{\partial}{\partial t}B_\theta = \frac{\partial}{\partial r}\eta_{NC}\left[\frac{1}{\mu_0}\frac{1}{r}\frac{\partial}{\partial r}rB_\theta - J_{BS}\right]. \tag{24.8}$$

Heating Power. In the energy balance equations (24.6) and (24.7), P_{OH} denotes joule heating power, i.e., a resistive dissipation of parallel plasma current. P_{ie} is equipartition power between electrons and ions. P_{He} and P_{Hi} stand for additional heating powers for electrons and ions, respectively.

The ohmic heating power P_{OH} is calculated by the use of the neoclassical resistivity,

$$P_{OH} = \eta_{NC} J_\zeta (J_\zeta - J_{BS}). \tag{24.9}$$

The equipartition between ions and electrons, P_{ie}, is caused by Coulomb collisions and by turbulence as is discussed in chapter 4. However, it is often modelled by only the Coulomb collision terms. In this case, it is expressed as

$$P_{ie} = \nu_{ie} n (T_i - T_e) \tag{24.10}$$

where ν_{ie} is the collision frequency of ions with electrons. Owing to the large difference of masses between ions and electrons, the collision frequency ν_{ie} is much smaller than ion–ion collision frequency or electron–electron (ion) collision frequency.

Power deposition profiles of additional heating, P_{Hi} and P_{He}, depend on plasma profiles. They could be consistently calculated by use of plasma profiles. Indeed, in the cases of rf wave launching and injection of energetic neutral particles, they have been intensively studied [2.4, 24.3]. However, we here show results, for simplicity, based on an assumption that the profile of energy supply is treated as an independent parameter. It is fixed in time, and is expressed by a Gaussian profile as

$$P_H(r) = P_{HO}\exp\left\{-\frac{(r - r_H)^2}{r_{Hw}^2}\right\}. \tag{24.11}$$

Density profile. In a present example, several simplifications are employed. The density profile is fixed in time in the following form as

$$n_e(r) = (n_0 - n_s)\left\{1 - \left(\frac{r}{a}\right)^2\right\}^{\sigma} + n_s. \tag{24.12}$$

24.1.2 Transport Coefficients

Energy transport is caused either by the collisional process (thermal fluctuations) or by turbulence. In the weak gradient limit, the collisional one χ_{NC} dominates. In the strong gradient limit, the turbulent transport χ_{TB} takes over. The limiting formula χ_{TB} is discussed in preceding chapters. As an interpolation formula, it is often expressed as a sum of a turbulent term χ_{TB} and neoclassical term χ_{NC},

$$\chi = \chi_{TB} + \chi_{NC}. \tag{24.13}$$

The choice of transport coefficient closes a set of equations that describes transport and profile. The following expressions are used for thermal diffusivity. Here, a neoclassical formula χ_{NC} is employed according to [17.11]. A theoretical formula of anomalous thermal transport coefficient, which is derived based on self-sustained turbulence in preceding chapters, is used:

$$\chi_{TB} = \frac{1}{1 + h_1 \omega_{E1}^2} C_{TB} F(s, \alpha) \alpha^{3/2} \delta^2 \frac{v_A}{qR} \tag{24.14}$$

where R is the major radius, a is the minor radius, v_A is the Alfvén velocity, δ is the collisionless skin depth, α is the normalized pressure gradient, $\alpha = -q^2 R\beta'$, $\beta = 2\mu_0(p_e + p_i)B^{-2}$, s is magnetic shear, $s = rq'/q$ and q is the safety factor. Electron and ion diffusivities are not separated in this turbulent transport model: we here assume that they have the same magnitude. $F(s, \alpha)$ represents the effect of magnetic shear and Shafranov shift as is given in chapter 10. The coefficient $(1 + h_1\omega_{E1}^2)^{-1}$ stands for the influence of the radial electric field. The parameter $\omega_{E1} = (\tau_{Ap} a/rsB)\,\mathrm{d}E_r/\mathrm{d}r$ denotes radial electric field gradient. Structural formation associated with a radial electric field is discussed later in this chapter. One numerical coefficient, C_{TB}, is an adjustment parameter from the comparison with experimental observations.

Apart from a factor $(1 + h_1\omega_{E1}^2)^{-1} F(s, \alpha)$, the gradient–flux relation has a dependence like

$$q_r \propto n^{-3/2} |\nabla p|^{3/2} \nabla T. \tag{24.15}$$

Heat flux has a higher order dependence on temperature gradient as is illustrated in figure 24.2. It is noted that χ_{NC} does not have a gradient dependence. As a result of this, a linear gradient–flux relation is obtained in the weak gradient limit (i.e., below a critical gradient of G_* in chapter 12). Figure 10.6 illustrates the contour of the transport coefficient.

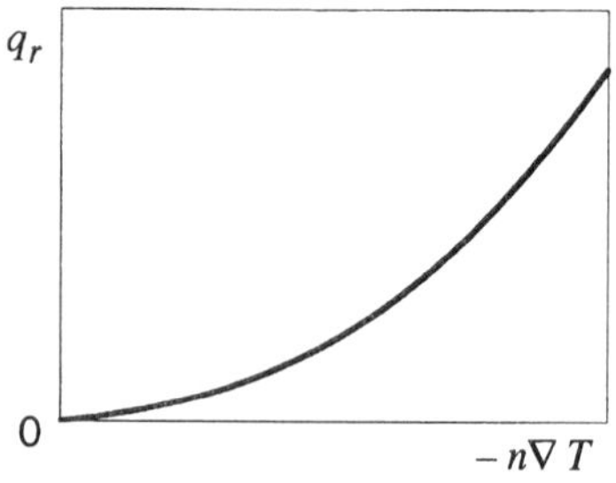

Figure 24.2. Nonlinear gradient–flux relation for the inhomogeneous plasmas.

In equation (24.14), coefficient $F(s,\alpha)$ represents an influence of a current profile (shape of magnetic surface) on anomalous transport. It is shown that turbulent transport coefficient is reduced if the magnetic shear is weak or negative [10.5, 24.4]. This is because an effective toroidal drift is reduced under such a circumstance as is discussed in chapter 10. The numerical solution, figures 10.5 and 10.6, is fitted to an interpolation formula as [24.4]

$$F(s,\alpha) = \frac{1}{\sqrt{2(1-2\hat{s})(1-2\hat{s}+3\hat{s}^2)}} \qquad (\hat{s} < 0) \qquad (24.16\text{-}1)$$

$$F(s,\alpha) = \frac{1+9\sqrt{2}\,\hat{s}^{5/2}}{\sqrt{2}\,(1-2\hat{s}+3\hat{s}^2+2\hat{s}^3)} \qquad (\hat{s} > 0) \qquad (24.16\text{-}2)$$

$(\hat{s} = s - \alpha)$.

It should be noted that expression (24.14) is employed as an example of the study of structural formation of plasmas. As is discussed in preceding chapters, there could exist other types of plasma turbulence in addition to current diffusive ballooning mode turbulence. When other turbulent modes become stronger, the explicit formula of the transport coefficient takes a different form. A survey of theoretical models is given in, e.g., [24.5], and a test of various models against observed plasma profiles has been reported in [24.6]. The characteristic nature shown in figure 24.2 is common to various models. The study of the model (24.14) provides a simple but deep and widely applicable insight into the problem.

24.2 Basic Feature of Confinement

We choose standard plasma parameters in section 24.2 as: $R = 3m$, $a = 1.2m$, toroidal magnetic field $B = 3$ T, plasma current $I_p = 3$ MA, central electron density $n_e(0) = 5 \times 10^{19}$ m^{-3}, heating power $P_{heat} = 10$ MW with $P_{He} = P_{Hi}$, and deuterium plasma. For these standard parameters, equations (24.6)–(24.8) are solved, and the solution is compared to experimental observations on JET tokamak. By this comparison, the numerical factor C_{TB} is chosen as

$$C_{TB} \simeq 8. \tag{24.17}$$

This numerical factor C_{TB}, which is determined from one set of device parameters, is applied to various conditions on the same device and on other tokamaks [24.6].

24.2.1 L-Mode plasmas

The model of χ is examined from the viewpoint of a global scaling of τ_E first, and then applied to a study of the radial profile $\chi(r)$.

Zero-dimensional analysis. A global energy confinement time τ_E can be estimated from a dimensional analysis $\tau_E \propto a^2\chi^{-1}$. Substituting equation (24.14) into this relation, one has

$$\tau_E \propto F^{-1} R^{1.5} B_p^2 T^{-1.5} A_i^{0.5} \ell_p^{1.5} \tag{24.18}$$

where A_i is the ion mass number, and ℓ_p is the scale-length of the pressure gradient. The gradient–flux relation, $q_r = -n\chi\nabla T$, is approximated by a zero-dimensional energy balance as

$$\frac{\chi n T}{a} \sim \frac{P_{tot}}{aR} \tag{24.19}$$

where P_{tot} is a total heating power. By use of equation (24.19), T is eliminated from equation (24.18), and a dimensional dependence is obtained as

$$\tau_E \propto a^{0.4} R^{1.2} I_p^{0.8} P_{tot}^{-0.6} A_i^{0.2} F^{-0.4} n_e^{0.6} \ell_p^{0.6} \tag{24.20}$$

Provided that the parameter dependence of $F(s, \alpha)$ is weak, we see that this scaling result (24.20) is close to experimental observations on the L-mode plasmas, for which a dependence like $\tau_E \propto P_{tot}^{-1/2}$ has been found empirically.

Spatial profiles of plasma parameter and transport coefficient. Simulation results are shown for the cases of ohmic heating and additional heating with $P_{heat} = 20$ MW. Differential equations (24.6)–(24.8) with equation (4.32-3) and equations (24.9)–(24.14) are solved. The density profile is fixed, and total plasma current I_p is kept constant. Time evolution is studied as follows: (1) the transport equation is solved without an additional heating power P_H; if a stationary profile is realized, then (2) an additional heating of the absorption profile (24.11) is applied and the response to it is solved. The schematic time trace is illustrated in figure 24.3. In the first phase of temporal evolution, the plasma temperature is sustained by the Joule heating term (24.9). The temperature profile and current

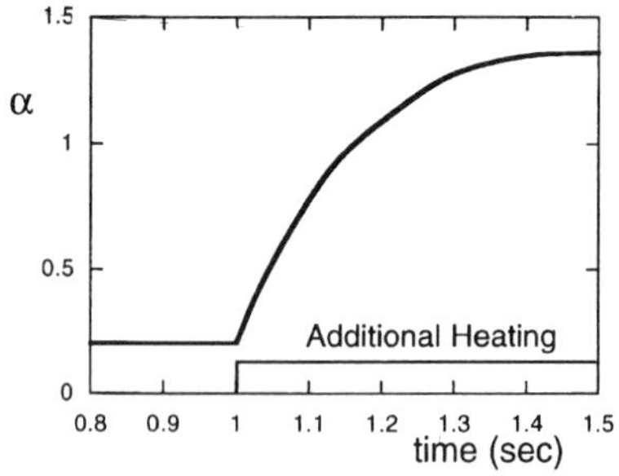

Figure 24.3. Time trace of plasma heating and the response of the plasma. (Pressure gradient at half radius is shown as an example.)

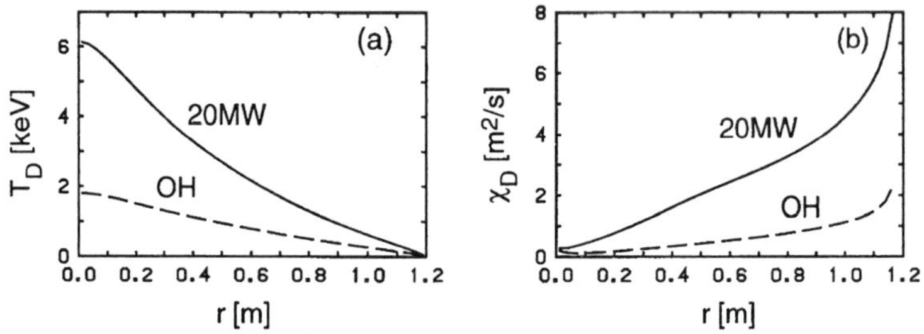

Figure 24.4. Radial profile of the temperature and thermal conductivity for the case of the Ohmic heating (denoted by OH) and the additional heating (denoted by 20MW).

profile are coupled to each other, and each profile reaches the stationary state. This state is called an ohmic plasma. With additional heating, the temperature starts to increase, and the plasma profile saturates at a new stationary state.

Figure 24.4 illustrates radial profiles of ion temperature and ion thermal diffusivity in two stationary states. In the ohmic heating case, for which the total heating power is a few MW, the ion temperature and the thermal conductivity are lower than those in the case of strong heating. It is found that χ is an increasing function of minor radius in both cases of ohmic heating and high power heating of $P_{heat} = 20$ MW; χ is increasing towards the edge. In the two cases in figure 24.4, the amounts of heating power supplied into plasma are different by nearly an order of magnitude. Nevertheless, the difference in plasma temperature (energy) does not appear so much. When an additional heating power of 20 MW is applied, the ion temperature is increased by a factor of about three, while the thermal diffusivity is enhanced by a factor of five. An increase of pressure gradient causes a rise of χ in the whole plasma column.

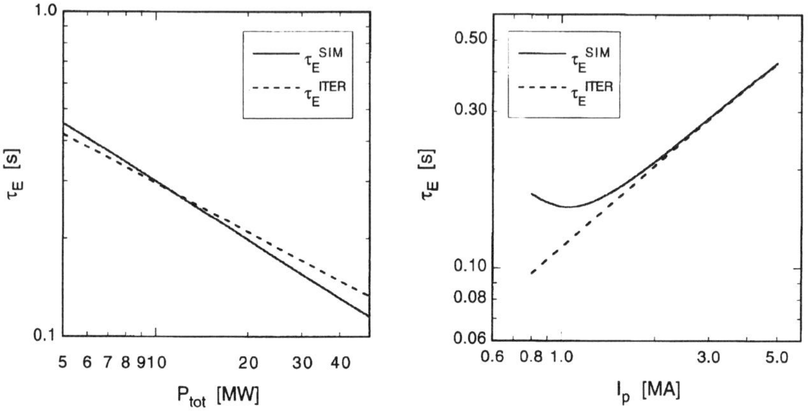

Figure 24.5. Dependence of the energy confinement time τ_E on the heating power (left) and that on the plasma current I_p (right). Dashed line indicates the empirical fitting of the experimental data in [24.8].

It is noted that radial profiles of temperature and thermal conductivity do not show large differences between the ohmic heating plasma and the one with intense additional heating. A similarity in the profile is a consequence of the fact that the plasma structure and transport coefficient interfere and determine each other.

Parameter dependence of τ_E. The parameter dependence of energy confinement time is studied from the solutions of the transport equation. Figure 24.5 (left) illustrates the power dependence of energy confinement time τ_E for standard parameters. The solid line and the dashed line indicate the simulation result and an empirical scaling, $\tau_E \propto P_{tot}^{-0.5}$ [24.7, 24.8], respectively. Over a wide power range, a so-called power degradation characteristic, $\tau_E \propto P_{tot}^{-0.6}$, is nearly confirmed. This behaviour is expected from the dimensional analysis of $\chi \propto T^{1.5}$, as is shown in equation (24.20).

The magnetic field dependence is studied. Other parameters (plasma radius, total toroidal current, heating power and density) are fixed as standard values as in figure 24.5 (left), for a fixed heating power of $P_{heat} = 10$ MW. Very weak dependence is obtained. This is because parameters s, α and v_A/qR are independent of B, and χ_{TB} depends on B only through an average curvature κ.

The dependence of τ_E on the total plasma current is strong. Figure 24.5 (right) demonstrates that τ_E asymptotically scales as $\tau_E \propto I_p^{0.81}$ in the high current limit. The analytic estimate of I_p dependence, $\tau_E \propto I_p^{0.8}$, comes from the q^2 dependence of χ. It is noticed that the energy confinement time deviates from this power dependence and takes the larger value in the low current regime

($\beta_p > 1$). This prediction of an improvement is pertinent to a high β_p plasma [24.9]. In this simulation study, it is caused by the reduced value of the coefficient $F(s, \alpha)$ in equation (24.14). The mechanism of reduction and the physics of improved confinement are discussed in detail in the next section.

24.3 Role of Current Profile

The result in the low current limit of figure 24.5 (right) suggests an important role of the configuration factor F for the global confinement time. In this section, we consider the case of intense heating with low toroidal plasma current, where bootstrap current significantly modifies the profile of the pitch of the magnetic field lines [24.4]. Namely, plasma pressure modifies the magnetic configuration. The impact of reduced or negative shear is elucidated.

The improved confinement in the low current case in figure 24.5 (right) is related to the β_p value of the plasma. For various values of I_p and P_{tot}, τ_E is analysed. It is found that the confinement improvement is observed when the input power exceeds a certain critical value of β_p.

When the value of β_p is less than unity, the relation

$$\beta_p \propto P_{tot}^{0.4} \tag{24.21}$$

is observed, representing the power degradation, $\tau_E \propto P_{tot}^{-0.6}$. However, if β_p exceeds a critical value close to unity, we observe

$$\beta_p \propto P_{tot}^{1.4} \tag{24.22}$$

A dramatic change in the confinement scaling is seen in the power dependence.

24.3.1 Internal Transport Barrier

The improvement of global confinement is related to the establishment of a barrier for plasma loss flux. A self-organized structure is found in the transport structure, which is represented by an internal transport barrier formation. The plasma profile with an internal transport barrier, which is observed in high β_p plasmas, is called the 'high β_p mode'. A theoretical model of the mechanism is discussed in the following.

We examine the radial profiles of a high β_p plasma based on a theoretical model of turbulent transport. A typical set of parameters, $I_p = 1$ MA and $P_{heat} = 20$ MW, is chosen. (Other parameters are the same as those for the case of figure 24.4.) In this case, a transport code study provides the solution with $\beta_p = 3.2$ and $I_{BS}/I_p = 0.935$. (I_{BS} is the total bootstrap current.) The radial profiles of electron temperature $T_e(r)$, normalized pressure gradient $\alpha(r)$, magnetic shear $s(r)$ and heat diffusivity $\chi(r)$ are calculated and shown in figure 24.6(a)–(d), respectively. A large pressure (temperature) gradient is found

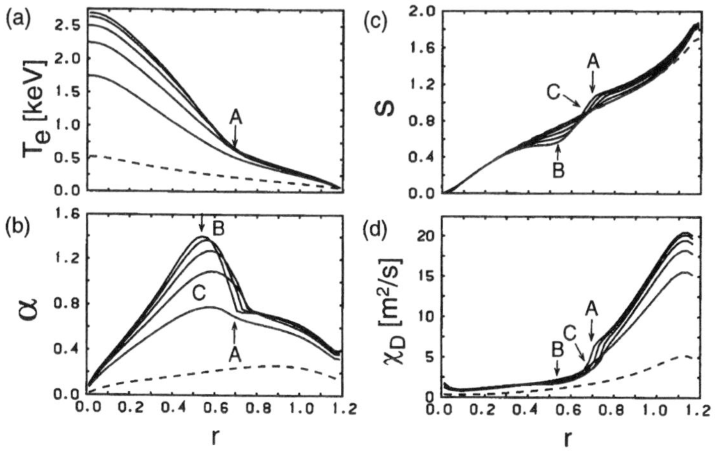

Figure 24.6. Plasma profile with an internal transport barrier formation. Temporal evolutions of temperature (a), normalized pressure gradient α (b), magnetic shear s (c), and thermal conductivity (d) are illustrated.

to be formed and sustained in the plasma core region

$$r < r_* \simeq 2a/3.$$

A sudden change of thermal conductivity near this surface occurs and the steep gradient is spontaneously formed which we attribute to the formation of an *internal transport barrier.*

The internal transport barrier is self-organized. Associated with the steep pressure gradient, the ratio of the bootstrap current to the total plasma current becomes high in this region. As a result, a hump of the current density profile appears in a region between $r = 0.6$ and 0.8 m, which results in a flattening of the $q(r)$ profile, i.e., reduction of shear $s(r)$ (figure 24.6(c)). In the region $r < r_* \simeq 2a/3$, a transport barrier is formed (figure 24.6(d)). In this region the strength of magnetic shear s is reduced and the value of α further increases, which cooperatively reduces the factor $F(s, \alpha)$ to a very small value as is given by equation (24.16). The reduction of $F(s, \alpha)$ cancels the increment of the term $\alpha^{3/2}$ in the formula of thermal diffusivity of equation (24.14). In spite of a large pressure gradient, thermal diffusivity is reduced.

The radial dependences of s and α, in figure 24.6(c) and (d), are traced on an s–α diagram in figure 24.7. The origin in radius $r = 0$ corresponds to the origin in this (s, α) diagram. In the core region, where an improved confinement is realized, magnetic shear is not strong, $s < 0.75$, and pressure gradient α is large. We see that χ does not increase so much in the region $r < r_*$, even if

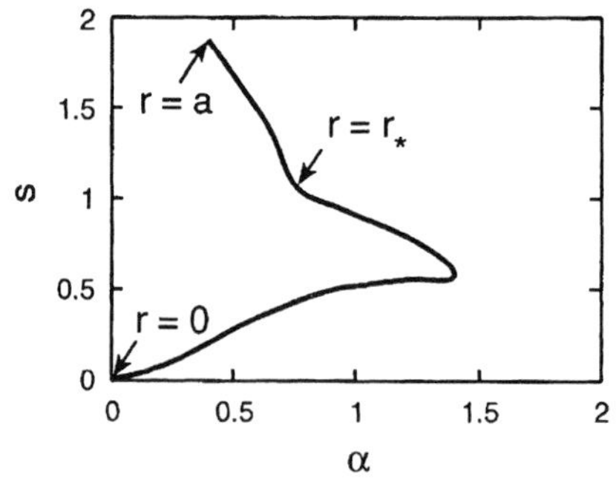

Figure 24.7. Relation of pressure gradient α and magnetic shear s in a plasma with an internal transport barrier. Establishment of a high pressure gradient in the low magnetic shear region $r \le r_*$ is shown.

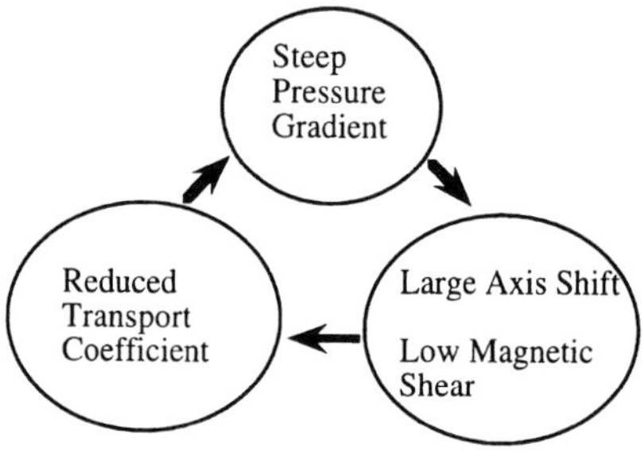

Figure 24.8. Nonlinear link between gradient, axis shift, low magnetic shear and reduced thermal conductivity.

the pressure gradient, which is in proportion to α, increases. The enhancement of pressure gradient is expected in the weak magnetic shear region, also shown in figure 10.6.

The improved confinement of a high β_p plasma is considered to take place as a result of a cooperative phenomenon between current profile and pressure profile. An increment of pressure gradient causes the larger shift of the toroidal axis. This shift increases an average magnetic well. The plasma current profile is also modified owing to the bootstrap current, and the magnetic shear is reduced. By these effects, the turbulent transport coefficient is decreased. The smaller thermal conductivity leads to a further increment of pressure gradient. These processes continue until they reach a state of meta-equilibrium balance. The nonlinear link is illustrated in figure 24.8.

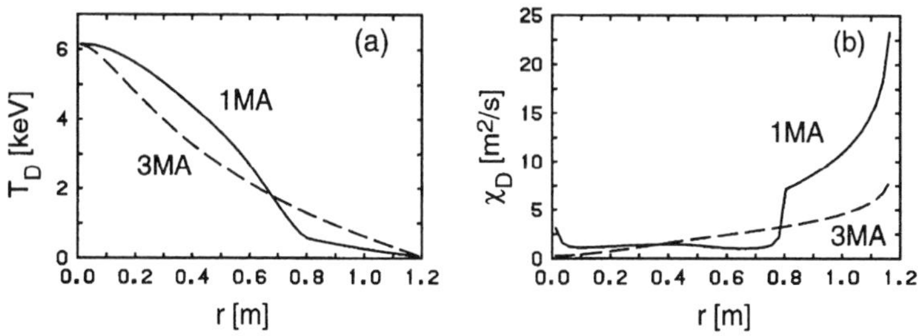

Figure 24.9. Radial profiles $T_D(r)$ and $\chi_D(r)$ in the presence of (solid line), and absence of (dashed line) internal transport barrier. The former is low current case, and the latter is high current case.

The location of the internal transport barrier responds to the magnitude of total plasma current. In figure 24.9, we compare radial profiles of $T_D(r)$ and $\chi_D(r)$ for various values of I_p with a fixed value of $P_{heat} = 20$ MW. The ratio of bootstrap current to total current I_p varies. The pressure profile has a kink point in the core region of the plasma column. The location of the barrier in the low I_p case is further in than that of the case of high I_p. The thermal diffusivities χ take similar values to each other near the centre. The increase of χ near magnetic axis is due to the neoclassical diffusion which scales as q^2.

24.4 Structural Formation of Density and Radial Electric Field

Structural formation in plasmas is caused by interference between particle and momentum transports as well. In this section, structural formation related to density and radial electric field is discussed. A simulation study on density and velocity profiles is shown. Velocities of ions and electrons are separately solved. Mean flow velocity and current density are deduced from the profiles of n_e, $\boldsymbol{V}_e$ and $\boldsymbol{V}_i$.

24.4.1 Simulation Model

Transport equations for density and flow velocity are solved by a numerical simulation. The transport equations of electrons and ions ($j = i, e$)

$$\frac{\partial}{\partial t} n_j + \boldsymbol{\nabla} \cdot \boldsymbol{\Gamma}_j = S_p \qquad (24.23\text{-}1)$$

$$m_j \frac{\mathrm{d}}{\mathrm{d}t} n_j \boldsymbol{V}_j = n_j e_j (\boldsymbol{E} + \boldsymbol{V}_j \times \boldsymbol{B}) - \boldsymbol{\nabla} p_j + \boldsymbol{\nabla} \cdot \mu_j \nabla \boldsymbol{V}_j + \boldsymbol{F}_j \qquad (24.23\text{-}2)$$

are solved with the Maxwell equations

$$\varepsilon_0 \frac{1}{r}\frac{\partial}{\partial r}(rE_r) = \sum_j e_j n_j \tag{24.24-1}$$

$$\frac{\partial}{\partial t}B_\theta = \frac{\partial}{\partial r}E_\zeta \tag{24.24-2}$$

$$\frac{1}{r}\frac{\partial}{\partial r}(rB_\theta) = \frac{1}{c^2}\frac{\partial}{\partial t}E_\zeta + \mu_0 \sum_j e_j n_j V_{j\zeta}. \tag{24.24-3}$$

The energy balance equation is not solved here and a temperature profile is given to elucidate the topics. This simplification clarifies a physical process concerning the density profile and the plasma flow [24.10].

The 'force' term $\boldsymbol{F}_j$ includes Coulomb collisions (collisional momentum transfer $\boldsymbol{F}_j^C$ or neoclassical bulk viscosity $\boldsymbol{F}_j^{NC}$), interaction force with a turbulence field $\boldsymbol{F}_j^W$, charge exchange force $\boldsymbol{F}_j^{CX}$ or force from a direct loss $\boldsymbol{F}_j^L$ etc. The physical pictures of these terms are discussed in preceding chapters, and explicit forms are listed in the appendix.

A particle source is given from ambient neutral particles and from the injected beam, which are denoted by suffixes 0 and b, respectively. The densities of neutral particles and beam ions are solved together.

A set of equations (24.23) and (24.24) is solved numerically with the following boundary conditions. On the magnetic axis, $r = 0$, we set

$$n_j V_{jr} = 0 \qquad n_j V_{j\theta} = 0 \qquad \frac{\partial}{\partial r}(n_j V_{j\zeta}) = 0,\ E_r = 0,\ B_\theta = 0 \qquad \frac{\partial}{\partial r}n_0 = 0. \tag{24.25-1}$$

On the wall surface, $r = b$, we impose

$$n_j V_{jr} = 0 \qquad n_j V_{j\theta} = 0 \qquad n_j V_{j\zeta} = 0 \qquad B_\theta = \frac{\mu_0 I_p}{2\pi b} \qquad \frac{\partial}{\partial r}n_0 = \frac{1}{D_0}\Gamma_0. \tag{24.25-2}$$

Parameters I_p, D_0 and Γ_0 are considered to be given. The surface of the main plasma is taken at $r = a$. The region between the plasma and wall surface, $a < r < b$, is called a scrape-off layer, and an additional loss mechanism exists in this region. In the scrape-off layer, plasma loss along magnetic field lines occurs. An additional loss is superimposed on the right-hand side of equations (24.23).

Force and flow are directly related as is discussed in the preceding chapters. The radial particle flux of electrons is expressed from equation (24.23-2) as

$$n_e V_{er} = -\frac{1}{1+c_e}\frac{\bar{\nu}_e + \nu_{ei}}{m_e \Omega_{e\zeta}^2}\left(\frac{\mathrm{d}p_e}{\mathrm{d}r} + \frac{\mathrm{d}p_i}{\mathrm{d}r}\right) - c_e n_e \frac{E_\zeta}{B_\theta} + \frac{1}{1+c_e}\frac{\bar{\nu}_e}{\Omega_{e\zeta}} n_e V_{i\theta}$$
$$+ c_e n_e \frac{B_\zeta}{B_\theta}\frac{\nu_{eb}}{\Omega_{e\zeta}}(V_{e\zeta} - V_{i\zeta}) + \frac{F_{e\theta}^W}{m_e \Omega_{e\zeta}} \tag{24.26}$$

where

$$c_e \equiv \frac{\nu_e + \nu_{ei}}{\nu_{ei} + \nu_{eb}} \frac{B_\theta^2}{B_\zeta^2} \qquad \bar{\nu}_e \equiv \sqrt{\pi}\, q^2 \frac{v_{the}}{qR} \frac{1.78\nu_{*e}}{1 + 1.78\nu_{*e}} \qquad \nu_* \equiv \frac{\nu_j q R}{\varepsilon^{3/2} v_{thj}}$$

and $\nu_{eb} = (n_b m_b / n_e m_e)\nu_{be}$. The damping rate $\bar{\nu}_e$ comes from the neoclassical bulk viscosity, and the factor c_e represents its contribution to the toroidal direction. This simple analytic expression is derived in the absence of a shear viscosity term in equation (24.23-2).

The first to fourth terms of the right-hand side of equation (24.26) represent the fluxes that are caused by Coulombic collisions. The first and the second terms stand for neoclassical diffusion and Ware pinch, respectively. The third term is a flux due to neoclassical bulk viscosity. Owing to the difference in mass between electrons and ions, this flux is usually small for electrons. The fourth is caused by a collisional drag force from beam ions. This is effective when a beam injection is set. The fifth term is a turbulent diffusion term. It is sometimes expressed as a diffusion term $F_{e\theta} = -m_e \Omega_e D_a (\mathrm{d}/\mathrm{d}r) n_e$. Here we write

$$F_{i\theta}^W = -F_{e\theta}^W = -\frac{Z^2 e^2 B_\zeta^2}{T_e} n_i (V_{i\theta} - \langle \omega r^{-1} k_\theta^{-1} \rangle r) \qquad (24.27)$$

where $\langle \omega / r k_\theta \rangle$ is the average rotation frequency of turbulence. The quantity $\langle \omega / r k_\theta \rangle$ could affect a momentum exchange between particles and fluctuations, but is set to be zero in this monograph for simplicity.

The ion orbit loss, when it is kept in a simulation, causes an additional force

$$\boldsymbol{F}^{lc} = e_i \Gamma_i^{lc} (\hat{\boldsymbol{r}} \times \boldsymbol{B}). \qquad (24.28)$$

24.4.2 Inward Pinch

Evaluation equation (24.26) shows that there is a variety of mechanisms that cause particle flux. Some are common for ions and electrons (e.g., flux by electron–ion collisions or a local part of turbulent-driven flux) but others are not (e.g., flux by bulk viscosity or by turbulent-driven viscosity). Thus profiles of density, radial electric field and flow velocity are coupled with each other [24.11]. As is discussed in chapter 2, an interaction between density profile and dynamics of flow (i.e., momentum transport) is one characteristic feature of far-nonequilibrium transport.

An example is taken for a comparison study with experimental results of the JFT-2M tokamak: the plasma major radius is $R = 1.3$ m, the minor radius is $a = 0.35$ m, the wall radius is $b = 0.4$ m, the toroidal magnetic field $B_\zeta = 1.3$ T, plasma current $I_p = 150$ kA with hydrogen plasmas. Neutral beam injection energy is 35 keV and the power is 600 kW. The plasma density at the wall is given as $n_b = 10^{18}$ m^{-3}.

In a simulation study of particle transport, a particle source is specified. In this simulation, a particle source is given by ionization of neutral particles.

The source profile of particles through ionization is calculated by use of an ionization cross-section that depends on plasma temperature, electron density and velocity of neutral particles. Three kinds of neutral particle are considered. (i) *Neutral particles* are injected as a neutral beam injection. Parameters of injection energy of 35 keV and power of 600 kW correspond to a particle source of the rate $S_{0.NBI} \simeq 10^{20}$ (particles per second). (ii) *A neutral particle supply by gas puffing.* The source rate is chosen as $S_0 = 2 \times 10^{19}$ (particles per second) with the particle velocity $v_0 = 1.5$ km s^{-1}. (iii) *Recycled particles.* Plasma particles which interact with the wall leave the wall and enter the plasma as recycled neutrals. The recycling rate on the wall, i.e., the ratio of neutral particles leaving the wall to the influx onto the wall, is chosen as $\gamma_0 = 0.8$ in this simulation.

A simple model of the diffusion coefficient is employed as

$$D_i(r) = (1 + 9r^2a^{-2})D_{i0}. \qquad (24.29)$$

Absolute values are chosen as $D_{i0} = 0.5$ m^2 s^{-1} and $\mu_{i0} = 1.5$ m^2 s^{-1}. Temperature profiles are given as a parabola, $T_e(r) = \{T_e(0) - T_e(a)\}\{1 - (r/a)^2\} + T_e(a)$ and $T_e = T_i$. We set $T(0) = 700$ eV and $T(a) = 50$ eV.

Variations of density and electric field owing to the beam injection are also simulated. Figure 24.10 illustrates the setup of neutral beam injection.

The simulation follows an experimental procedure, whose scheme is as follows. (i) First, plasma particles are sustained by gas puffing and current profile is sustained by ohmic current drive. A stationary state is realized. (ii) Then a neutral beam injection in the co-direction to the toroidal plasma current is applied for the period of 200 ms. (iii) Finally, the direction of the neutral beam is reversed and is injected in the counter direction to the plasma current. The duration of counter-injection is 200 ms.

The simulation result is shown in figure 24.11. The stationary plasma is first sustained without an external momentum source. At the time of $t = 0.5$ s of figure 24.11, the neutral beam (co-injection) is onset. Forces in the toroidal direction $F_{j,\zeta}$ are set in equation (24.23-2). After a duration of 200 ms, the

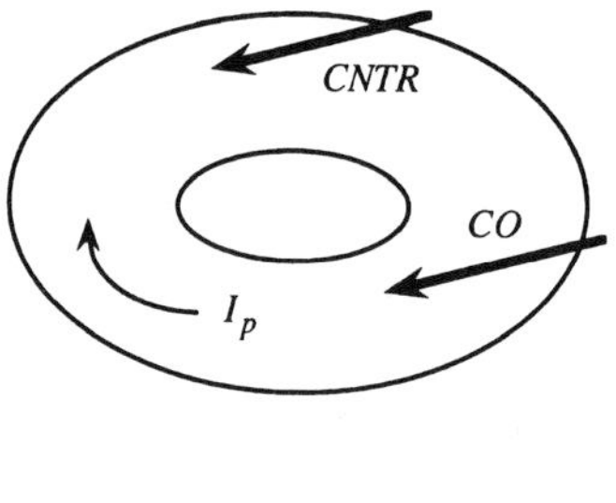

Figure 24.10. Neutral beam injection. Co-injection to the plasma current direction (co-injection) and injection counter to the direction of plasma current (counter-injection).

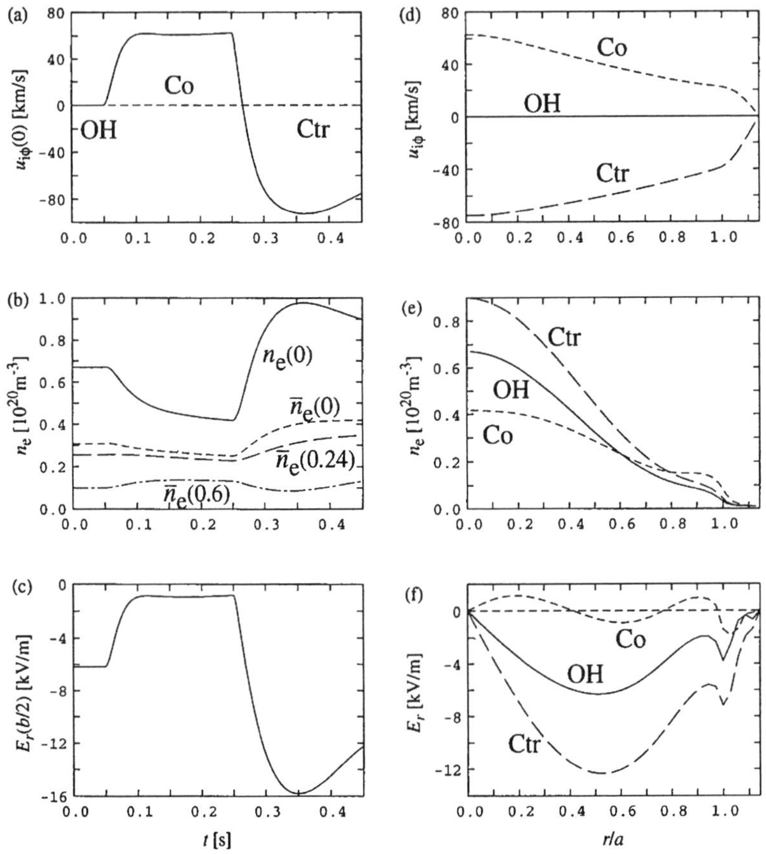

Figure 24.11. Response of the plasma profile against the external injection of the momentum. (Co-injection during the period of 0.05 s $< t <$ 0.25 s and counter-injection for 0.25 s $< t <$ 0.45 s.) Temporal evolution (a)–(c) and radial profiles (d)–(f). In (d)–(f) solid lines, short dashed lines and long dashed lines are for ohmic plasma, co-injection and counter-injection, respectively.

direction is reversed, and counter-injection is applied. (Note that a typical time scale of momentum confinement, $a^2/4\mu_i$, is about 16 ms for this simulation. The choice of 200 ms allows us to establish a stationary state of velocity profile.) Time evolutions are shown for a central toroidal velocity (a), electron density at various radial locations (b: solid line for the centre), and radial electric field near half radius (c), respectively. Typical profiles are illustrated for toroidal velocity (d), density (e) and radial electric field (f), respectively. In (d)–(f), solid lines are for initial profiles where the beam source is absent, short dashed lines are at the time of $t = 0.25$ s (end of co-injection) and long dashed lines are at the time of $t = 0.45$ s (end of counter-injection), respectively. It is shown that the density is expelled from the inside of the plasma column, when co-injection is imposed. The density profile becomes flatter, and the radial electric field is reduced. In contrast, when counter-injection is applied, the density increases, even though the same supply rate of neutral particles is kept. In the case of counter-injection, the density profile is peaked. The radial electric field becomes more negative.

This example clearly demonstrates that the supply of momentum influences the transport of plasma particles. Interference between fluxes of particles (scalar quantity) and momentum (vector quantity) occurs in confined plasmas. This is due to a long range interaction of self-induced radial electric field, and is characteristic of the turbulent transport of nonuniform plasmas.

24.5 Transport Barrier of H-Mode Plasmas

In chapter 21, formation of the interface is discussed. A steep gradient could be established at an interface, the thickness of which is in proportion to $\sqrt{\mu_i}$. Turbulent viscosity could be reduced when the gradient of radial electric field becomes strong enough. The reduced viscosity enhances the gradient of velocity and electric field. This link sustains the sharp gradient of radial electric field and reduces the turbulent transport. A model is discussed in chapter 21. An example of transport simulation is shown [24.12].

The transport model of turbulent viscosity is employed based on a theory of the current diffusive ballooning mode as

$$\mu_i = C_\mu \frac{F(s, \alpha)}{1 + h_1 \omega_{E1}^2} \alpha^{3/2} \frac{c^2}{\omega_p^2} \frac{v_A}{qR}. \tag{24.30}$$

Particle diffusivity is not obtained by the present theory of CDBM turbulence. We simply assume that the particle diffusivity has the same dependence as μ_i, that is,

$$D = C_n \frac{F(s, \alpha)}{1 + h_1 \omega_{E1}^2} \alpha^{3/2} \frac{c^2}{\omega_p^2} \frac{v_A}{qR}. \tag{24.31}$$

Coefficient h_1 has a parameter dependence [18.14], but is taken as a constant here. A typical value is $h_1 = 24$ for $r/R = 1/3$, $q = 3$, $s = 0.5$ and $\alpha = 0.3$. Coefficients C_μ and C_n are numerical constants.

As in section 24.4, the temperature profile is set, and is given as a parabola, $T_e(r) = \{T_e(0) - T_e(a)\}\{1 - (r/a)^2\} + T_e(a)$, $T_e = T_i$ and $T(0) = 700$ eV. The edge temperature is a parameter to be varied for the study of transport barrier formation. The momentum source is set to be zero. Other parameters are the same as those in the case of figure 24.11. Bifurcation associated with an evolution of radial electric field is shown in chapters 19–22. In this simulation, bifurcation is not addressed, but the spatial structure of reduced transport, i.e., a transport barrier, is illustrated.

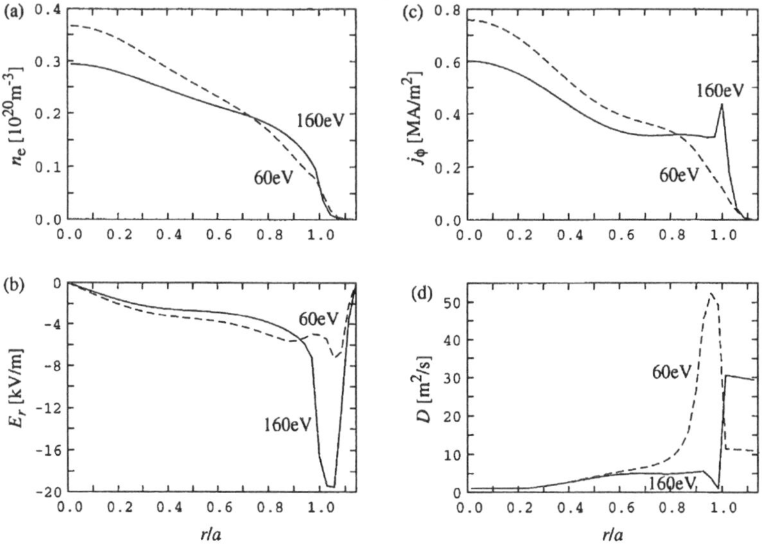

Figure 24.12. Radial profiles of the density (a), the radial electric field (b), the toroidal current (c) and the turbulent diffusivity (d). The solid line is for the case of high edge temperature, $T_{edge} = 160$ eV, and the dashed lines are for the low edge temperature case, $T_{edge} = 60$ eV.

Figure 24.12 illustrates the radial profile of density (a), that of radial electric field (b), the toroidal current density profile (c) and the profile of anomalous diffusion coefficient (d). Solid lines are for the case of $T(a) = 160$ eV and dashed lines are for a low edge temperature case, $T(a) = 60$ eV. When edge temperature is low, radial electric field profile depends approximately linearly on minor radius, and $\boldsymbol{E} \times \boldsymbol{B}$ rotation is seen to be nearly rigid body. The gradient of $\boldsymbol{E} \times \boldsymbol{B}$ rotation is weak. Turbulent diffusion coefficients μ and D are increasing towards the edge. When edge temperature increases, electric field gradient becomes strong near the edge. As the gradient of radial electric field increases, the transport coefficient is reduced, and the edge density gradient also becomes steeper. The transport coefficient has a dip at the plasma surface, and an edge transport barrier is formed. The reduction of transport near the edge, relative to the low edge temperature case, has two noticeable factors. First, a sharp reduction is observed in the location of $0.9 < r/a < 1$. Second, a moderate reduction of transport remains in the region of $0.7 < r/a < 0.9$.

One transport barrier is related to the structure of the electric field. A very strong reduction of transport at the edge is caused by strong electric field inhomogeneity. The electric field gradient extends into a scrape-off layer region, $0.95 < r/a < 1.05$. The strong gradient of radial electric field is a primary reason for transport barrier formation. In addition to this, the influence of modification of magnetic shear is also observed in figure 24.12. Another transport reduction of the inside region, $0.7 < r/a < 0.9$, is caused by a reduction of magnetic shear. The pressure gradient is increased by the onset of a transport barrier at the edge. Therefore a bootstrap current is also increased near the edge. This causes, under the condition of constant toroidal current I_p, the flattening of toroidal current. The mechanism which is discussed in section 24.3 is effective to decrease the turbulence-driven transport. Through these processes, the transport barrier is formed at the edge and could extend into the core region. The penetration of the transport barrier into the core is also discussed in [24.13].

24.6 On Experimental Observations

Before closing this chapter, some characteristic observations in experiments are visited.

24.6.1 Profile Resilience in L-Mode Plasmas

In a wide range of plasma parameters, energy confinement time has been observed to be a decreasing function of average plasma temperature. Such plasmas, being called L-mode plasmas, are characterized by (i) a degradation of the confinement time when heating power is increased [24.7], (ii) the shape of the electron temperature profile, $T_e(r)/T_e(0)$, is insensitive to the shape of the heating power profile, $P_H(r)/P_H(0)$, and (iii) the profile of thermal diffusivity increases towards the edge [24.14]. These features have been observed very widely, and are general properties.

These properties are attributed to the dependence of thermal diffusivity on the pressure gradient and on the pitch of magnetic field line, q^{-1}. As is explained in chapter 3, the poloidal magnetic field, B_p, is essential for the toroidal confinement. This important role is also applied to transport properties: the turbulent transport coefficient is governed by B_p, not B. Such conditions must be satisfied by theoretical models of turbulent transport coefficient. A particular model of turbulent thermal conductivity equation (24.14) has the property that χ is an increasing function of $|\nabla T|$ and q. This model predicts a correlation between an increment of χ and the access of parameter α to the stability boundary of the ideal MHD ballooning mode, α_c(MHD), as is shown in figure 24.3. An empirical correlation between χ and ratio α/α_c(MHD) has been pointed out [24.15].

24.6.2 Internal Transport Barrier

Formation of an internal transport barrier has been observed in various experimental circumstances and the profile of ion temperature can have a steep

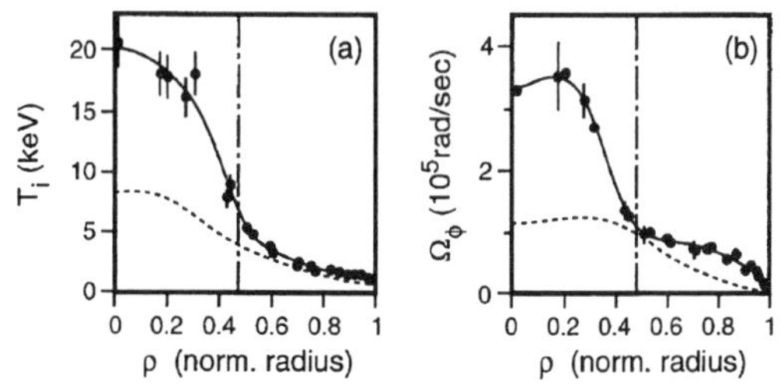

Figure 24.13. Experimental observation of an internal transport barrier. The cases with internal transport barrier (solid lines) and without (dashed lines) are shown. (Quoted from [24.19].)

gradient inside plasmas [24.16]. For instance, when plasma current I_p is small and poloidal beta value β_p is high [24.17], an internal transport barrier is formed. By deforming the current profile, i.e., by inverting the sign of q' in the core region, the transport reduction in the central region becomes very clear [24.18–20]. Observed profiles of the ion temperature and toroidal rotation velocity are illustrated in figure 24.13.

24.6.3 Particle Pinch

At the moment, a systematic study has not yet been developed in this aspect. One important observation is the improved ohmic confinement (IOC) [24.21]; an enhancement of energy confinement was realized by a reduction of gas influx. It has been observed that the energy confinement time in ohmic heating plasma increases with the increment of density, $\tau_E \propto \bar{n}_e$, in a low density regime [24.22]. In a high density region, the proportionality $\tau_E \propto \bar{n}_e$ is lost, and τ_E weakly depends on the density. When the gas puffing, which is used to sustain the high density in ohmic plasma, is suddenly reduced, then the plasma density profile shows a spontaneous peaking. Associated with the peaking, τ_E increases so as to recover the proportionality $\tau_E \propto \bar{n}_e$.

This kind of density peaking with an improved confinement has also been observed under the control of the injection direction of a neutral beam [24.23, 24.24]. An experimental observation on the JFT-2M tokamak is shown in figure 24.14. The change of direction of neutral beam, from co-direction to counter-direction, reverses the rotation velocity. Associated with this the radial electric field profile changes and the density peaking starts. This result is to be compared with theoretical analysis (figure 24.11).

24.6.4 H-Mode Transport Barrier

Intense experimental study has been performed on H-mode plasmas. Progress has been reported in [2.21, 24.25–24.27].

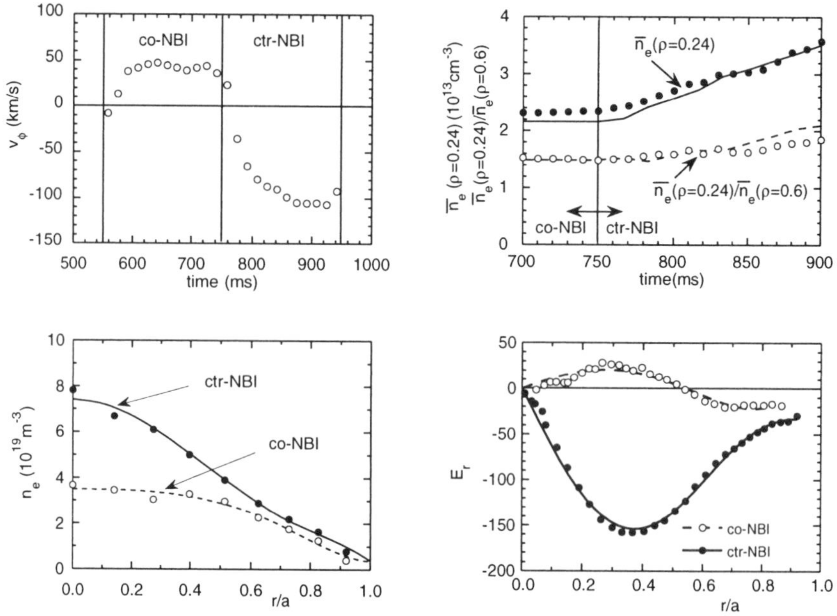

Figure 24.14. Density peaking by counter-injection on JFT-2M. Temporal evolution of toroidal velocity at plasma centre (top left), and those of central density and density peaking factor (top right) are shown. At the bottom, radial profiles of electron density and radial electric field are illustrated. (Based on JFT-2M experiment, [24.24].)

One important issue is the causality of radial electric field for an improved confinement in the H-mode. It has a fairly firm experimental basis for the improved confinement. Other fundamental issues are (i) a trigger of bifurcation, and (ii) spatial structure of the transport barrier.

The temporal evolution of H-mode transition has been observed. A rapid change of electric field at the onset of spontaneous transition has been observed, which suggests a bifurcation. The existence of electric field bifurcation has been clearly demonstrated in other experiments where the H-mode transition is induced by a bias current across the plasma surface [24.28]. A nonlinear relation between the radial electric field and radial current has been observed as is illustrated in figure 24.15. A jump of current is observed at a critical voltage, above which current is a decreasing function of the voltage [24.29]. A spontaneous electric bifurcation has also been observed in toroidal helical plasmas [2.23]. (See figure 2.11.) There is now no doubt that a plasma has a feature of electric field bifurcation.

The spatial profile of the transport barrier of the H-mode has also been

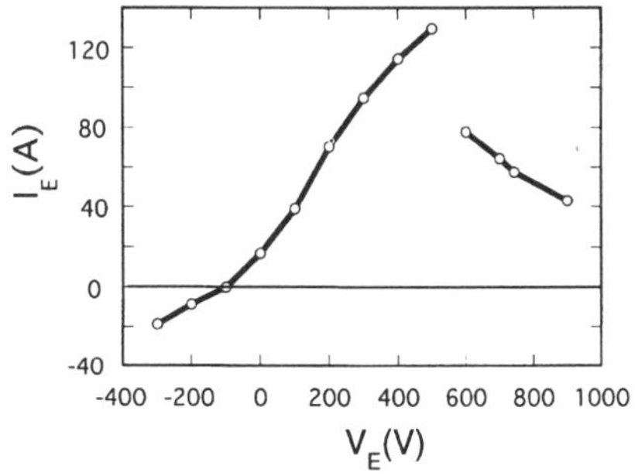

Figure 24.15. Current across magnetic surfaces (between electrode and limiter) against electrode voltage. (Quoted from [24.29].)

studied. Figure 24.16 shows the observed change of gradient near the edge in the JFT-2M tokamak [24.27]. The detailed structure is measured. The dependence of the width of the transport barrier is discussed and was found to have a weak dependence on poloidal gyro-radius of ions [24.30].

24.6.5 Fluctuations

Observation of fluctuations is a key element for the understanding of turbulent transport in plasmas. Due to progress of diagnostic methods, a prominent achievement has been seen in the experimental study of fluctuations [24.31]. In particular, observations of the fluctuations with longer wavelength have piled up in recent years.

It should be emphasized that fluctuations in the range of collisionless skin depth have been observed quite recently. For this range of fluctuations, the theory of self-sustained turbulence is developed in this monograph. The potential importance of such fluctuations has been recognized [24.32] but the existence of the mode has remained a theoretical hypothesis. Recently, a measurement was made on a short wavelength mode, the wavelength of which is in the range of collisionless skin depth, by use of a TFTR microwave scattering experiment [24.33]. The existence of fluctuations in the range of collisionless skin depth is demonstrated experimentally without doubt. Correlation of fluctuation level with electron anomalous transport is also suggested in the experiments.

Based on experimental observations, one could say that theories on turbulent transport and the role of electric field in structural formation are no longer a hypothesis but capture some essential nature of real plasmas. Nevertheless, the theoretical study is far from satisfactory and future study is necessary. Experimental progress and deepening of theoretical modelling are expected in the near future.

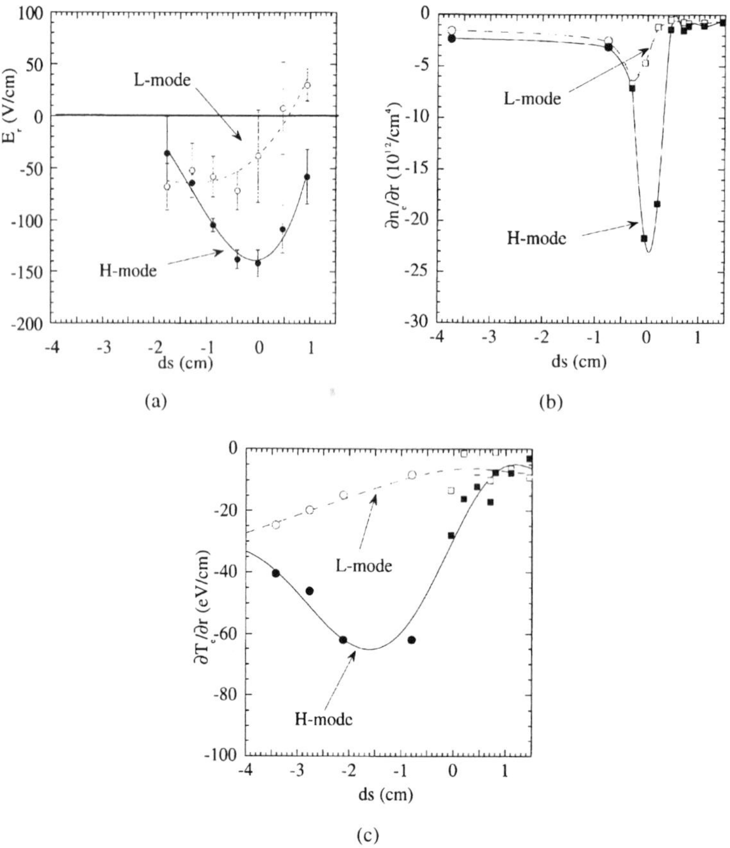

Figure 24.16. Experimental observation of the transport barrier. Radial electric field (a), density gradient profile (b) and electron temperature gradient (c) and are shown. A distance d*s* is measured from the plasma surface, and d*s* is negative for inside. Solid lines (H-mode) and dashed lines (L-mode) are drawn for an eye-guide. (Quoted from JFT-2M experiments [24.27].)

Appendix 24A Models of Forces in Simulation Model

Details of the model expressions, that are employed in numerical simulation, are described.

24A.1 Injection Rate of Toroidal Momentum by NBI

The momentum injection associated with the neutral beam injection (NBI) is expressed in terms of the power deposition density P_b and the energy of the beam particle W_b as

$$F_{b\zeta} = m_b v_b \frac{P_b}{W_b}.$$

The slowing down rate of the beam ions is given by

$$\nu_b = \frac{3}{\ln(1 + v_b^3 v_c^{-3})} \nu_{be}$$

where

$$\nu_{be} = \frac{n_e Z_b^2 e^4 \ln \Lambda}{6\pi \sqrt{2\pi}\, \varepsilon_0^2 m_b} \left(\frac{T_e}{m_e}\right)^{3/2}$$

is the beam–electron collision frequency and the critical energy of the beam particles is given by

$$v_c^3 = 3\sqrt{\frac{\pi}{2}} \frac{n_i Z_i^2}{n_e} \frac{m_e}{m_i} \left(\frac{T_e}{m_e}\right)^{3/2}$$

24A.2 Charge Exchange

Charge exchange collision between ions and neutral particles induces momentum loss of ions. The force on ions associated with the charge exchange is expressed as

$$\begin{aligned} F_{i\theta}^{CX} &= -m_i n_i n_0 \langle \sigma_{CX} v \rangle V_{i\theta} \\ F_{i\zeta}^{CX} &= -m_i n_i n_0 \langle \sigma_{CX} v \rangle V_{i\zeta} \end{aligned}$$

where the charge exchange cross-section is approximately given, for the parameters $T_e > 50$ eV, as

$$\langle \sigma_{CX} v \rangle = 1.57 \times 10^{-16} \sqrt{T_{i.[eV]}}\, (z^2 - 14.63z + 53.65).$$

In this expression, the argument z is defined as $z = \log_{10}(\max\{T_{i.[eV]}, 50\})$ where $T_{i.[eV]}$ is the ion temperature which is measured in the unit of eV.

24A.3 Collisional Relaxation

Collisional momentum exchange between plasma species is given as

$$\begin{aligned} F_{ei.\theta}^{C} &= -F_{ie.\theta}^{C} = -\nu_{ei} m_e n_e (V_{e\theta} - V_{i\theta}) \\ F_{ei.\zeta}^{C} &= -F_{ie.\zeta}^{C} = -\nu_{ei} m_e n_e (V_{e\zeta} - V_{i\zeta}) \end{aligned}$$

where

$$\nu_{ei} = \frac{Z_{eff} e^4 \ln \Lambda}{6\pi \sqrt{2\pi}\, \varepsilon_0^2 m_e^{1/2} T_e^{3/2}}$$

is the electron–ion collision frequency, $Z_{eff} = n_e^{-1} \sum_i Z_i n_i$ is the effective charge of ions, $\ln\Lambda = 15 - \frac{1}{2}\ln(n_{e.[20]}) + \ln(T_{e.[keV]})$ is the Coulomb logarithmic constant, and $n_{e.[20]}$ and $T_{e.[keV]}$ are the electron density (measured in the unit of 10^{20} m^{-3}) and electron temperature (measured in the unit of keV), respectively.

The collisional force between beam particle and bulk plasma particles is given as

$$F^C_{eb.\zeta} = -F^C_{be.\zeta} = -\nu_{be} m_b n_b (V_{e\zeta} - V_{b\zeta})$$
$$F^C_{ib.\zeta} = -F^C_{bi.\zeta} = -\nu_{bi} m_b n_b (V_{i\zeta} - V_{b\zeta})$$

where

$$\nu_{bi} = \frac{n_i Z_i^2 Z_b^2 e^4 \ln\Lambda}{4\pi\varepsilon_0^2 m_b}\left(\frac{1}{m_i} + \frac{1}{m_b}\right)\frac{1}{|v_b^3| + (3\sqrt{\pi}/4)(3T_i m_i^{-1})^{3/2}}$$

is the beam–ion collision frequency.

24A.4 Neoclassical Viscosity

The neoclassical bulk viscosity is in proportion to the collision frequency and is simplified in this case as

$$F^{NC}_{j\theta} = -\sqrt{\pi}\, q^2 m_j n_j \frac{v_{thj}}{qR}\frac{1.78\nu_j^*}{1 + 1.78\nu_j^*} V_{j\theta}$$

where $\nu_j^* \equiv \nu_j q R/\varepsilon^{3/2} v_{thi}$ is the normalized collision frequency. In order to simplify the analysis, the suppression factor owing to the strong radial electric field is not taken into account.

REFERENCES

[24.1] Example of the representative study of transport simulations are
Düchs D F, Post D E and Rutherford P H 1977 *Nucl. Fusion* **17** 565
Dnestrovskii Y N and Kostomarov D P 1985 *Numerical Simulation of Plasmas* (Berlin: Springer)

[24.2] Kikuchi H and Azumi M 1995 *Plasma Phys. Control. Fusion* **37** 1215

[24.3] Stix T H 1962 *Theory of Plasma Waves* (New York: McGraw-Hill)
Takamura S 1986 *Introduction to Plasma Heating* (Nagoya: Nagoya University Press) (in Japanese)
Fukuyama A, Itoh K and Itoh S-I 1986 *Computor Physics Rep.* **4** 137
Morishita T, Fukuyama A, Hamamatsu K, Itoh S-I and Itoh K 1987 *Nucl. Fusion* **27** 1291

[24.4] Fukuyama A, Itoh K, Itoh S-I, Yagi M and Azumi M 1995 *Plasma Phys. Control. Fusion* **37** 611

[24.5] Connor J W 1995 *Plasma Phys. Control. Fusion* **37** A119

[24.6] Connor J W, Alexander M, Attenberger S E *et al* 1997 *Fusion Energy 1996 (16th Conf. Proc.)* vol 2 (Montreal: IAEA) p 935

[24.7] Goldston R J 1984 *Plasma Phys. Control. Fusion* **26** 87
Kaye S M 1985 *Phys. Fluids* **28** 2327
Takizuka T 1992 *Proc. 19th Eur. Conf. on Controlled Fusion and Plasma Heating (Innsbruck, 1992)* vol 16C, part I, p 51
Stroth U, Kaiser M, Ryter F and Wagner F 1995 *Nucl. Fusion* **35** 131

[24.8] Yushmanov P N, Takizuka T, Riedel K S, Kardaun O J W F *et al* 1990 *Nucl. Fusion* **30** 1999

[24.9] Fukuyama A, Itoh K, Itoh S-I, Yagi M and Azumi M 1994 *Plasma Phys. Control. Fusion* **36** 1385

[24.10] Fuji Y 1997 Self-consistent modelling of particle transport in tokamaks *PhD Thesis* Okayama University

[24.11] Itoh S-I 1990 *J. Phys. Soc. Japan* **59** 3431

[24.12] Fukuyama A, Fuji Y, Itoh S-I, Yagi M and Itoh K 1996 *Plasma Phys. Control. Fusion* **38** 1319

[24.13] Hinton F L and Staebler G M 1993 *Phys. Fluids* B **5** 1281
Hinton F L, Staebler G M and Kim Y-B 1994 *Plasma Phys. Control. Fusion* **36** A237

[24.14] See, e.g., Perkins F W 1984 *Proc. 4th Int. Symp. on Heating in Toroidal Plasmas (Rome)* vol 2, p 977

[24.15] Connor J W, Taylor J B and Turner M 1984 *Nucl. Fusion* **24** 256

[24.16] Keilhacker M and the JET Team 1993 *Plasma Physics and Controlled Nuclear Fusion Research 1992* vol 1 (Vienna: IAEA) p 15

[24.17] Koide Y, Kikuchi M, Mori M, Tsuji S, Ishida S *et al* 1994 *Phys. Rev. Lett.* **72** 3662

[24.18] Levinton F M, Zarnstorff M C, Batha S H, Bell M *et al* 1995 *Phys. Rev. Lett.* **75** 4417

[24.19] Strait E J, Lao L L, Mauel M E, Rice B W, Taylor T S *et al* 1995 *Phys. Rev. Lett.* **75** 4421

[24.20] Koide Y and the JT-60 Team 1997 *Phys. Plasmas* **4** 1623
Equipe TORE SUPRA 1994 *Plasma Physics and Controlled Nuclear Fusion Research 1994* vol 1 (Vienna: IAEA) p 105

[24.21] Soldner F X, Muller E R, Wagner F *et al* 1988 *Phys. Rev. Lett.* **61** 1105

[24.22] Greenwald M, Gwinn D, Milora S *et al* 1984 *Phys. Rev. Lett.* **53** 352

[24.23] Gehre O, Gruber O, Murmann H D, Robert D E, Wagner F *et al* 1988 *Phys. Rev. Lett.* **60** 1502
Fussmann G, Gruber O, Niedermeyer H *et al* 1989 *Plasma Physics and Controlled Nuclear Fusion Research 1988* vol 1 (Vienna: IAEA) p 145

[24.24] Ida K, Itoh S-I, Itoh K, Hidekuma S, Miura Y *et al* 1992 *Phys. Rev. Lett.* **68** 182

[24.25] Groebner R J, Burrell K H and Seraydarian R P 1990 *Phys. Rev. Lett.* **64** 3015

[24.26] Burrell K H 1994 *Plasma Phys. Control. Fusion* **36** A291
Burrell K H 1997 *Phys. Plasmas* **4** 1499
Ida K 1998 *Plasma Phys. Control. Fusion* **40** 1429

[24.27] Ida K, Hidekuma S, Miura Y, Fujita T, Mori M *et al* 1990 *Phys. Rev. Lett.* **65** 1364

[24.28] Taylor R J, Brown M L, Fried B D *et al* 1989 *Phys. Rev. Lett.* **63** 2365
Weynants R R, Van Oost G, Bertschinger G *et al* 1992 *Nucl. Fusion* **32** 837

[24.29] Cornelis J, Sporken R, Van Oost G and Weynants R R 1994 *Nucl. Fusion* **34** 171

[24.30] Ida K, Itoh K, Itoh S-I, Miura M *et al* 1994 *Phys. Plasmas* **1** 116

[24.31] Fonck R J, Bretz N L, Cosby G *et al* 1992 *Plasma Phys. Control. Fusion* **34** 1993

Wootton A, Tsui H Y W and Prager S 1992 *Plasma Phys. Control. Fusion* **34** 2030

[24.32] Ohkawa T 1978 *Phys. Lett.* **67A** 35

Parail V V and Pogutse O P 1980 *JETP Lett.* **32** 384

Kadomtsev B B and Pogutse O P 1985 *Plasma Physics and Controlled Nuclear Fusion Research 1984* vol 2 (Vienna: IAEA) p 69

[24.33] Wong K-L, Bretz N L, Hahm T S and Synakowski E 1997 *Phys. Lett.* A **236** 339

Index